SpringerWienNewYork

CISM COURSES AND LECTURES

The series presents lecture notes, monographs, edited works and proceedings in the field of Mechanics, Engineering, Computer Scienc and Applied Mathematics.
Purpose of the series is to make known in the international scientific and technical community results obtained in some of the activities organized by CISM, the International Centre for Mechanical Science

TERNATIONAL CENTRE FOR MECHANICAL SCIENCES

COURSES AND LECTURES - No. 503

CLASSICAL AND ADVANCED THEORIES OF THIN STRUCTURES

MECHANICAL AND MATHEMATICAL ASPECTS

EDITED BY

ANTONINO MORASSI
UNIVERSITY OF UDINE, ITALY

ROBERTO PARONI
UNIVERSITY OF SASSARI, ITALY

SpringerWienNewYork

This volume contains 32 illustrations

Printed in Italy

ISBN 978-3-211-99940-0
e-ISBN 978-3-211-85430-3

PREFACE

Thin structures are frequently used in several areas of civil, mechanical and aeronautical engineering. Slender beams, plates and shells, just to mention some of the most common examples, are especially advantageous in structural design applications since they are able to ensure a high ratio between "strength" and weight.

Approximate analytical models for thin structures are hundreds of years old and go back to the pioneering works by Euler, D. Bernoulli, Navier and Kirchhoff, to mention a few. These classical theories are based on the introduction of some a-priori assumptions motivated by the smallness of certain dimensions with respect to others, on the stress field or on the deformation of the thin body. In the last few decades a considerable amount of work has been done in order to rigorously justify these a-priori assumptions. In particular, approaches based on rigorous asymptotic expansion (mainly due to the French school) or inspired by the Gamma-convergence of energy functionals (proposed by E. De Giorgi in 1979) have been successfully used in deriving one or two-dimensional classical mechanical models for thin structures, in linear and non-linear elasticity. Besides these asymptotic methods, another approach, called the method of internal constraints, was developed. This method is based on viewing the a-priori kinematical assumptions as internal constraints, which generate reactive stresses, and to obtain the lower dimensional theory by integration over the "small" dimensions. Asymptotic methods and the method of internal constraints could be seen as complementary: on one hand, asymptotic methods provide a rigorous justification of the starting point of the method of internal constraints, on the other hand, this latter method provides an intuitive and consistent mechanical deduction of lower dimensional theories without an extensive use of deep mathematical tools.

The aim of the CISM course entitled "Classical and Advanced Theories of Thin Structures: Mechanical and Mathematical Aspects", held in Udine on June 5-9 2006, was to present an up-to-date overview of the general aspects and applications of the theories for thin structures, through the interaction of several topics, ranging from non-linear thin-films, shells, beams of different materials and in different contexts (elasticity, plasticity, etc.).

The course was addressed to PhD students and researchers in the fields of continuum mechanics, structural engineering and applied mathematics. The plan of the course and the lectures have taken into account the different background of the audience.

The first two chapters by Morassi and Paroni introduce the main notions of continuum mechanics: the concepts of stress, deformation and constitutive equations. Moreover, in discussing the existence of the solution for the elasticity problem, several mathematical tools such as Sobolev spaces, weak convergence, Lax-Milgram lemma and Korn's inequalities are recalled. As anticipated earlier, the notion of Gamma-convergence plays a fundamental role in the deduction of lower dimensional models. Percivale in the third chapter introduces the notion and describes the main properties of Gamma-convergence. Besides presenting some illustrative examples, he also discusses the properties of the limit problem for an elastic-plastic beam. In the fourth chapter Podio-Guidugli presents a unified approach to classic rod and plate theories by means of a procedure which is an outgrowth of the method of internal constraints. He also discusses how to approximate the stress field in a linearly elastic structure-like body. The next chapter by Ciarlet is devoted to a detailed description of the nonlinear and linear equations proposed by W.T. Koiter for modelling thin elastic shells. The existence, uniqueness and regularity of solutions to the linear Koiter equations is then established, thanks to a fundamental Korn inequality on surfaces. At the beginning of the chapter the basic notions about differential geometry for surfaces are introduced. The book ends with the lectures by Fonseca. In the first two she applies Gamma-convergence techniques to deduce the energy for brittle and non-brittle thin films. The last four lectures are dedicated to the study of various interesting aspects of nonlinear membranes, including the deduction for two nonlinear membrane models: one based on the so called Cosserat vector and the other written in terms of Young measures. This latter model is appropriate for the study of phase transitions.

Antonino Morassi and Roberto Paroni

CONTENTS

Strain, Stress and Linearized Elasticity

Antonino Morassi
Dipartimento di Georisorse e Territorio, Università degli Studi di Udine

Abstract In these notes I introduce the concepts of strain and stress, and I present the linearized elasticity problem.

1 Introduction

In these notes I present some basic elements of Continuum Mechanics. The first two chapters introduce the concepts of strain and stress, respectively. The third chapter is devoted to the derivation of the linearized elasticity and to the presentation of some related existence and uniqueness results.

The manuscript reproduces more or less the pattern followed in the notes of the three lectures delivered at CISM. The presentation of the arguments is deliberately simplified and there is no claim to general and exhaustive treatment. Indeed, the aim of this first series of lectures, and of those by Paroni (2008), was to introduce the audience of the course to the principles of Continuum Mechanics and to the mathematical tools adequate for the correct formalization of the problems.

The reader interested to further investigate the issues outlined in these notes is invited to consult the classical treatises of Continuum Mechanics and Theory of Elasticity, an essential list of which can be found in the references.

2 Strain

2.1 Points, vectors and tensors

Let us denote by $\mathbb{R}^3$ the cartesian space formed by the usual vectorial space of all ordered triplets, or *points*, $X = (X_1, X_2, X_3)$, $X_i \in \mathbb{R}$ for $i = 1, 2, 3$, endowed with an euclidean structure induced by the distance

$$d(X,Y) = (\sum_{i=1}^{3}(X_i - Y_i)^2)^{1/2} \quad (1)$$

between any two points X and $Y = (Y_1, Y_2, Y_3)$ of $\mathbb{R}^3$. Here, X_1, X_2, X_3 are the *orthogonal cartesian coordinates* of X in $\mathbb{R}^3$.

We shall denote by V the vector space associated with $\mathbb{R}^3$. The elements of V are called *vectors*. Every vector $\mathbf{v}$ may be interpreted as difference between two points of $\mathbb{R}^3$, namely the head, Y, and the end, X, of the pointed arrow used to denote the vector $\mathbf{v}$, e.g. $\mathbf{v} = Y - X$. The euclidean distance of $\mathbb{R}^3$ endows V with its natural metric structure: for every $\mathbf{v} \in V$, the norm $|\mathbf{v}|$ of $\mathbf{v}$ is the length of the arrow which represents $\mathbf{v}$. In particular, $|\mathbf{v}| = (\mathbf{v} \cdot \mathbf{v})^{1/2}$, where $\cdot$ is the usual inner product of V.

We think V equipped with a fixed cartesian frame $\{O; \mathbf{e}_1, \mathbf{e}_2, \mathbf{e}_3\}$, where O is a chosen point of $\mathbb{R}^3$, called *origin*, and $\{\mathbf{e}_i\}_{i=1}^3$ is an orthonormal basis of V, e.g. $\mathbf{e}_i \cdot \mathbf{e}_j = \delta_{ij}$, where δ_{ij} is the Kronecker symbol. Therefore, the (cartesian) components of a vector $\mathbf{v}$ are given by

$$v_i = \mathbf{v} \cdot \mathbf{e}_i, \tag{2}$$

$i = 1, 2, 3$, so that

$$\mathbf{v} \cdot \mathbf{u} = \sum_{i=1}^{3} v_i u_i. \tag{3}$$

Similarly, if $\{\mathbf{e}_i\}_{i=1}^3$ are the versors of the axes of the cartesian space $\mathbb{R}^3$, recalling that $\mathbf{X} = X - O$, the ith coordinate of the point X is $X_i = \mathbf{X} \cdot \mathbf{e}_i$, $i = 1, 2, 3$.

The set of all the linear transformations of V into itself is denoted by Lin. Each element of Lin is called second-order tensor. Equipped with the usual operations of addition and scalar multiplication, Lin is a vector space. Given an orthonormal basis $\{\mathbf{e}_i\}_{i=1}^3$ of V, the components A_{ij} of $\mathbf{A} \in \text{Lin}$ with respect to this basis are defined by

$$A_{ij} = \mathbf{e}_i \cdot \mathbf{A}\mathbf{e}_j, \quad i, j = 1, 2, 3. \tag{4}$$

2.2 Bodies and deformations

For us, a *body* B is a set of particles. These particles can be considered as primitive elements of Mechanics. The body B is assumed to be smooth, that is the particles $X \in B$ can be set into one-to-one correspondence with a bounded *domain*[1] in $\mathbb{R}^3$ with regular boundary, and the mapping is assumed to be differentiable as many times as desired (usually two or three times).

Bodies are available to us only in their *configurations*, that is the regions they happen to occupy in $\mathbb{R}^3$. It is often convenient to select one particular configuration and refer everything concerning the body to that

[1] Let Ω be an open and connected set of $\mathbb{R}^3$. The closure of Ω, $\overline{\Omega} = \Omega \cup \partial\Omega$, is called here *domain*.

configuration. Let $\boldsymbol{\kappa}$ be such configuration

$$\begin{array}{rccl} \boldsymbol{\kappa}: & B & \rightarrow & \mathbb{R}^3 \\ & X & \mapsto & \boldsymbol{\kappa}(X) = \mathbf{X}\,, \end{array} \tag{5}$$

where $\mathbf{X}$ is the place occupied by the particle X in the configuration $\boldsymbol{\kappa}$. $\boldsymbol{\kappa}$ is assumed to be smooth, in particular $X = \boldsymbol{\kappa}^{-1}(\mathbf{X})$ is well-defined. $\boldsymbol{\kappa}(B)$ denotes the set of all the reference placements of the body; it will be called *reference configuration* and denoted by $\overline{\Omega}$.

The deformation of the body will be described as a mapping acting on the reference configuration $\boldsymbol{\kappa}$ onto the *actual configuration* $\boldsymbol{\chi}$. Therefore, a deformation of the body

$$\begin{array}{rccl} \boldsymbol{\chi}: & \overline{\Omega} & \rightarrow & \mathbb{R}^3 \\ & \mathbf{X} & \mapsto & \boldsymbol{\chi}(\mathbf{X}) = \mathbf{x} \end{array} \tag{6}$$

gives the actual placement $\mathbf{x}$ occupied by the particle $\mathbf{X}$ in the reference configuration. We indicate with $\boldsymbol{\chi}(\overline{\Omega})$ the image of $\overline{\Omega}$ under the map $\boldsymbol{\chi}$ or, briefly, the *deformation* of $\overline{\Omega}$ under $\boldsymbol{\chi}$.

We accept the following a priori assumptions on $\boldsymbol{\chi}$:

i) There is a bijection between $\overline{\Omega}$ and $\boldsymbol{\chi}(\overline{\Omega})$.

ii) $\boldsymbol{\chi} \in C^1(\overline{\Omega})$.

iii) The orientation preserving condition

$$\det \nabla\boldsymbol{\chi}(\mathbf{X}) > 0 \tag{7}$$

holds in $\overline{\Omega}$. Here, ∇ is the gradient operator[2] and det is the determinant operator.[3]

By above assumptions, the map $\boldsymbol{\chi}$ is globally invertible in $\overline{\Omega}$, e.g. there exists the inverse of $\boldsymbol{\chi}$, say $\boldsymbol{\chi}^{-1}$, such that

$$\boldsymbol{\chi}^{-1}(\mathbf{x}) = \mathbf{X} \quad \text{in } \boldsymbol{\chi}(\overline{\Omega}),$$

[2]Let $\boldsymbol{\chi} : \Omega \rightarrow \mathbb{R}^3$ be a function of class C^1 in Ω. The *gradient* of $\boldsymbol{\chi}$ at $\mathbf{X} \in \Omega$ is the unique element $\nabla\boldsymbol{\chi}$ such that

$$\lim_{|\mathbf{h}|\to 0} \frac{|\boldsymbol{\chi}(\mathbf{X}+\mathbf{h}) - \boldsymbol{\chi}(\mathbf{X}) - \nabla\boldsymbol{\chi}(\mathbf{X})\mathbf{h}|}{|\mathbf{h}|} = 0.$$

If $\boldsymbol{\chi}$ is of C^1-class up to the (smooth) boundary $\partial\Omega$ of Ω, then the gradient of $\boldsymbol{\chi}$ at $\mathbf{X} \in \partial\Omega$ is well-defined and it can be evaluated as limit of $\nabla\boldsymbol{\chi}(\mathbf{Y})$, $\mathbf{Y} \in \Omega$, as $\mathbf{Y} \rightarrow \mathbf{X}$.

[3]Let $\mathbf{A} \in \mathrm{Lin}$; $\det \mathbf{A} \in \mathbb{R}$ is defined as

$$(\det \mathbf{A})\mathbf{a} \times \mathbf{b} \cdot \mathbf{c} = (\mathbf{A}\mathbf{a} \times \mathbf{A}\mathbf{b}) \cdot \mathbf{A}\mathbf{c} \quad \text{for every } \mathbf{a}, \mathbf{b}, \mathbf{c} \in V \text{ such that } \mathbf{a} \times \mathbf{b} \cdot \mathbf{c} \neq 0.$$

Here $\times$ is the vectorial product in $V \times V$.

and it still satisfies conditions i)-iii).

The *displacement field* $\mathbf{u}(\mathbf{X})$ associated with the deformation χ is defined as follows:

$$\begin{array}{rccl} \mathbf{u}: & \overline{\Omega} & \rightarrow & \mathbb{R}^3 \\ & \mathbf{X} & \mapsto & \mathbf{u}(\mathbf{X}) = \chi(\mathbf{X}) - \mathbf{X} \end{array} \tag{8}$$

The vector $\mathbf{u}(\mathbf{X})$ is the displacement from the reference configuration to the deformed shape at the point $\mathbf{X}$. The *displacement gradient* is given by

$$\nabla \mathbf{u} = \nabla \chi - \mathbf{1} \,, \tag{9}$$

where $\mathbf{1} \in \text{Lin}$ is the identity tensor.

Following a standard notation, we shall denote by $\mathbf{F}$ and $\mathbf{H}$ the deformation gradient and the displacement gradient, respectively, i.e.

$$\begin{array}{l} \nabla \chi(\mathbf{X}_0) = \mathbf{F}(\mathbf{X}_0) \,, \\ \nabla \mathbf{u}(\mathbf{X}_0) = \mathbf{H}(\mathbf{X}_0), \quad \mathbf{X}_0 \in \overline{\Omega}. \end{array} \tag{10}$$

The values $\mathbf{F}(\mathbf{X}_0)$, $\mathbf{H}(\mathbf{X}_0)$, $\mathbf{X}_0 \in \overline{\Omega}$, are elements of Lin. In particular, the values $\mathbf{F}(\mathbf{X}_0)$ are elements of the set of positive linear transformations Lin^+:

$$\text{Lin}^+ = \{\mathbf{A} \in \text{Lin} \mid \det \mathbf{A} > 0\} \tag{11}$$

2.3 Deformation: examples

i) *Rigid deformation*

We say that a deformation is rigid if it leaves the distance between any pair of points of $\overline{\Omega}$ unchanged, i.e.

$$|\chi(\mathbf{X}) - \chi(\mathbf{Y})| = |\mathbf{X} - \mathbf{Y}| \quad \text{for every } \mathbf{X},\ \mathbf{Y} \in \overline{\Omega} \,.$$

One can show that

$$\chi(\mathbf{X}) = \chi(\mathbf{X}_0) + \mathbf{Q}(\mathbf{X} - \mathbf{X}_0) \,, \tag{12}$$

where $\mathbf{X}_0$ is any point of $\overline{\Omega}$ and $\mathbf{Q}$ is a constant *orthogonal tensor*[4] with $\det \mathbf{Q} = 1$. Then, we have

$$\mathbf{F}(\mathbf{X}) = \mathbf{Q} \quad \text{in } \overline{\Omega}$$

[4] A second-order tensor $\mathbf{Q} \in \text{Lin}$ is called *orthogonal* if $\mathbf{Q}\mathbf{Q}^T = \mathbf{1}$. We shall denote by Orth the collection of all second-order orthogonal tensors. We call *rotation* an orthogonal tensor $\mathbf{Q}$ with $\det \mathbf{Q} = 1$ and we shall indicate

$$\text{Orth}^+ = \{\mathbf{Q} \in \text{Orth} \mid \det \mathbf{Q} = 1\}.$$

and the *rigid displacement field* is given by

$$\mathbf{u}(\mathbf{X}) = \mathbf{u}(\mathbf{X}_0) + (\mathbf{Q} - \mathbf{1})(\mathbf{X} - \mathbf{X}_0) \quad \text{in } \overline{\Omega} .$$

A rigid deformation is a translation if $\mathbf{Q} = \mathbf{1}$, a rotation (about $\mathbf{X}_0$) if $\mathbf{u}(\mathbf{X}_0) = \mathbf{0}$, cf. Ex. 3.

ii) *Simple extension*

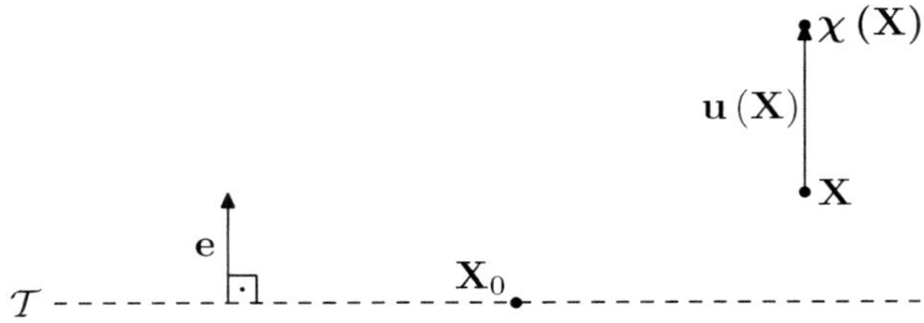

Figure 1. Simple extension in the direction $\mathbf{e}$.

Let $\mathbf{X}_0$ be a point on $\overline{\Omega}$ which remains fixed during deformation. Let $\mathcal{T}$ the plane passing through $\mathbf{X}_0$ with normal $\mathbf{e}$, $|\mathbf{e}| = 1$. The simple extension, of amount α in the direction $\mathbf{e}$, is such that the displacement $\mathbf{u}(\mathbf{X})$ of point $\mathbf{X}$ is parallel to $\mathbf{e}$-direction and its modulus is proportional to the distance of $\mathbf{X}$ from $\mathcal{T}$. That is

$$\chi(\mathbf{X}) = \mathbf{X} + \alpha[(\mathbf{X} - \mathbf{X}_0) \cdot \mathbf{e}]\mathbf{e} = \mathbf{X} + \alpha(\mathbf{e} \otimes \mathbf{e})(\mathbf{X} - \mathbf{X}_0) , \tag{13}$$

where, for any two vectors $\mathbf{a}, \mathbf{b} \in V$, $\mathbf{a} \otimes \mathbf{b}$ denotes the element of Lin such that $(\mathbf{a} \otimes \mathbf{b})\mathbf{v} = (\mathbf{b} \cdot \mathbf{v})\mathbf{a}$ for every $\mathbf{v} \in V$. We have

$$\mathbf{u}(\mathbf{X}) = \alpha((\mathbf{X} - \mathbf{X}_0) \cdot \mathbf{e})\mathbf{e} ,$$

$$\mathbf{F}(\mathbf{X}) = \mathbf{1} + \alpha(\mathbf{e} \otimes \mathbf{e}), \quad \mathbf{H}(\mathbf{X}) = \alpha(\mathbf{e} \otimes \mathbf{e}) ,$$

$$\det \mathbf{F} > 0 \Longleftrightarrow \alpha > -1 .$$

iii) *Simple shear*

Let $\mathbf{e}_1 \cdot \mathbf{e}_2 = 0$, $|\mathbf{e}_1| = |\mathbf{e}_2| = 1$. With the notation of Figure 2, the deformation of simple shear, of amount γ in the plane $(\mathbf{e}_1, \mathbf{e}_2)$, is such that

$$\chi(\mathbf{X}) = \mathbf{X} + \gamma((\mathbf{X} - \mathbf{X}_0) \cdot \mathbf{e}_2)\mathbf{e}_1 = \mathbf{X} + \gamma(\mathbf{e}_1 \otimes \mathbf{e}_2)(\mathbf{X} - \mathbf{X}_0) \tag{14}$$

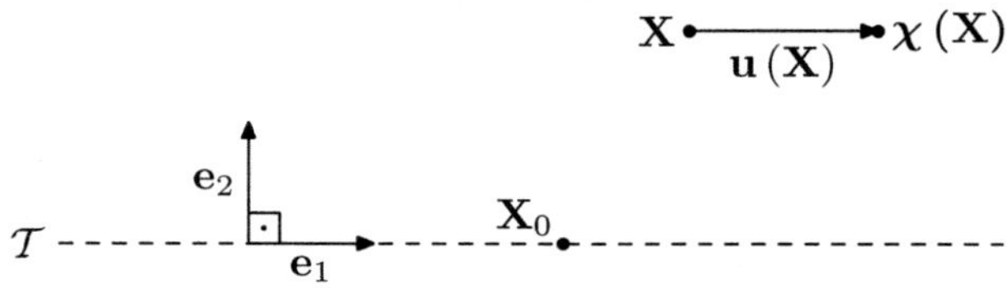

Figure 2. Simple shear in the plane $(\mathbf{e}_1, \mathbf{e}_2)$.

and then

$$\mathbf{u}(\mathbf{X}) = \gamma(\mathbf{e}_1 \otimes \mathbf{e}_2)(\mathbf{X} - \mathbf{X}_0),$$

$$\mathbf{F}(\mathbf{X}) = \mathbf{1} + \gamma(\mathbf{e}_1 \otimes \mathbf{e}_2), \quad \mathbf{H}(\mathbf{X}) = \gamma(\mathbf{e}_1 \otimes \mathbf{e}_2),$$

$$\det \mathbf{F} = 1.$$

These three cases are examples of *homogeneous deformations*, that is deformations with constant gradient in $\overline{\Omega}$:

$$\boldsymbol{\chi}(\mathbf{X}) = \boldsymbol{\chi}(\mathbf{X}_0) + \mathbf{F}(\mathbf{X} - \mathbf{X}_0), \quad \mathbf{X}_0 \in \overline{\Omega}, \quad \text{in } \overline{\Omega}, \tag{15}$$

with $\mathbf{F} = const$ in $\overline{\Omega}$.

2.4 Changes in length, area, volume, angle. Strain measures

The knowledge of the gradient $\mathbf{F}$ at $\mathbf{X}_0 \in \Omega$ allows to characterize locally the deformation. In fact, by the regularity of $\boldsymbol{\chi}$ we have

$$\boldsymbol{\chi}(\mathbf{X}_0 + d\mathbf{X}) = \boldsymbol{\chi}(\mathbf{X}_0) + \mathbf{F}(\mathbf{X}_0)d\mathbf{X} + o(d\mathbf{X};\ \mathbf{X}_0)\ , \tag{16}$$

with $\displaystyle\lim_{|d\mathbf{X}|\to 0} \frac{|o(d\mathbf{X};\ \mathbf{X}_0)|}{|d\mathbf{X}|} = 0$. When $|d\mathbf{X}|$ is *sufficiently small*, then the deformation can be locally approximated by

$$\boldsymbol{\chi}(\mathbf{X}_0 + d\mathbf{X}) = \boldsymbol{\chi}(\mathbf{X}_0) + \mathbf{F}(\mathbf{X}_0)d\mathbf{X}\ . \tag{17}$$

Denoting by

$$d\mathbf{x} = \boldsymbol{\chi}(\mathbf{X}_0 + d\mathbf{X}) - \boldsymbol{\chi}(\mathbf{X}_0), \tag{18}$$

we have

$$d\mathbf{x} = \mathbf{F}(\mathbf{X}_0)d\mathbf{X}. \tag{19}$$

This equation can be regarded as a local description of the deformation in a neighborhood of $\mathbf{X}_0$. In this description the vector $d\mathbf{X}$ is mapped linearly into $\mathbf{F}(\mathbf{X}_0)d\mathbf{X}$.

If $d\mathbf{X}$ in Eq. (19) is identified with an *infinitesimal* oriented line element (or *material fiber*) through $\mathbf{X}_0$, then Eq. (19) can be used for evaluating the local changes in length, area, etc.

Change in length

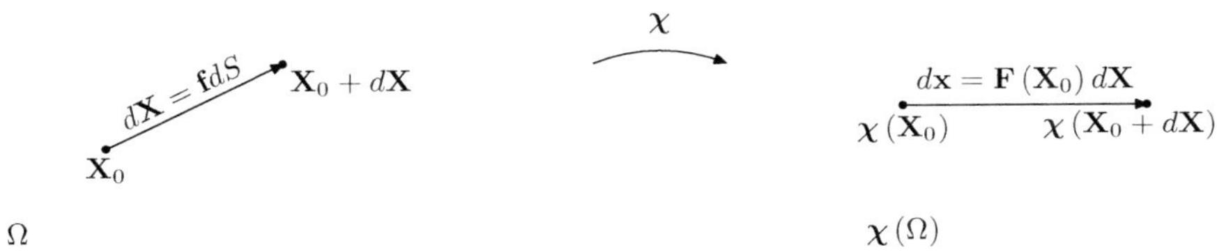

Figure 3. Local change in length.

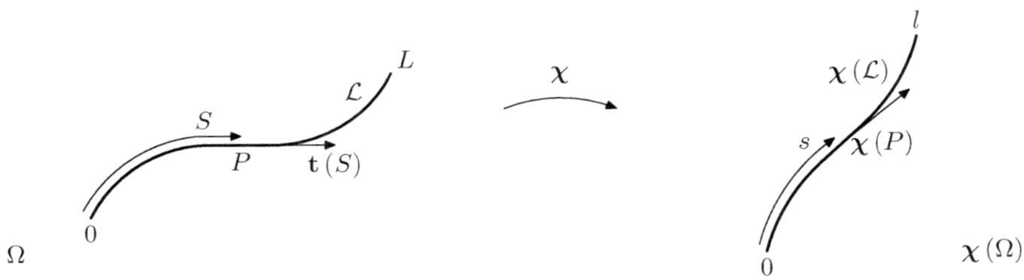

Figure 4. Change in length of finite curves.

With reference to Figure 3, let $d\mathbf{X} = dS\mathbf{f}$ with $|\mathbf{f}| = 1$. The length of the material fiber $d\mathbf{X}$ after the deformation is (here $\mathbf{F} = \mathbf{F}(\mathbf{X}_0)$)

$$ds = |d\mathbf{x}| = \sqrt{\mathbf{F}^T\mathbf{F}\mathbf{f} \cdot \mathbf{f}}\, dS$$

and, therefore, the local change in length is controlled by the *Cauchy-Green right tensor* $\mathbf{C} = \mathbf{F}^T\mathbf{F}$[5]:

$$\frac{(ds)^2 - (dS)^2}{(dS)^2} = (\mathbf{C} - \mathbf{1})\mathbf{f} \cdot \mathbf{f}. \tag{20}$$

$\mathbf{C}$ is a symmetric positive tensor.[6]

We can use the above result for estimating the change in length of finite (simple and regular) curves $\mathcal{L} \subset \overline{\Omega}$, see Figure 4. Let S and s be the arch length on $\mathcal{L}$ and the corresponding on $\chi(\mathcal{L})$, respectively. The length of the deformed curve is given by $l = \int_{0(\chi(\mathcal{L}))}^{l} ds$. By making the change of variables $s = s(S)$ we have $l = \int_{0(\mathcal{L})}^{L} (\frac{ds}{dS}) dS$ and $\frac{ds}{dS} = \sqrt{\mathbf{C}(\mathbf{S})\boldsymbol{\tau} \cdot \boldsymbol{\tau}}$, where $\boldsymbol{\tau} = \boldsymbol{\tau}\,(S)$ is the unit tangent vector to $\mathcal{L}$ at the point of abscissa S.

Change in area

Consider two material fibers through $\mathbf{X}_0$, say $d\mathbf{X}_1 = dS_1\mathbf{f}_1$ and $d\mathbf{X}_2 = dS_2\mathbf{f}_2$, with $|\mathbf{f}_1| = |\mathbf{f}_2| = 1$ and $\mathbf{f}_1 \times \mathbf{f}_2 \neq \mathbf{0}$. Denote by $d\mathbf{A} = d\mathbf{X}_1 \times d\mathbf{X}_2$ the *infinitesimal oriented surface* in the referential configuration $\overline{\Omega}$. Under the deformation χ the images of $d\mathbf{X}_1$, $d\mathbf{X}_2$ are respectively $d\mathbf{x}_1 = \mathbf{F}d\mathbf{X}_1$, $d\mathbf{x}_2 = \mathbf{F}d\mathbf{X}_2$, where $\mathbf{F} = \mathbf{F}(\mathbf{X}_0)$. Therefore, the image of $d\mathbf{A}$ is $d\mathbf{a} = \mathbf{F}d\mathbf{X}_1 \times \mathbf{F}d\mathbf{X}_2$, which corresponds to the oriented infinitesimal surface area in the actual configuration $\chi(\overline{\Omega})$, see Figure 5.

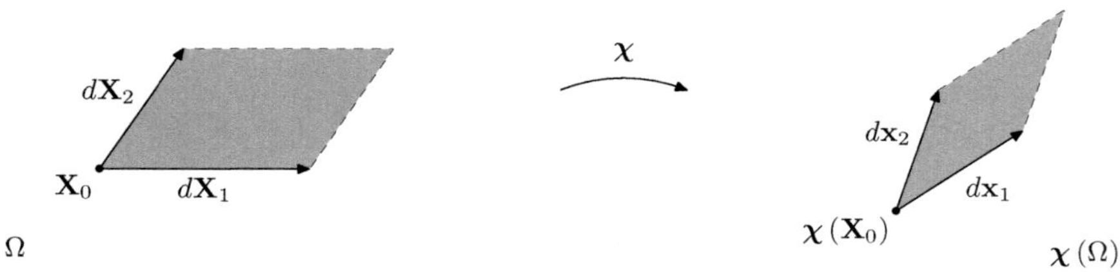

Figure 5. Local change in area.

We have

$$dA = |d\mathbf{A}| = |d\mathbf{X}_1 \times d\mathbf{X}_2|,$$

[5] For each $\mathbf{A} \in$ Lin, the transpose $\mathbf{A}^T$ of $\mathbf{A}$ is the unique tensor $\mathbf{A}^T\mathbf{a} \cdot \mathbf{b} = \mathbf{a} \cdot \mathbf{A}\mathbf{b}$ for every $\mathbf{a}, \mathbf{b} \in V$.

[6] A tensor $\mathbf{A} \in$ Lin is *symmetric* if $\mathbf{A} = \mathbf{A}^T$. The set of second-order symmetric tensors is denoted by Sym. $\mathbf{A} \in$ Sym is *positive definite*, and we write $\mathbf{A} \in \mathrm{Sym}^+$, if $\mathbf{A}\mathbf{a} \cdot \mathbf{a} \geq 0$ for every $\mathbf{a} \in V$ and $\mathbf{A}\mathbf{a} \cdot \mathbf{a} = 0$ implies $\mathbf{a} = \mathbf{0}$.

$$da = |d\mathbf{a}| = |d\mathbf{x}_1 \times d\mathbf{x}_2|$$

and, therefore, using the vectorial identities $\mathbf{a} \cdot (\mathbf{b} \times \mathbf{c}) = \mathbf{b} \cdot (\mathbf{c} \times \mathbf{a})$ and $\mathbf{a} \times (\mathbf{b} \times \mathbf{c}) = (\mathbf{c} \cdot \mathbf{a})\mathbf{b} - (\mathbf{a} \cdot \mathbf{b})\mathbf{c}$, we obtain

$$\frac{da}{dA} = \left\{ \frac{(\mathbf{C}\mathbf{f}_1 \cdot \mathbf{f}_1)(\mathbf{C}\mathbf{f}_2 \cdot \mathbf{f}_2) - (\mathbf{C}\mathbf{f}_1 \cdot \mathbf{f}_1)^2}{(\mathbf{f}_1 \cdot \mathbf{f}_1)(\mathbf{f}_2 \cdot \mathbf{f}_2) - (\mathbf{f}_1 \cdot \mathbf{f}_2)^2} \right\}^{1/2} . \tag{21}$$

The change in area of a finite regular surface $\mathcal{S}$ can be evaluated as follows. Let $\mathbf{X} = \mathbf{X}(\vartheta_1, \vartheta_2)$ be a representation of the surface $\mathcal{S}$, where the two parameters ϑ_α, $\alpha = 1, 2$, belong to some bounded domain D of $\mathbb{R}^2$, see Figure 6. The tangent space $\mathcal{T}(\mathcal{S}, \mathbf{X}_0)$ of $\mathcal{S}$ at the point $\mathbf{X}_0$ is spanned by the vectors $\mathbf{A}_\alpha = \frac{\partial \mathbf{X}}{\partial \vartheta_\alpha}(\mathbf{X}_0)$. Similarly, the vectors $\mathbf{a}_\alpha = \frac{\partial \boldsymbol{\chi}}{\partial \vartheta_\alpha} = \mathbf{F}(\mathbf{X}_0)\mathbf{A}_\alpha$, $\alpha = 1, 2$, constitute a basis of the tangent space $\mathcal{T}(\boldsymbol{\chi}(\mathcal{S}), \boldsymbol{\chi}(\mathbf{X}_0))$.

The initial area of $\mathcal{S}$ is $A = \int_{\mathcal{S}} dA$, whereas the area after the deformation is

$$a = \int_{\boldsymbol{\chi}(\mathcal{S})} da = \int_{\mathcal{S}} (\frac{da}{dA}) dA ,$$

and $\frac{da}{dA}$ can be evaluated as in Eq. (21) with $\mathbf{f}_\alpha$ replaced by $\mathbf{a}_\alpha$, $\alpha = 1, 2$.

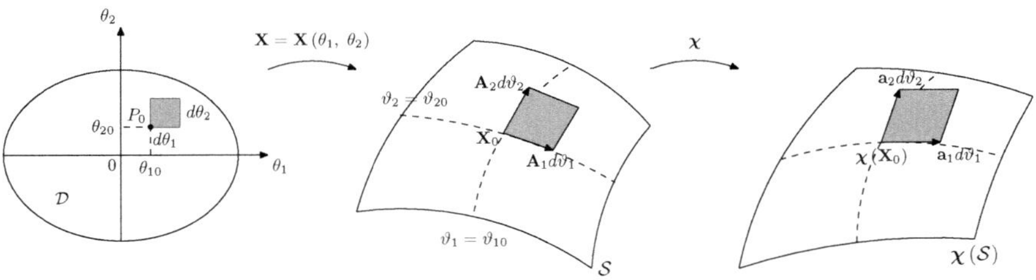

Figure 6. Change in area of a finite surface $\mathcal{S}$.

Change in volume

Let $d\mathbf{X}_i$, $i = 1, 2, 3$, be three non coplanar fibers along edges through $\mathbf{X}_0$ of a parallelepiped volume element in the reference configuration $\overline{\Omega}$. The local change in volume is equal to

$$\frac{dv - dV}{dV} = \frac{\mathbf{F}d\mathbf{X}_1 \times \mathbf{F}d\mathbf{X}_2 \cdot \mathbf{F}d\mathbf{X}_3 - d\mathbf{X}_1 \times d\mathbf{X}_2 \cdot d\mathbf{X}_3}{d\mathbf{X}_1 \times d\mathbf{X}_2 \cdot d\mathbf{X}_3}, \tag{22}$$

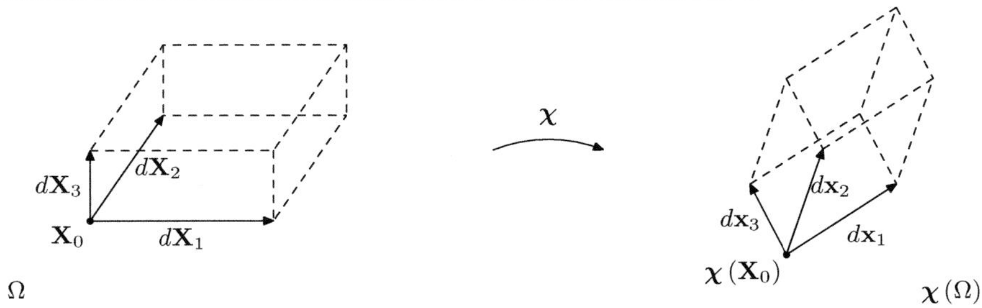

Figure 7. Local change in volume.

where $\mathbf{F}$ is the gradient at $\mathbf{X}_0$. Recalling the definition of determinant of a tensor, see footnote 3, we have

$$\frac{dv}{dV} = \det \mathbf{F},$$

which is a positive number in view of condition (7). Therefore, the volume after the deformation can be calculated as

$$v = \int_{\chi(\Omega)} dv = \int_{\Omega} (\frac{dv}{dV}) dV = \int_{\Omega} \det \mathbf{F} dV \; .$$

Change in angle

To fix the ideas, let $d\mathbf{X}_1$ and $d\mathbf{X}_2$ be two orthogonal fibers through $\mathbf{X}_0$ in the reference configuration $\overline{\Omega}$. Let $d\mathbf{X}_\alpha = dS_\alpha \mathbf{f}_\alpha$, $\alpha = 1, 2$. By definition of scalar product we have

$$\cos \varphi_{(\mathbf{f}_1,\, \mathbf{f}_2)} = \frac{d\mathbf{x}_1 \cdot d\mathbf{x}_2}{|d\mathbf{x}_1||d\mathbf{x}_2|} = \frac{\mathbf{C}\mathbf{f}_1 \cdot \mathbf{f}_2}{(\mathbf{C}\mathbf{f}_1 \cdot \mathbf{f}_1)^{1/2} (\mathbf{C}\mathbf{f}_2 \cdot \mathbf{f}_2)^{1/2}} \tag{23}$$

and, since cosine function is uniquely invertible on $[0, \pi]$, the change in angle between the directions of $\mathbf{f}_1$ and $\mathbf{f}_2$ is equal to

$$\gamma_{(\mathbf{f}_1,\, \mathbf{f}_2)} = \frac{\pi}{2} - \arccos \left(\frac{\mathbf{C}\mathbf{f}_1 \cdot \mathbf{f}_2}{(\mathbf{C}\mathbf{f}_1 \cdot \mathbf{f}_1)^{1/2} (\mathbf{C}\mathbf{f}_2 \cdot \mathbf{f}_2)^{1/2}} \right) \; . \tag{24}$$

The above analysis (local and global) shows that the changes in length, area, volume and angle, or the so-called *essential* changes in geometry, are controlled by the tensor $\mathbf{C} = \mathbf{F}^T \mathbf{F}$. This suggests how to measure strain.

A reasonable requirement in the definition of a strain measure is that its value should be constant over the set of all rigid deformations. In fact, it is easy to verify that $\mathbf{C} = \mathbf{1}$ for any rigid deformation. Therefore, $\mathbf{C}$ can be chosen as a local strain measure. There are other strain measures, for example that provided by the tensor

$$\mathbf{E} = \frac{1}{2}(\mathbf{C} - \mathbf{1}), \tag{25}$$

which is identically equal to $\mathbf{0}_{\mathrm{Lin}}$ at any rigid deformation.

2.5 Polar decomposition theorem

This theorem states that there exists a unique symmetric, positive definite tensor $\mathbf{U}$ and a unique rotation $\mathbf{R}$ such that the gradient deformation $\mathbf{F}$ can be expressed as

$$\mathbf{F} = \mathbf{R}\mathbf{U}\,. \tag{26}$$

The tensor $\mathbf{U}$ is called the (right) *stretch tensor* and $\mathbf{R}$ is the *rotation tensor* in the polar decomposition (26). This factorization is a well-known linear algebra result which holds for every non singular tensor (e.g., with non zero determinant) and tell us that the deformation correponding locally to $\mathbf{F}$ may be obtained by applying first the stretch $\mathbf{U}$, followed by the rotation $\mathbf{R}$.

To make this statement more expressive, let $\mathbf{C} = \sum_{i=1}^{3} u_i^2 \boldsymbol{\eta}_i \otimes \boldsymbol{\eta}_i$, where $u_i^2 > 0$ are the eigenvalues of $\mathbf{C}$ and $\{\boldsymbol{\eta}_i\}_{i=1}^{3}$, $|\boldsymbol{\eta}_i| = 1$ and $\boldsymbol{\eta}_i \cdot \boldsymbol{\eta}_j = \delta_{ij}$, are the corresponding eigenvectors. Therefore, since $\mathbf{U}^2 = \mathbf{C}$, $\mathbf{U}$ can be represented as $\mathbf{U} = \sum_{i=1}^{3} u_i \boldsymbol{\eta}_i \otimes \boldsymbol{\eta}_i$, $u_i > 0$, $i = 1, 2, 3$. The quantities u_i are the *principal stretches* and $\boldsymbol{\eta}_i$ are the *principal directions of strain* in the reference configuration. Therefore, $\mathbf{F}$ may be obtained by affecting pure stretches of amounts, say, u_i, along the principal directions of strain $\boldsymbol{\eta}_i$, $i = 1, 2, 3$, followed by a rotation of those directions of amount $\mathbf{R}$.

2.6 Exercises

1. Prove that the following conditions are equivalent:

i) $\boldsymbol{\chi}$ is a rigid deformation.

ii) There exists $\mathbf{Q} \in \mathrm{Orth}^+$, $\mathbf{Q}$ constant, and there exists $\mathbf{a} \in V$, $\mathbf{a}$ constant, such that $\boldsymbol{\chi}(\mathbf{X}) = \mathbf{Q}\mathbf{X} + \mathbf{a}$.

iii) $\mathbf{C}(\mathbf{X}) = \mathbf{1}$ in $\overline{\Omega}$.

2. Let $\mathbf{A} \in \mathrm{Lin}$. We define $\mathrm{cof}\mathbf{A} \in \mathrm{Lin}$ as follows:

$$\mathrm{cof}\mathbf{A}(\mathbf{a} \times \mathbf{b}) := \mathbf{A}\mathbf{a} \times \mathbf{A}\mathbf{b}, \quad \text{for every } \mathbf{a},\, \mathbf{b} \in V\,.$$

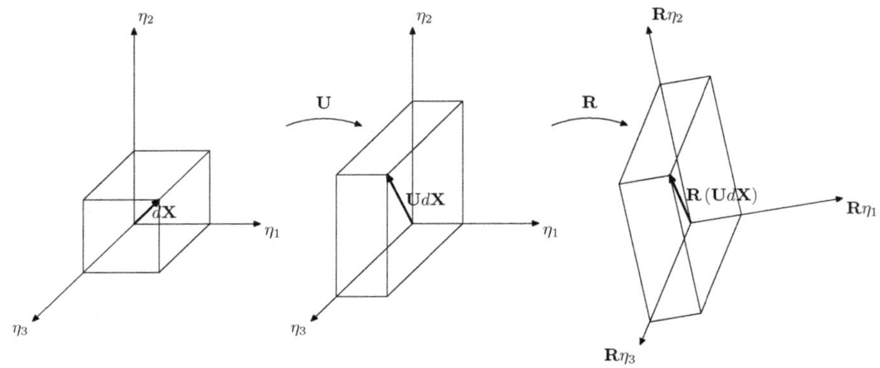

Figure 8. Polar decomposition of $\mathbf{F}$.

Prove that $(\det \mathbf{A})\mathbf{1} = \mathbf{A}^T \mathrm{cof}\mathbf{A}$. Use the above identity to prove that the normal $\mathbf{n}_{\boldsymbol{\kappa}} = \frac{\mathbf{e}_1 \times \mathbf{e}_2}{|\mathbf{e}_1 \times \mathbf{e}_2|}$ of the oriented surface $d\mathbf{A} = \mathbf{e}_1 \times \mathbf{e}_2 dS_1 dS_2$ has (unitary) image under the deformation $\boldsymbol{\chi}$ equal to $\mathbf{n} = \frac{\mathrm{cof}\mathbf{F}\mathbf{n}}{|\mathrm{cof}\mathbf{F}\mathbf{n}|}$.

3. Prove that a rotation around an axis defined by the unit vector $\mathbf{w} \in V$ of an angle θ can be represented as

$$\mathbf{Q}\,(\mathbf{w},\ \theta) = \mathbf{1} + \sin\theta \mathbf{W} + (1 - \cos\theta)\,\mathbf{W}^2\ ,$$

where $\mathbf{W}\mathbf{a} = \mathbf{w} \times \mathbf{a}$ for every $\mathbf{a} \in V$.

4. Prove the polar decomposition theorem.

5. Let $\boldsymbol{\chi} : \overline{\Omega} \to \mathbb{R}^3$ be a deformation. Denote by $\{O; \mathbf{e}_1, \mathbf{e}_2, \mathbf{e}_3\}$ a Cartesian frame. Prove that

$$\nabla\boldsymbol{\chi} = \sum_{i=1}^{3} \frac{\partial \boldsymbol{\chi}}{\partial X_i} \otimes \mathbf{e}_i,$$

where X_i, $i = 1, 2, 3$, are the cartesian coordinates of a point X. Compute the components of $\nabla\boldsymbol{\chi}$ in cylindrical and spherical coordinates.

6. Let Ω be a regular domain in $\mathbb{R}^3$ and let $\mathbf{v} = \mathbf{v}(\mathbf{X})$ be a regular vector field on $\overline{\Omega}$. Prove the following identities:

$$\int_\Omega \nabla\mathbf{v} = \int_{\partial\Omega} \mathbf{v} \otimes \mathbf{n}\ ,$$

$$\int_\Omega \mathrm{Div}\,\mathbf{v} = \int_{\partial\Omega} \mathbf{v} \cdot \mathbf{n}\ ,$$

where $\mathbf{n}$ is the unit outer normal to $\partial\Omega$ and $\text{Div}\,\mathbf{v} = \text{tr}(\nabla\mathbf{v})$.[7]

Let $\mathbf{A} = \mathbf{A}(\mathbf{X})$ a regular second order tensor field over Ω. Prove that

$$\int_\Omega \text{Div}\mathbf{A} = \int_{\partial\Omega} \mathbf{A}\mathbf{n}\,,$$

where the divergence of a tensor field is defined as $\text{Div}\left(\mathbf{A}^T\mathbf{a}\right) = \text{Div}\mathbf{A}\cdot\mathbf{a}$ for every $\mathbf{a} \in V$ constant.

3 Stress

3.1 Forces and Moments

Forces are primitive elements of Mechanics and they express one of the most simple models describing the interactions between a body and the surrounding environment or between different parts of the same body.

We will introduce this notion with reference to a body B undergoing a deformation $\boldsymbol{\chi}$. Let p be a part of the body B and let $\boldsymbol{\chi}(p)$ be the deformation of p under $\boldsymbol{\chi}$. We shall postulate the presence of two kinds of forces acting on p in $\boldsymbol{\chi}(p)$:

- a body force $\mathbf{f}_b(p)$, or a volume force, caused, for example, by the gravity field or by electrostatic effects;

- a contact force $\mathbf{f}_c(p)$, due, for example, to the interaction of the body and the environment by means of the *external* boundary of the body or, more interestingly, caused by the interaction between different parts of the same body exerted through their common boundary.

We shall assume that $\mathbf{f}_b(p)$ is an absolutely continuous function of the volume of $\boldsymbol{\chi}(p)$, that is

$$\mathbf{f}_b(p) = \int_{\boldsymbol{\chi}(p)} \mathbf{b}\,dv, \tag{27}$$

where the function $\mathbf{b} : \boldsymbol{\chi}(\Omega) \to V$ is the body-force density (per unit volume). Moreover, also $\mathbf{f}_c(p)$ is assumed to be absolutely continuous function of the surface area of the boundary $\partial(\boldsymbol{\chi}(p))$ of $\boldsymbol{\chi}(p)$, that is

$$\mathbf{f}_c(p) = \int_{\partial\boldsymbol{\chi}(p)} \mathbf{t}\,da, \tag{28}$$

[7]The *trace* is the linear operator that assigns to each tensor $\mathbf{A}$ a scalar $\text{tr}(\mathbf{A})$ and satisfies

$$\text{tr}(\mathbf{a}\otimes\mathbf{b}) = \mathbf{a}\cdot\mathbf{b},$$

for all vectors $\mathbf{a}$ and $\mathbf{b}$.

where the function $\mathbf{t}$ defined in $\chi(\overline{\Omega})$ is the contact-force density (per unit area). $\mathbf{t}$ will be called *traction* in the sequel.

The *resultant force* $\mathbf{r}(p)$ acting on p in χ is given by

$$\mathbf{r}(p) = \mathbf{f}_b(p) + \mathbf{f}_c(p) = \int_{\chi(p)} \mathbf{b}\, dv + \int_{\partial\chi(p)} \mathbf{t}\, da. \tag{29}$$

The *resultant moment* $\mathbf{m}(p; \mathbf{x}_0)$ acting on p with respect to $\mathbf{x}_0$, $\mathbf{x}_0 \in \mathbb{R}^3$, is defined as

$$\mathbf{m}(p; \mathbf{x}_0) = \int_{\chi(p)} (\mathbf{x} - \mathbf{x}_0) \times \mathbf{b}\, dv + \int_{\partial\chi(p)} (\mathbf{x} - \mathbf{x}_0) \times \mathbf{t}\, da, \tag{30}$$

where we recall that $\mathbf{x} = \chi(\mathbf{X})$ denotes the actual placement of the particle $\mathbf{X} \in \boldsymbol{\kappa}(B)$. We have, for any $\mathbf{x}_0$, $\mathbf{x}_0' \in \mathbb{R}^3$,

$$\mathbf{m}(p; \mathbf{x}_0') = \mathbf{m}(p; \mathbf{x}_0) + (\mathbf{x}_0 - \mathbf{x}_0') \times \mathbf{r}(p) \tag{31}$$

and $\mathbf{m}(p; \mathbf{x}_0)$ does not depend on $\mathbf{x}_0$ if and only if $\mathbf{r}(p) = \mathbf{0}$.

Remark 3.1. The expression given for $\mathbf{m}(p; \mathbf{x}_0)$ in Eq. (30) implicitly assumes that the *contact-couple density* is identically equal to zero in $\chi(\Omega)$. This assumption can be weakened, with deep consequences on the remaining of the theory.

3.2 Euler's Laws

We can now formulate the *Euler's laws of equilibrium.* We shall say that the body B undergoing a deformation χ is in *equilibrium*, under a system of contact and body forces, if the resultant force and the resultant moment on each part of B vanish:

$$\mathbf{r}(p) = \mathbf{0} \quad \text{for every } \chi(p) \subset \chi(\overline{\Omega}) \quad \textit{(force balance)}, \tag{32}$$

$$\mathbf{m}(p; \mathbf{x}_0) = \mathbf{0} \quad \text{for every } \chi(p) \subset \chi(\overline{\Omega}) \quad \textit{(couple balance)}, \tag{33}$$

where $\mathbf{x}_0$ is a given point of $\mathbb{R}^3$. Under the above conditions, we say that χ is an *equilibrium deformation* and that $\chi\left(\overline{\Omega}\right)$ is an *equilibrium shape* for $\overline{\Omega}$.

3.3 The Euler-Cauchy Stress Principle

The body-force density $\mathbf{b}$ and the contact-force density $\mathbf{t}$ may have very general expression and they may depend on $\mathbf{x}$, p, B, etc. In the sequel we shall restrict our attention to the so-called *external* body forces

$$\mathbf{b} = \mathbf{b}(\mathbf{x}). \tag{34}$$

Concerning the contact-force density, we shall make the following assumptions. Let $\mathbf{x}$ be a point of $\chi(\Omega)$ and consider an oriented, regular surface Σ through $\mathbf{x}$, see Figure 9. Let $\mathbf{n}$ be the unit normal to Σ in $\mathbf{x}$. We indicate with $\mathbf{t}$ the contact-force density exerted on $\mathbf{x} \in \Sigma$ through Σ by the surrounding particles of the body which are on the portion of the body, adjacent to Σ, toward which the normal vector $\mathbf{n}$ points.

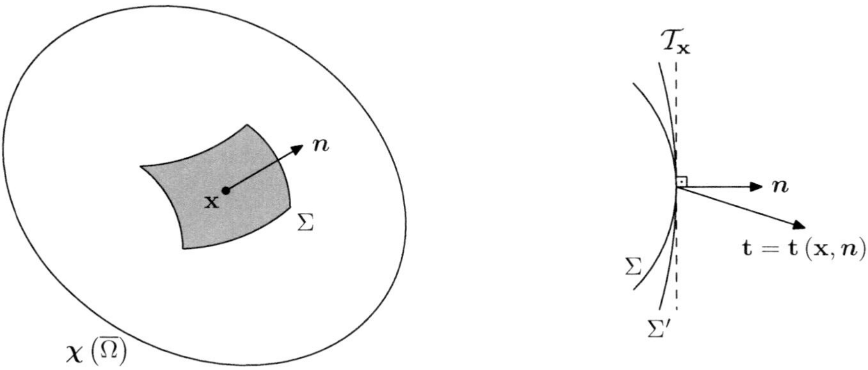

Figure 9. Traction on an oriented surface Σ through $\mathbf{x} \in \chi(\Omega)$.

We shall assume that the traction $\mathbf{t}$ at $\mathbf{x}$ has a common value for all those surfaces having a common tangent plane $\mathcal{T}_{\mathbf{x}}$ at $\mathbf{x}$, that is

$$\mathbf{t} = \mathbf{t}(\mathbf{x}, \mathbf{n}), \qquad \mathbf{t} : \chi(\overline{\Omega}) \times S_1 \to \mathbb{R}^3, \tag{35}$$

$S_1 = \{\mathbf{v} \in V \mid |\mathbf{v}| = 1\}$, see Figure 9. Such tractions are called *simple.* The notion of simple traction can be extended also to points $\mathbf{x}$ belonging to the *external* boundary of $\chi(\Omega)$. The assumption embodied in the expression of the contact-force $\mathbf{f}_c(p)$ and in Eq. (35) is the so-called *Euler-Cauchy Stress Principle.*

3.4 Consequences of the Euler-Cauchy Stress Principle

Cauchy's fundamental lemma

Assume that $\mathbf{b} \in C^0(\chi(\Omega))$ and $\mathbf{t}(\cdot, \mathbf{n}) \in C^0(\chi(\overline{\Omega}))$ (continuity with respect to the spatial variable). Then, the force balance equation (32) implies that there exists a continuous second-order tensor field $\mathbf{T}(\mathbf{x}) \in \mathrm{Lin}$, called *stress tensor* or *Cauchy tensor*, such that

$$\mathbf{t}(\mathbf{x}, \mathbf{n}) = \mathbf{T}(\mathbf{x})\mathbf{n} \qquad \text{in } \overline{\Omega}. \tag{36}$$

Equation (36) says that the traction $\mathbf{t}$ is in fact a linear homogeneous function of the unit normal $\mathbf{n}$ of the surface on which $\mathbf{t}$ is acting. More precisely, it can be shown that $\mathbf{T}(\mathbf{x})$ has the following representation

$$\mathbf{T}(\mathbf{x}) = \sum_{i=1}^{3} \mathbf{t}(\mathbf{x}, \mathbf{e}_i) \otimes \mathbf{e}_i, \tag{37}$$

where $\{\mathbf{e}_1, \mathbf{e}_2, \mathbf{e}_3\}$ is the usual orthonormal basis of $\mathbb{R}^3$ and $\mathbf{t}(\mathbf{x}, \mathbf{e}_i)$ is the traction acting on a plane surface through $\mathbf{x}$, with normal $\mathbf{e}_i$, $i = 1, 2, 3$.

Local form of equilibrium equations

Let the assumptions of the Cauchy's fundamental lemma be satisfied and let $\mathbf{T} \in C^1(\chi(\Omega)) \cap C^0(\partial\chi(\Omega))$. Then, the force balance equation (32) and the couple balance equation (33) imply that

$$\mathrm{div}\mathbf{T} + \mathbf{b} = \mathbf{0} \quad \text{in } \chi(\Omega)^8 \tag{38}$$

and

$$\mathbf{T} = \mathbf{T}^T \quad \text{in } \chi(\overline{\Omega}). \tag{39}$$

In fact, by the force balance equation and the Cauchy's representation of simple tractions, we have

$$\int_{\chi(p)} \mathbf{b}\, dv + \int_{\partial\chi(p)} \mathbf{T}\mathbf{n}\, da = \mathbf{0} \quad \text{for every } \chi(p) \subset \chi(\Omega).$$

By the Divergence Theorem (see Ex. 6 of Section 2.6) we can write

$$\int_{\chi(p)} (\mathbf{b} + \mathrm{div}\mathbf{T})\, dv = \mathbf{0} \quad \text{for every } \chi(p) \subset \chi(\Omega).$$

By the continuity of the integrand function and by the arbitrariness of $\chi(p)$, we obtain equation (38).

By the moment balance equation (33) and the Cauchy's Lemma, we have (let $\mathbf{x}_0 = \mathbf{0}$)

$$\int_{\chi(p)} \mathbf{W}\mathbf{b} dv + \int_{\partial\chi(p)} \mathbf{W}\mathbf{T}\mathbf{n} da = \mathbf{0} \quad \text{for every } \chi(p) \subset \chi(\Omega),$$

where $\mathbf{W} \in \mathrm{Skw} = \{\mathbf{A} \in \mathrm{Lin} \mid \mathbf{A}^T = -\mathbf{A}\}$ is the skew-symmetric tensor associated with the position vector $\mathbf{x}$, i.e. $\mathbf{W}\mathbf{v} = \mathbf{x} \times \mathbf{v}$ for every $\mathbf{v} \in V$.

[8]Note that $\mathrm{div}\mathbf{T}$ is calculated on $\chi(\Omega)$.

By the Divergence Theorem and by continuity of the integrand function we find

$$\mathbf{W}\mathbf{b} + \mathrm{div}(\mathbf{W}\mathbf{T}) = \mathbf{0} \qquad \text{in } \chi(\Omega). \tag{40}$$

We now compute $\mathrm{div}(\mathbf{W}\mathbf{T})$. Let $\mathbf{a}$ any constant vector. By the definition of divergence of a tensor field, we have

$$\mathrm{div}(\mathbf{W}\mathbf{T}) \cdot \mathbf{a} = \mathrm{div}(\mathbf{T}^T\mathbf{W}^T\mathbf{a})^9 = \mathbf{W}\mathrm{div}\mathbf{T} \cdot \mathbf{a} + \mathbf{T} \cdot \nabla(\mathbf{W}^T\mathbf{a}).$$

Since $\mathbf{W} \in \mathrm{Skw}$, $\mathbf{W}^T\mathbf{a} = \mathbf{a} \times \mathbf{x} = \mathbf{A}\mathbf{x}$, where $\mathbf{A}$ is the constant skew-symmetric tensor associated to $\mathbf{a}$. Therefore:

$$\nabla(\mathbf{W}^T\mathbf{a}) = \nabla(\mathbf{A}\mathbf{x}) = \mathbf{A}. \tag{41}$$

Finally, by multiplying the differential equation (40) by $\mathbf{a}$ and recalling (41), we have

$$\mathbf{W}(\mathbf{b} + \mathrm{div}\mathbf{T}) \cdot \mathbf{a} + \mathbf{T} \cdot \mathbf{A} = 0 \quad \text{in } \chi(\Omega),$$

for every constant vector $\mathbf{a}$. By the force balance equation and by the arbitrariness of $\mathbf{A}$, $\mathbf{A} \in \mathrm{Skw}$, the thesis follows.

3.5 Boundary conditions of traction and equilibrium problem

An appropriate boundary condition of traction is difficult to formulate. We recall that our main goal is to find the deformation caused by applying given forces to a given body in a reference configuration $\boldsymbol{\kappa}$. We wish to specify the tractions acting on $\partial\chi(\Omega)$ but this configuration in unknown. Even if we try to assign the tractions on $\partial\boldsymbol{\kappa}(\Omega)$, the deformation produced will move the point of application and deform the surface, and still we find difficulties to impose given forces.

Therefore, it is customarily assumed that the boundary condition of traction in the actual state takes the form

$$\mathbf{t}(\mathbf{x}, \mathbf{n}) = \text{ given function of } \mathbf{x} \quad \text{on } \partial\chi(\Omega),$$

that is

$$\mathbf{T}(\mathbf{x})\mathbf{n}(\mathbf{x}) = \boldsymbol{\varphi}(\mathbf{x}) \quad \text{on } \partial\chi(\Omega), \tag{42}$$

where $\boldsymbol{\varphi} : \partial\chi(\Omega) \to V$ is a (regular) given surface traction field.

[9] The space Lin has a natural inner product $\mathbf{A} \cdot \mathbf{B} = \mathrm{tr}(\mathbf{A}^T\mathbf{B})$, which in cartesian components reads as $\mathbf{A} \cdot \mathbf{B} = \sum_{i,j=1}^{3} A_{ij}B_{ij}$. Let $\mathbf{v}(\mathbf{x})$ and $\mathbf{A}(\mathbf{x})$ a regular vector field and a regular tensor field defined on $\chi(\Omega)$, respectively. We have: $\mathrm{div}(\mathbf{A}^T\mathbf{v}) = \mathrm{div}\mathbf{A} \cdot \mathbf{v} + \mathbf{A} \cdot \nabla\mathbf{v}$.

Summarizing, the equilibrium of the body B in configuration χ is described by the following boundary value problem on the stress tensor $\mathbf{T} \in C^1(\chi(\Omega)) \cap C^0(\partial\chi(\Omega))$:

$$\begin{cases} \operatorname{div}\mathbf{T} + \mathbf{b} = \mathbf{0} & \text{in } \chi(\Omega), \\ \mathbf{T} = \mathbf{T}^T & \text{in } \chi(\overline{\Omega}), \\ \mathbf{T}\mathbf{n} = \boldsymbol{\varphi} & \text{on } \partial\chi(\Omega), \end{cases} \tag{43}$$

where $\mathbf{b} \in C^0(\chi(\Omega))$ and $\boldsymbol{\varphi} \in C^0(\partial\chi(\Omega))$ are the given body-force and surface-traction densities and $\mathbf{n}$ is the unit outer normal to $\partial\chi(\Omega)$. Note that $\boldsymbol{\varphi}$ and $\mathbf{b}$ must satisfy the necessary global equilibrium conditions $\mathbf{r}(\Omega) = \mathbf{0}$ and $\mathbf{m}(\Omega; \mathbf{x}_0) = \mathbf{0}$ for some $\mathbf{x}_0 \in \mathbb{R}^3$.

Problem (43) is largely undetermined and, accordingly, can have many solutions.

3.6 Virtual work theorem

There are situations in which it is useful to rewrite the equilibrium problem (43) in an integral form. Let $\mathbf{v} \in C^1(\chi(\overline{\Omega}))$ be a vector field on $\chi(\overline{\Omega})$. Scalar multiplication of the differential equilibrium equation by $\mathbf{v}$ and integration over $\chi(\overline{\Omega})$ yield

$$\int_{\chi(\Omega)} (\operatorname{div}\mathbf{T} \cdot \mathbf{v} + \mathbf{b} \cdot \mathbf{v})dv = 0.$$

Recalling that $\operatorname{div}(\mathbf{T}^T\mathbf{v}) = \operatorname{div}\mathbf{T} \cdot \mathbf{v} + \mathbf{T} \cdot \nabla\mathbf{v}$, by Divergence Theorem, by the boundary condition on $\partial\chi(\Omega)$ and by $\mathbf{T} = \mathbf{T}^T$, we have

$$\int_{\chi(\Omega)} \mathbf{T} \cdot \nabla\mathbf{v} dv = \int_{\chi(\Omega)} \mathbf{b} \cdot \mathbf{v} dv + \int_{\partial\chi(\Omega)} \boldsymbol{\varphi} \cdot \mathbf{v} da \quad \text{for every } \mathbf{v} \in C^1(\chi(\Omega)), \tag{44}$$

which is the usual form of the so-called *Virtual Work Theorem* on the equilibrium shape $\chi(\Omega)$ ($(\text{VWT})_\chi$). The field $\mathbf{v}$ is usually called *virtual displacement field.*

By using the arbitrariness of $\mathbf{v}$ and assuming enough regularity of the fields involved in (44), one can prove that (44) implies the equilibrium equations (43).

3.7 Reference description

A disadvantage of the equilibrium equations (43) over the deformed configuration (or, equivalently, of the $(\text{VWT})_\chi$ in equation (44)) is that the equilibrium problem is formulated in terms of the unknown actual placement $\chi(\Omega)$ of the body. To circumvent this difficulty, one can rewrite these equations in lagrangian form, that is in terms of the reference configuration $\overline{\Omega}$ of the body. This transformation involves the introduction of the so-called *Piola-Kirchhoff stress tensor.*

Let us assume that the body B, represented by the domain $\overline{\Omega}$ in its reference configuration $\boldsymbol{\kappa}$, is in equilibrium under the body forces $\mathbf{b}$ and the surface forces φ in the deformed configuration $\boldsymbol{\chi}(\Omega)$. We introduce the (regular) change of variables $\mathbf{x} = \boldsymbol{\chi}(\mathbf{X})$ within the $(\text{VWT})_{\boldsymbol{\chi}}$ and we define

$$\mathbf{w}(\mathbf{X}) = (\mathbf{v} \circ \boldsymbol{\chi})(\mathbf{X}).$$

Then, the $(\text{VWT})_{\boldsymbol{\chi}}$ expression becomes

$$\int_\Omega \nabla\mathbf{w} \cdot \mathbf{T} \quad ((\frac{\partial \boldsymbol{\chi}}{\partial \mathbf{X}})^{-1})^T \det \frac{\partial \boldsymbol{\chi}}{\partial \mathbf{X}}\, dV = \\ = \int_\Omega \mathbf{w} \cdot \mathbf{b}(\boldsymbol{\chi}(\mathbf{X})) \det \frac{\partial \boldsymbol{\chi}}{\partial \mathbf{X}}\, dV + \int_{\partial\Omega} \mathbf{w} \cdot \varphi(\boldsymbol{\chi}(\mathbf{X}))(\frac{da}{dA})dA\ ,$$

or, using the standard notation $\frac{\partial \boldsymbol{\chi}}{\partial \mathbf{X}} = \mathbf{F}$,

$$\int_\Omega \nabla\mathbf{w} \cdot \mathbf{T} \quad (\mathbf{F}^{-1})^T \det \mathbf{F}\, dV = \\ = \int_\Omega \mathbf{w} \cdot \mathbf{b}(\boldsymbol{\chi}(\mathbf{X})) \det \mathbf{F}\, dV + \int_{\partial\Omega} \mathbf{w} \cdot \varphi(\boldsymbol{\chi}(\mathbf{X}))(\frac{da}{dA})dA\ . \tag{45}$$

Define

$$\mathbf{S}(\mathbf{X}) = \mathbf{T}(\boldsymbol{\chi}(\mathbf{X}))(\mathbf{F}^{-1})^T \det \mathbf{F}, \tag{46}$$

$$\mathbf{b}_{\boldsymbol{\kappa}}(\mathbf{X}) = \mathbf{b}(\boldsymbol{\chi}(\mathbf{X})) \det \mathbf{F}, \tag{47}$$

$$\varphi_{\boldsymbol{\kappa}}(\mathbf{X}) = \varphi(\boldsymbol{\chi}(\mathbf{X}))\frac{da}{dA}\ . \tag{48}$$

Then, Eq. (45) can be written as

$$\int_\Omega \nabla\mathbf{w} \cdot \mathbf{S}\, dV = \int_\Omega \mathbf{w} \cdot \mathbf{b}_{\mathbf{k}}\, dV + \int_{\partial\Omega} \mathbf{w} \cdot \varphi_{\mathbf{k}}\, dA, \tag{49}$$

which expresses the Virtual Work Theorem written on the reference configuration $\boldsymbol{\kappa}$ from which the deformation $\boldsymbol{\chi}$ is applied $((\text{VWT})_{\boldsymbol{\kappa}})$.

The body-force density $\mathbf{b}_{\boldsymbol{\kappa}}$, $\mathbf{b}_{\boldsymbol{\kappa}} : \Omega \to \mathbb{R}^3$, measures the applied body force per unit volume in the reference configuration, and it depends on the deformation $\boldsymbol{\chi}$. The vector field $\varphi_{\boldsymbol{\kappa}}$ is the density of the applied surface force per unit area on the deformation $\boldsymbol{\chi}$. The tensor $\mathbf{S}$ appearing in Eq. (46) is called the *Piola-Kirchhoff stress tensor.*

It is worth noticing that while $\mathbf{T}$ and $\mathbf{S}$ serve to determine the traction upon every part of the body, neither determines the other unless the deformation gradient $\mathbf{F}$ from $\boldsymbol{\kappa}$ to the present configuration $\boldsymbol{\chi}$ be specified. The Cauchy stress tensor $\mathbf{T}$ gives the present tractions alone and completely on $\boldsymbol{\chi}(\overline{\Omega})$. The Piola-Kirchhoff stress tensor $\mathbf{S}$ delivers those same tractions only if $\mathbf{F}$ is explicitly specified.

There is another, perhaps more expressive, way to introduce $\mathbf{S}$. Making reference to Figure 10, let $d\mathbf{A}$ be a small plane portion of surface through $\mathbf{X}$, $\mathbf{X} \in \Omega$, of normal $\mathbf{n}_{\boldsymbol{\kappa}}$, and denote by $d\mathbf{a}$ its image under the deformation $\boldsymbol{\chi}$. The vector $\mathbf{t}_{\boldsymbol{\kappa}}$ is simply the density of the contact force acting on $d\mathbf{a}$

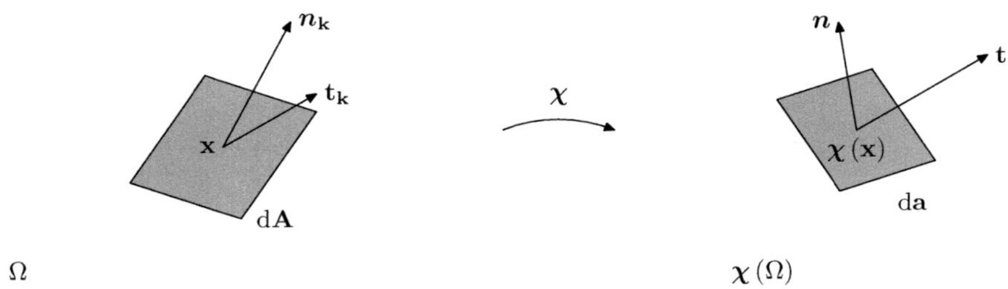

Figure 10. Actual, $\mathbf{t}$, and referential, $\mathbf{t_k}$, contact force density.

expressed in terms of the reference area dA, that is

$$\mathbf{t}_{\boldsymbol{\kappa}} = \mathbf{t}\frac{da}{dA}. \tag{50}$$

By Cauchy's fundamental lemma applied on the reference configuration there exists a tensor $\mathbf{S} = \mathbf{S}(\mathbf{X})$ such that

$$\mathbf{t}_{\boldsymbol{k}} = \mathbf{S}\mathbf{n}_{\boldsymbol{k}} = \mathbf{T}\mathbf{n}\frac{da}{dA}, \tag{51}$$

where $\mathbf{n}_{\boldsymbol{k}}$ is the unit outer normal at $\mathbf{X}$ in $\boldsymbol{\kappa}$. We know that (see Ex. 2, Section 2.6)

$$\mathbf{n} = (\mathbf{F}^{-1})^T \mathbf{n}_{\boldsymbol{\kappa}} \det \mathbf{F}\frac{dA}{da}. \tag{52}$$

Therefore

$$\mathbf{S}\mathbf{n}_{\boldsymbol{\kappa}} = \mathbf{T}(\mathbf{F}^{-1})^T \mathbf{n}_{\boldsymbol{\kappa}} \det \mathbf{F} \tag{53}$$

and we obtain the expression (46) of $\mathbf{S}$.

Finally, we conclude with the reference form of the equilibrium equations. By the $(\mathrm{VWT})_{\boldsymbol{\kappa}}$ (equation (49)), under the usual regularity assumptions on the quantities involved and by the arbitrariness of $\mathbf{w}$ we obtain

$$\begin{cases} \mathrm{Div}\mathbf{S} + \mathbf{b}_{\boldsymbol{\kappa}} = \mathbf{0} & \text{in } \Omega, \\ \mathbf{S}\mathbf{n}_{\boldsymbol{\kappa}} = \varphi_{\boldsymbol{\kappa}} & \text{on } \partial\Omega, \\ \mathbf{S}\mathbf{F}^T = (\mathbf{S}\mathbf{F}^T)^T & \text{in } \overline{\Omega}. \end{cases} \tag{54}$$

Note that $\mathbf{S}$ is not, in general, a symmetric tensor and that $\mathrm{Div}\mathbf{S}$ means divergence of the tensor field $\mathbf{S}$ with respect to the $\mathbf{X}$ variable.

3.8 Exercises

1. Show that

$$\begin{cases} \mathbf{r}(p) = \mathbf{0}, \\ \mathbf{m}(p;\mathbf{x}_0) = \mathbf{0} & \text{for every } \chi(p) \subset \chi(\overline{\Omega}), \end{cases}$$

if and only if the $(\mathrm{VWT})_{\boldsymbol{\chi}}$ holds for all sub-domains of the body (in $\boldsymbol{\chi}(\overline{\boldsymbol{\Omega}})$) and all vector fields $\mathbf{r} : \chi(\overline{\Omega}) \to \mathbb{R}^3$ of the form $\mathbf{r}(\mathbf{x}) = \mathbf{a}+\mathbf{b}\times\mathbf{x}$, where $\mathbf{a}$ and $\mathbf{b}$ are constant vectors. The field $\mathbf{r}$ is called *infinitesimal rigid displacement field.*

2. Prove that if $\mathbf{S} \in \mathrm{Sym}$ and $\mathbf{W} \in \mathrm{Skw}$, then $\mathbf{S}\cdot\mathbf{W} = 0$. Show that if $\mathbf{T}\cdot\mathbf{W} = 0$ for every $\mathbf{W} \in \mathrm{Skw}$, then $\mathbf{T} \in \mathrm{Sym}$.

3. Prove that $\mathbf{t}(\mathbf{x}, -\mathbf{n}) = -\mathbf{t}(\mathbf{x}, \mathbf{n})$, where $\mathbf{n}$ is the unit normal to a regular surface Σ through $\mathbf{x}$, $\Sigma \subset \chi(\overline{\Omega})$ (Cauchy's Reciprocal Theorem).

4. Prove the Cauchy's fundamental lemma (36).

5. Prove that the $(\mathrm{VWT})_{\boldsymbol{\chi}}$ of Eq. (44) implies the equilibrium equations (43) (for regular enough $\mathbf{T}$).

4 Linearized Elasticity

In this section we shall formulate the equilibrium problem for a body made of hyperelastic material in the case of infinitesimal deformation. Then, we conclude with some existence and uniqueness results for linearized elasticity.

Let the reference configuration of a continuous body be represented by a bounded domain $\Omega = \boldsymbol{\kappa}(B)$ with smooth boundary $\partial\Omega$. Assume that $\partial_t\Omega$, $\partial_d\Omega$ are two subsets of $\partial\Omega$ such that $\partial_t\Omega \cup \partial_d\Omega = \partial\Omega$ and $\partial_t\Omega \cap \partial_d\Omega = \emptyset$. Kinematical conditions are imposed on $\partial_d\Omega$, e.g. $\mathbf{u} = \mathbf{g}$ with $\mathbf{g}$ smooth, whereas a smooth vector force field $\varphi_{\boldsymbol{\kappa}}$ is assigned on $\partial_t\Omega$. A field of volume loads $\mathbf{b}_{\boldsymbol{\kappa}}$ is prescribed over Ω.

Our starting point is the differential form of the equilibrium problem in finite elasticity (see Paroni (2008))

$$\begin{cases} \text{Div}\mathbf{S} + \mathbf{b}_{\boldsymbol{\kappa}} = \mathbf{0} & \text{in } \Omega, \\ \mathbf{S}\mathbf{n}_{\boldsymbol{\kappa}} = \boldsymbol{\varphi}_{\boldsymbol{\kappa}} & \text{on } \partial_t\Omega, \\ \mathbf{u} = \mathbf{g} & \text{on } \partial_d\Omega, \\ \mathbf{S}\mathbf{F}^T = (\mathbf{S}\mathbf{F}^T)^T & \text{in } \overline{\Omega}, \\ \mathbf{S} = \mathcal{F}_{\boldsymbol{\kappa}}(\mathbf{X},\mathbf{F}) = D_{\mathbf{F}}w_{\boldsymbol{\kappa}}(\mathbf{X},\mathbf{F}) & \text{in } \overline{\Omega}, \end{cases} \tag{55}$$

where $w_{\boldsymbol{\kappa}} : \overline{\Omega} \times \text{Lin}^+ \to \mathbb{R}$ is the stored-energy function of the hyperelastic material, $\mathbf{u}(\mathbf{X}) = \boldsymbol{\chi}(\mathbf{X}) - \mathbf{X}$ is the displacement field, $\mathbf{F} = \nabla\boldsymbol{\chi}$, $\nabla\mathbf{u} := \mathbf{H} = \mathbf{F} - \mathbf{1}$ and $\mathbf{n}_{\boldsymbol{\kappa}}$ denotes the unit outward normal to $\partial\Omega$.

We recall that the stored-energy function $w_{\boldsymbol{\kappa}}$ satisfies the *Principle of Material Frame-Indifference* (see Paroni (2008))

$$w_{\boldsymbol{\kappa}}(\mathbf{X},\mathbf{F}) = w_{\boldsymbol{\kappa}}(\mathbf{X},\mathbf{QF}) \quad \text{for every } \mathbf{F} \in \text{Lin}^+ \text{ and for every } \mathbf{Q} \in \text{Orth}^+ \ . \tag{56}$$

4.1 Infinitesimal deformation

Let us consider the case when the displacement field and its gradient are small in some sense. More precisely, corresponding to any given displacement field $\widetilde{\mathbf{u}}(\mathbf{X})$ consider the family of displacements

$$\mathbf{u}_\varepsilon(\mathbf{X}) = \varepsilon\widetilde{\mathbf{u}}(\mathbf{X}) \qquad \text{as } \varepsilon \to 0, \tag{57}$$

and denote quantities defined from it with a subscript ε. Here, ε is an adimensional parameter and $O(\varepsilon^2)$ denotes a term of order ε^2. The value of $O(\varepsilon^2)$ may change from line to line in the sequel. We have

$$\mathbf{H}_\varepsilon = \nabla(\varepsilon\widetilde{\mathbf{u}}) = \varepsilon\widetilde{\mathbf{H}}, \tag{58}$$

with $\widetilde{\mathbf{H}} = \nabla\widetilde{\mathbf{u}}$, and

$$\mathbf{C}_\varepsilon = \mathbf{F}_\varepsilon^T\mathbf{F}_\varepsilon = \mathbf{1} + \varepsilon(\widetilde{\mathbf{H}} + \widetilde{\mathbf{H}}^T) + O(\varepsilon^2). \tag{59}$$

Then

$$\mathbf{U}_\varepsilon = \mathbf{1} + \frac{\varepsilon}{2}(\widetilde{\mathbf{H}} + \widetilde{\mathbf{H}}^T) + O(\varepsilon^2) \tag{60}$$

and from the polar decomposition theorem $\mathbf{F}_\varepsilon = \mathbf{R}_\varepsilon\mathbf{U}_\varepsilon$ we have

$$\mathbf{R}_\varepsilon = \mathbf{1} + \frac{\varepsilon}{2}(\widetilde{\mathbf{H}} - \widetilde{\mathbf{H}}^T) + O(\varepsilon^2). \tag{61}$$

Define

$$\frac{1}{2}(\widetilde{\mathbf{H}} + \widetilde{\mathbf{H}}^T) = \widetilde{\mathbf{E}} \in \text{Sym}, \tag{62}$$

$$\frac{1}{2}(\widetilde{\mathbf{H}} - \widetilde{\mathbf{H}}^T) = \widetilde{\mathbf{W}} \in \mathrm{Skw}. \tag{63}$$

To within $O(\varepsilon^2)$ terms, we have

$$\mathbf{U}_\varepsilon = \mathbf{1} + \mathbf{E}_\varepsilon, \quad \mathbf{R}_\varepsilon = \mathbf{1} + \mathbf{W}_\varepsilon, \tag{64}$$

where

$$\mathbf{E}_\varepsilon = \varepsilon\widetilde{\mathbf{E}}, \quad \mathbf{W}_\varepsilon = \varepsilon\widetilde{\mathbf{W}}, \tag{65}$$

are called the *infinitesimal strain* and *infinitesimal rotation*, respectively.

The above results show that when ε is sufficiently small, the exact polar decomposition $\mathbf{F}_\varepsilon = \mathbf{R}_\varepsilon\mathbf{U}_\varepsilon$ is replaced by the additional decomposition of $\mathbf{H}_\varepsilon$ into its symmetric and skew parts: $\mathbf{H}_\varepsilon = \mathbf{E}_\varepsilon + \mathbf{W}_\varepsilon$.

A theory in which only the terms of lowest non-trivial order in ε are retained is described as a theory of *infinitesimal deformations.*

For infinitesimal deformation theory, $\mathbf{E}_\varepsilon$ determines the change in length of a fiber, the change in area, the change in volume and the change in angle between two fibers. In fact, by using the notation of the section 2.4, the local *change in length* can be evaluated as

$$\begin{aligned}\frac{ds}{dS} &= \sqrt{\mathbf{C}_\varepsilon\mathbf{f}\cdot\mathbf{f}} = \sqrt{(\mathbf{1} + 2\mathbf{E}_\varepsilon + O(\varepsilon^2))\mathbf{f}\cdot\mathbf{f}} = \\ &= 1 + \mathbf{E}_\varepsilon\mathbf{f}\cdot\mathbf{f} + O(\varepsilon^2),\end{aligned}$$

so that

$$\frac{ds - dS}{dS} = \mathbf{E}_\varepsilon\mathbf{f}\cdot\mathbf{f}, \tag{66}$$

to within terms of order $O(\varepsilon^2)$, compare with Eq. (20).

As to the *change in area*, by the exact formula (21) and under the assumption $\mathbf{f}_1\cdot\mathbf{f}_2 = 0$, $|\mathbf{f}_i| = 1$, $i = 1, 2$, we have

$$\frac{da}{dA} = 1 + \mathbf{E}_\varepsilon\mathbf{f}_1\cdot\mathbf{f}_1 + \mathbf{E}_\varepsilon\mathbf{f}_2\cdot\mathbf{f}_2 + O(\varepsilon^2), \tag{67}$$

that is, to within terms of order $O(\varepsilon^2)$,

$$\frac{da - dA}{dA} = \mathbf{E}_\varepsilon\mathbf{f}_1\cdot\mathbf{f}_1 + \mathbf{E}_\varepsilon\mathbf{f}_2\cdot\mathbf{f}_2. \tag{68}$$

Similarly, by linearization of Eq. (24), the *change in angle* of two orthogonal fibers $d\mathbf{X}_\alpha = dS_\alpha\mathbf{f}_\alpha$, $\alpha = 1, 2$, is given by

$$\gamma_{(\mathbf{f}_1,\mathbf{f}_2)} = 2\mathbf{E}_\epsilon\mathbf{f}_1\cdot\mathbf{f}_2\ . \tag{69}$$

Concerning the *change of volume*, let $\{\mathbf{e}_i\}_{i=1}^3$ be a positively oriented triplet such that $\mathbf{e}_i \cdot \mathbf{e}_j = \delta_{ij}$. By the polar decomposition theorem we have $\det \mathbf{F}_\varepsilon = \det(\mathbf{R}_\varepsilon \mathbf{U}_\varepsilon) = \det \mathbf{R}_\varepsilon \det \mathbf{U}_\varepsilon = \det \mathbf{U}_\varepsilon$. Recalling the definition of determinant we obtain

$$\begin{aligned}\det \mathbf{U}_\varepsilon &\simeq \det(\mathbf{1} + \mathbf{E}_\varepsilon) = (\mathbf{1} + \mathbf{E}_\varepsilon)\,\mathbf{e}_1 \times (\mathbf{1} + \mathbf{E}_\varepsilon)\,\mathbf{e}_2 \cdot (\mathbf{1} + \mathbf{E}_\varepsilon)\,\mathbf{e}_3 \\ &\simeq 1 + \mathbf{E}_\varepsilon \cdot \sum_{i=1}^{3} \mathbf{e}_i \otimes \mathbf{e}_i = 1 + \mathrm{tr}\mathbf{E}_\varepsilon.\end{aligned}$$

Therefore, the change in volume is determined by the trace of $\mathbf{E}_\varepsilon$:

$$\frac{dv - dV}{dV} = \mathrm{tr}\mathbf{E}_\varepsilon. \tag{70}$$

4.2 Constitutive equations for infinitesimal deformation

Aim of this section is to linearize the condition $(55)_5$ for infinitesimal deformations.

To simplify the notation, in this part we omit to indicate the dependance on $\mathbf{X}$ and we write w instead of $w_{\boldsymbol{\kappa}}$. By assuming w regular enough, one can prove the following two identities:

$$Dw(\mathbf{F}) = \mathbf{Q}^T Dw(\mathbf{QF}) \quad \text{for every } \mathbf{Q} \in \mathrm{Orth} \text{ and for every } \mathbf{F} \in \mathrm{Lin}^+, \tag{71}$$

$$D^2 w(\mathbf{F})\mathbf{WF} = \mathbf{W} Dw(\mathbf{F}) \quad \text{for every } \mathbf{W} \in \mathrm{Skw} \text{ and for every } \mathbf{F} \in \mathrm{Lin}^+. \tag{72}$$

Here, given a function $f = f(z)$, we denote by $Df(z_0)$ the derivative of f evaluated at $z = z_0$, that is $Df(z_0) = \left.\frac{\partial f(z)}{\partial z}\right|_{z=z_0}$. A proof of (71), (72) will be given at the end of this section.

We formally develop the function w by Taylor series for the family of displacements $\varepsilon\widetilde{\mathbf{u}}$ as $\varepsilon \to 0$. We have:

$$w(\mathbf{F}_\varepsilon) = w(\mathbf{1} + \varepsilon\widetilde{\mathbf{H}}) = w(\mathbf{1}) + \varepsilon Dw(\mathbf{1}) \cdot \widetilde{\mathbf{H}} + \frac{\varepsilon^2}{2} D^2 w(\mathbf{1})\widetilde{\mathbf{H}} \cdot \widetilde{\mathbf{H}} + o(\varepsilon^2), \tag{73}$$

where $o(\varepsilon^2)$ is a term of higher order with respect to ε^2, e.g. $\lim_{\varepsilon \to 0} o(\varepsilon^2)\varepsilon^{-2} = 0$.

We shall denote by

$$\mathring{\mathbf{T}} = Dw(\mathbf{1}) = \mathcal{F}(\mathbf{1}) \tag{74}$$

the *initial stress tensor*, that is the stress tensor within the body when it occupies the reference configuration. Since $\mathbf{S}(\mathbf{F}) = \mathbf{T}(\mathbf{F}^{-1})^T \det \mathbf{F}$ and $\mathbf{S}(\mathbf{1}) = \mathring{\mathbf{T}}$, we have

$$\mathring{\mathbf{T}} \in \mathrm{Sym}. \tag{75}$$

Now, from the symmetry of the second order derivatives of w

$$\left(D^2 w\right)_{ijrs} = \frac{\partial}{\partial F_{ij}}\left(\frac{\partial w}{\partial F_{rs}}\right) = \frac{\partial}{\partial F_{rs}}\left(\frac{\partial w}{\partial F_{ij}}\right) = \left(D^2 w\right)_{rsij},$$

it turns out that $D^2 w(\mathbf{1})$ is a symmetric fourth-order tensor.[10]

Concerning the $O(\varepsilon^2)$-term in the Taylor development (73), by using the identity (72) with $\mathbf{F} = \mathbf{1}$ and $\mathbf{W} = \widetilde{\mathbf{W}}$, we have

$$\begin{aligned} & D^2 w(\mathbf{1})\widetilde{\mathbf{H}} \cdot \widetilde{\mathbf{H}} = D^2 w(\mathbf{1})(\widetilde{\mathbf{W}} + \widetilde{\mathbf{E}}) \cdot \widetilde{\mathbf{H}} = \\ & = \widetilde{\mathbf{W}}\mathring{\mathbf{T}} \cdot \widetilde{\mathbf{H}} + \widetilde{\mathbf{E}} \cdot \widetilde{\mathbf{W}}\mathring{\mathbf{T}} + D^2 w(\mathbf{1})\widetilde{\mathbf{E}} \cdot \widetilde{\mathbf{E}} = \\ & = \widetilde{\mathbf{H}}\mathring{\mathbf{T}} \cdot \widetilde{\mathbf{H}} - \widetilde{\mathbf{E}}\mathring{\mathbf{T}} \cdot \widetilde{\mathbf{H}} + \widetilde{\mathbf{E}} \cdot \widetilde{\mathbf{W}}\mathring{\mathbf{T}} + D^2 w(\mathbf{1})\widetilde{\mathbf{E}} \cdot \widetilde{\mathbf{E}} = \\ & = \widetilde{\mathbf{H}}\mathring{\mathbf{T}} \cdot \widetilde{\mathbf{H}} - \widetilde{\mathbf{E}}\mathring{\mathbf{T}} \cdot \widetilde{\mathbf{E}} + D^2 w(\mathbf{1})\widetilde{\mathbf{E}} \cdot \widetilde{\mathbf{E}}, \end{aligned} \tag{76}$$

where in the last line the identity $\mathbf{A} \cdot \mathbf{BC} = \mathbf{AC}^T \cdot \mathbf{B}$, for all $\mathbf{A}$, $\mathbf{B}$, $\mathbf{C} \in$ Lin, has been used. Therefore, we have

$$\begin{aligned} w(\mathbf{F}_\varepsilon) & = w(\mathbf{1}) + \varepsilon \mathring{\mathbf{T}} \cdot \widetilde{\mathbf{H}} + \frac{\varepsilon^2}{2}\widetilde{\mathbf{H}}\mathring{\mathbf{T}} \cdot \widetilde{\mathbf{H}} + \\ & + \frac{\varepsilon^2}{2}(D^2 w(\mathbf{1})\widetilde{\mathbf{E}} \cdot \widetilde{\mathbf{E}} - \widetilde{\mathbf{E}}\mathring{\mathbf{T}} \cdot \widetilde{\mathbf{E}}) + o(\varepsilon^2). \end{aligned} \tag{77}$$

We can now compute the linearized constitutive equation for the Piola-Kirchhoff tensor for small ε. By Eq. (77) we have

$$\begin{aligned} \mathbf{S} & = \mathcal{F}_{\boldsymbol{\kappa}}(\mathbf{F}_\varepsilon) = D_{\mathbf{F}_\varepsilon} w(\mathbf{F}_\varepsilon) = D_{\mathbf{H}_\varepsilon} w(\mathbf{1} + \mathbf{H}_\varepsilon) = \\ & = \mathring{\mathbf{T}} + \varepsilon \widetilde{\mathbf{H}}\mathring{\mathbf{T}} + \frac{\varepsilon^2}{2} D_{\mathbf{E}_\varepsilon}(D^2 w(\mathbf{1})\widetilde{\mathbf{E}} \cdot \widetilde{\mathbf{E}} - \widetilde{\mathbf{E}}\mathring{\mathbf{T}} \cdot \widetilde{\mathbf{E}}) + o(\varepsilon) = \\ & = \mathring{\mathbf{T}} + \varepsilon \widetilde{\mathbf{H}}\mathring{\mathbf{T}} + \varepsilon D^2 w(\mathbf{1})\widetilde{\mathbf{E}} - \varepsilon \widetilde{\mathbf{E}}\mathring{\mathbf{T}} + o(\varepsilon) = \\ & = \mathring{\mathbf{T}} + \varepsilon \widetilde{\mathbf{W}}\mathring{\mathbf{T}} + \varepsilon D^2 w(\mathbf{1})\widetilde{\mathbf{E}} + o(\varepsilon), \quad \text{as } \varepsilon \to 0. \end{aligned} \tag{78}$$

[10]We shall denote by $\mathbb{L}\mathrm{in}$ the linear space of all fourth-order tensors, that is the linear transformations of Lin into itself. The transpose of $\mathbb{A} \in \mathbb{L}\mathrm{in}$ is the element $\mathbb{A}^T \in \mathbb{L}\mathrm{in}$ defined by $\mathbb{A}\mathbf{A} \cdot \mathbf{B} = \mathbf{A} \cdot \mathbb{A}^T\mathbf{B}$ for all $\mathbf{A}$, $\mathbf{B} \in$Lin. $\mathbb{A} \in \mathbb{L}\mathrm{in}$ is symmetric if $\mathbb{A}^T = \mathbb{A}$. Given an orthonormal basis $\{\mathbf{e}_i\}_{i=1}^3$ of V, the components A_{ijkl} of $\mathbb{A} \in \mathbb{L}\mathrm{in}$ with respect to this basis can be calculated as

$$A_{ijkl} = \mathbf{e}_i \otimes \mathbf{e}_j \cdot \mathbb{A}(\mathbf{e}_k \otimes \mathbf{e}_l).$$

The symmetry (or *major symmetry*) of $\mathbb{A} \in \mathbb{L}\mathrm{in}$ can be expressed in components as $A_{ijkl} = A_{klij}$, $i, j, k, l = 1, 2, 3$.

Recalling Eq. (46), the Cauchy stress tensor is equal to

$$\begin{aligned}\mathbf{T} &= (1+\varepsilon\mathrm{tr}\widetilde{\mathbf{E}}+o(\varepsilon))^{-1}(\mathring{\mathbf{T}}+\varepsilon\widetilde{\mathbf{W}}\mathring{\mathbf{T}}+\varepsilon D^2w(\mathbf{1})\widetilde{\mathbf{E}}+o(\varepsilon))(\mathbf{1}+\mathbf{H}_\varepsilon)^T = \\ &= (1-\varepsilon\mathrm{tr}\widetilde{\mathbf{E}}+o(\varepsilon))(\mathring{\mathbf{T}}+\varepsilon\widetilde{\mathbf{W}}\mathring{\mathbf{T}}+\varepsilon D^2w(\mathbf{1})\widetilde{\mathbf{E}}+o(\varepsilon))(\mathbf{1}+\mathbf{H}_\varepsilon^T) = \\ &= (\mathring{\mathbf{T}}+\varepsilon\widetilde{\mathbf{W}}\mathring{\mathbf{T}}+\varepsilon D^2w(\mathbf{1})\widetilde{\mathbf{E}}-\varepsilon(\mathrm{tr}\widetilde{\mathbf{E}})\mathring{\mathbf{T}})(\mathbf{1}+\mathbf{H}_\varepsilon^T)+o(\varepsilon) = \\ &= \mathring{\mathbf{T}}+\varepsilon\mathring{\mathbf{T}}\widetilde{\mathbf{H}}^T+\varepsilon\widetilde{\mathbf{W}}\mathring{\mathbf{T}}+\varepsilon D^2w(\mathbf{1})\widetilde{\mathbf{E}}-\varepsilon(\mathrm{tr}\widetilde{\mathbf{E}})\mathring{\mathbf{T}}+o(\varepsilon) = \\ &= \mathring{\mathbf{T}}\left(1-\varepsilon\mathrm{tr}\widetilde{\mathbf{E}}\right)+2\varepsilon\left(\widetilde{\mathbf{H}}\mathring{\mathbf{T}}\right)^{\mathrm{sym}}-\varepsilon\widetilde{\mathbf{E}}\mathring{\mathbf{T}}+\varepsilon D^2w(\mathbf{1})\widetilde{\mathbf{E}}+o(\varepsilon).\end{aligned} \tag{79}$$

By comparing Eqs. (78) and (79) it is evident that if the reference configuration is not natural, that is $\mathring{\mathbf{T}} \neq \mathbf{0}$, then the Piola-Kirchhoff stress tensor is different from the Cauchy stress tensor. Moreover, by Eq. (79) it follows that if $\mathring{\mathbf{T}} \neq \mathbf{0}$ then the Cauchy stress tensor depends also on the skew-symmetric part of the displacement gradient.

If $\mathring{\mathbf{T}} = \mathbf{0}$, then Eqs.(78) and (79) reduce respectively to

$$\mathbf{S} = \mathbb{C}\mathbf{E}_\varepsilon + o(\varepsilon), \quad \mathbf{T} = \mathbb{C}\mathbf{E}_\varepsilon + o(\varepsilon) \quad \text{as } \varepsilon \to 0, \tag{80}$$

and $\mathbf{S} = \mathbf{T}$ to within higher order terms. Here, the fourth-order symmetric tensor

$$\mathbb{C} = D^2w(\mathbf{1}) \tag{81}$$

is the *elasticity tensor* of the material for the reference configuration $\boldsymbol{\kappa}$. The elasticity tensor satisfies the so-called *minor symmetries*

$$C_{ijrs} = C_{jirs} = C_{ijsr}, \quad i,j,r,s = 1,2,3. \tag{82}$$

We conclude the section with a proof of identities (71) and (72). The first identity follows by differentiating the condition of frame-indifference

$$w(\mathbf{F}) = w\left(\mathbf{QF}\right),$$

which holds for every $\mathbf{Q} \in \mathrm{Orth}^+$ and for every $\mathbf{F} \in \mathrm{Lin}^+$, with respect to $\mathbf{F}$. Let us introduce the function $\mathbf{Z}\left(\mathbf{F}\right) = \mathbf{QF}$. By using the derivation rule we obtain

$$[Dw\left(\mathbf{F}\right)]_{ij} = \frac{\partial w\left(\mathbf{Z}\left(\mathbf{F}\right)\right)}{\partial F_{ij}} = \left(\frac{\partial w\left(\mathbf{Z}\right)}{\partial Z_{rs}}\right)_{\mathbf{Z}=\mathbf{Z}(\mathbf{F})} \frac{\partial Z_{rs}\left(\mathbf{F}\right)}{\partial F_{ij}} = \left[\mathbf{Q}^T Dw\left(\mathbf{QF}\right)\right]_{ij}.$$

In order to prove identity (72) we use the following well-known result (see Ex. 3 of Section 2.6): for any $\mathbf{W} \in \mathrm{Skw}$ and for any $\vartheta \in [0, 2\pi]$,

$$\mathbf{Q}\left(\vartheta\right) = \mathbf{1} + \sin\vartheta\mathbf{W} + \left(1-\cos\vartheta\right)\mathbf{W}^2 \in \mathrm{Orth}^+. \tag{83}$$

By (71), we have

$$\mathbf{Q}Dw(\mathbf{F}) - Dw(\mathbf{F}) = Dw(\mathbf{QF}) - Dw(\mathbf{F}) \quad \text{for every } \mathbf{Q} \in \mathrm{Orth}^{+}.$$

Choose $\mathbf{Q} = \mathbf{Q}(\vartheta)$ as in Eq. (83), with $|\vartheta|$ small. Since $\mathbf{Q}(\vartheta) = \mathbf{1} + \vartheta\mathbf{W} + O(\vartheta^2)$, for any $\mathbf{W} \in \mathrm{Skw}$, we obtain

$$\left(\mathbf{1} + \vartheta\mathbf{W} + O\left(\vartheta^2\right)\right) Dw(\mathbf{F}) - Dw(\mathbf{F}) = Dw\left(\left(\mathbf{1} + \vartheta\mathbf{W} + O\left(\vartheta^2\right)\right)\mathbf{F}\right) - Dw(\mathbf{F}),$$

that is (assuming $w = w(\mathbf{X}, \cdot)$ regular enough)

$$\vartheta\mathbf{W}Dw(\mathbf{F}) = \vartheta D^2 w(\mathbf{F})\,\mathbf{WF} + O\left(\vartheta^2\right).$$

Dividing by ϑ and taking the limit as $\vartheta \to 0$, the thesis follows.

4.3 The boundary value problem of linearized elasticity

We shall linearize the equilibrium problem in finite elasticity for infinitesimal deformations assigned from a reference configuration which is a natural state for an hyper-elastic material. By Eq. (80) we have

$$\mathrm{Div}\mathbf{S} = \mathrm{Div}\left(\mathbb{C}\mathbf{E}_\varepsilon\right) + o(\varepsilon) \qquad \text{in } \Omega, \text{ as } \varepsilon \to 0. \tag{84}$$

We assume that the actual body-force density is small:

$$\mathbf{b}_\varepsilon(\mathbf{x}) \quad = \varepsilon\widetilde{\mathbf{b}}(\mathbf{x}) \qquad \text{in } \boldsymbol{\chi}(\Omega). \tag{85}$$

Therefore, by (57) we have

$$\mathbf{b}_{\boldsymbol{\kappa}\varepsilon}(\mathbf{X}) \quad = \mathbf{b}_\varepsilon(\mathbf{X}) + o(\varepsilon) \qquad \text{in } \Omega \text{ as } \varepsilon \to 0. \tag{86}$$

By inserting Eqs. (85), (86) in Eq. $(55)_1$, the equilibrium equation becomes

$$\mathrm{Div}\left(\mathbb{C}\widetilde{\mathbf{E}}_\varepsilon\right) + \widetilde{\mathbf{b}}_\varepsilon + o(\varepsilon) = 0 \qquad \text{in } \Omega \text{ as } \varepsilon \to 0$$

and, neglecting higher orded terms,

$$\mathrm{Div}\left(\mathbb{C}\mathbf{E}_\varepsilon\right) + \mathbf{b}_\varepsilon = \mathbf{0} \qquad\qquad \text{in } \Omega, \tag{87}$$

which is the well-known system of partial differential equations of linearized elasticity.

The boundary condition $(55)_2$ can be discussed and linearized similarly.

In conclusion, dropping the subscript ε and $\boldsymbol{\kappa}$, the boundary value problem of linearized elasticity can be formulated as follows: to find the displacement field $\mathbf{u}$, $\mathbf{u}: \overline{\Omega} \to \mathbb{R}^3$, $\mathbf{u}$ *regular enough*, solution to

$$(\mathrm{LE}) \begin{cases} \mathrm{Div}\,(\mathbb{C}\nabla \mathbf{u}) + \mathbf{b} = \mathbf{0} & \text{in } \Omega, \\ (\mathbb{C}\nabla \mathbf{u})\,\mathbf{n} = \boldsymbol{\varphi} & \text{on } \partial_t\Omega, \\ \mathbf{u} = \mathbf{g} & \text{on } \partial_d\Omega, \end{cases} \qquad (88)$$

where $\boldsymbol{\varphi}$ and $\mathbf{g}$ are the Neumann and Dirichlet (regular) data, respectively, and $\partial_t\Omega \cup \partial_d\Omega = \partial\Omega$.

In the next section we shall present some uniqueness and existence results for the problem (LE) defined in Eq. (88), see Ciarlet (2004), Fichera (1972), Valent (1988).

4.4 Some existence and uniqueness results for the linearized elasticity problem

In this section we shall mainly concerned with the (LE) problem (88) for homogeneous Dirichlet and Neumann boundary conditions:

$$(\mathrm{LE})_{\mathrm{Mix}} \begin{cases} \mathrm{Div}\,(\mathbb{C}\nabla \mathbf{u}) + \mathbf{b} = \mathbf{0} & \text{in } \Omega, \\ (\mathbb{C}\nabla \mathbf{u})\,\mathbf{n} = \mathbf{0} & \text{on } \partial_t\Omega, \\ \mathbf{u} = \mathbf{0} & \text{on } \partial_d\Omega, \end{cases} \qquad (89)$$

where $\partial_d\Omega \neq \emptyset$, $\partial_t\Omega \neq \emptyset$, $\overset{\circ}{\partial_d\Omega} \cup \overset{\circ}{\partial_t\Omega} = \emptyset$, $\partial_d\Omega \cap \partial_t\Omega = \partial\Omega$. We shall introduce the following a-priori assumptions:

i) (*on the domain* Ω) Ω is an open, bounded and connected subset of $\mathbb{R}^n$, $n = 3$, and it has the boundary $\partial\Omega$ of Lipschitz class with constant r_0, M_0[11]; moreover, $|\Omega| \leq M_1 r_0^n$, $M_1 > 0$.

ii) (*On the material*) The elasticity tensor $\mathbb{C}: \overline{\Omega} \to \mathbb{L}\mathrm{in}$ is symmetric ($\mathbb{C} = \mathbb{C}^T$ or, in components, $\mathbb{C}_{ijrs} = \mathbb{C}_{rsij}$, $i,j,r,s = 1,2,3$), it has the minor symmetries $\mathbb{C}_{ijrs} = \mathbb{C}_{jirs} = \mathbb{C}_{ijsr}$, $i,j,r,s = 1,...,n$; moreover, it is strongly convex in $\overline{\Omega}$: there exists $\xi_0 > 0$ such that $\mathbb{C}(\mathbf{X})\,\mathbf{A}\cdot\mathbf{A} \geq \xi_0\,|\mathbf{A}|^2$,

[11] Given the domain Ω, we shall say that $\partial\Omega$ is of Lipschitz class with constants $r_0 > 0$, $M_0 > 0$ if, for any $Q \in \partial\Omega$, there exists a rigid transformation of coordinates under which we have $Q = 0$ (origin) and $\Omega \cap B_{r_0}(0) = \{\mathbf{X} \in B_{r_0} | X_n > \Phi(\mathbf{X}')\}$, $\mathbf{X} = (\mathbf{X}', X_n)$, $\mathbf{X}' \in \mathbb{R}^{n-1}$, $X_n \in \mathbb{R}$, where Φ is a Lipschitz continuous function in $B_{r_0}(0) \subset \mathbb{R}^{n-1}$, satisfying $\Phi(\mathbf{0}) = 0$ and $\|\Phi\|_{C^{0,1}(B_{r_0}(0))} \leq M_0$. Here $\|\Phi\|_{C^{0,1}(B_{r_0}(0))} = \|\Phi\|_{L^\infty(B_{r_0}(0))} + \|\nabla\Phi\|_{L^\infty(B_{r_0}(0))}$.

for every $\mathbf{A} \in$ Sym a.e. in $\overline{\Omega}$; finally, $\mathbb{C} \in L^{\infty}(\Omega)$, that is $\mathbb{C}_{ijrs} \in L^{\infty}(\Omega)$, $i,j,r,s = 1,...,n$[12].

iii) (*On the data*) The body force density $\mathbf{b}$, $\mathbf{b} : \Omega \to \mathbb{R}^3$, is square integrable in Ω: $\|\mathbf{b}\|_{L^2(\Omega)} = \left(\sum_{i=1}^{n} \int_{\Omega} b_i^2 \, dV\right)^{1/2} < \infty$.

Moreover, we shall assume that the portion of the boundary $\partial_d\Omega$ on which the displacement is given (and equal to zero) is *large enough*. More precisely: there exists $P \in \partial_d\Omega$ such that

$$\partial\Omega \cap B_{r_0}(P) \subset \partial_d\Omega,$$

where $B_{r_0}(P) = \{Q \in \mathbb{R}^n | \, |Q - P| < r_0\}$.

We have the following well-known result.

Theorem 4.1. *(Existence and uniqueness for the mixed (LE) problem with homogeneous Dirichlet and Neumann data)*
Under the above assumptions, the problem (89) *has a unique solution* $\boldsymbol{u} \in H^1(\Omega)$. *Moreover*

$$\|\boldsymbol{u}\|_{H^1(\Omega)} \leq C\|\boldsymbol{b}\|_{L^2(\Omega)}, \tag{90}$$

where $C > 0$ *is a constant only depending on* r_0, M_0, M_1, ξ_0.

Remark 4.2. Here $H^1(\Omega)$ denotes the Hilbert space $W^{1,2}(\Omega)$, with norm $\|\mathbf{f}\|_{H^1(\Omega)} = \left(\|\mathbf{f}\|^2_{L^2(\Omega)} + \|\nabla\mathbf{f}\|^2_{L^2(\Omega)}\right)^{1/2}$.

Proof. We shall use the following Riesz representation theorem, see Brezis (1986).

Theorem 4.3. *Let* H *be an Hilbert space and let* $a(\cdot,\cdot) : H \times H \to \mathbb{R}$ *a bilinear, continuous, coercive form on* $H \times H$. *Let* $\ell(\cdot)$, $\ell : H \to \mathbb{R}$, *be a linear and continuous functional on* H. *Then, there exists a unique* $u \in H$ *such that*

$$a(u,v) = \ell(v) \qquad \textit{for every } v \in H. \tag{91}$$

We define $H = \left\{\mathbf{v} \in H^1(\Omega), \mathbf{v} = \mathbf{0} \text{ on } \partial_d\Omega\right\}$[13]. It turns out that H is an Hilbert space with norm $\|\cdot\|_{H^1(\Omega)}$. By multiplying the equilibrium equation

[12] Let $\mathbb{C}$ as before, $\mathbb{C} \in L^{\infty}(\Omega)$. We define, for $\mathbf{X} \in \overline{\Omega}$, $\|\mathbb{C}(\mathbf{X})\|_{\infty} = \sup_{|\mathbf{A}|} |\mathbb{C}(\mathbf{X})\mathbf{A}|$ for every $\mathbf{A} \in$ Sym .

[13] The functions of H are essentially the functions $\mathbf{v}$ of $H^1(\Omega)$ which vanish on $\partial_d\Omega$. It is not easy to give a precise meaning to the condition $\mathbf{v} = \mathbf{0}$ on $\partial_d\Omega$, because functions in $H^1(\Omega)$ are defined only a.e. and do not have continuous representants. For a rigorous definition one can use the *theory of traces*.

by $\mathbf{v} \in H$, integrating by parts and taking the boundary conditions of $(89)_1$, $(89)_2$ into account, we obtain Eq. (91) with $a(\cdot,\cdot) : H \times H \rightarrow \mathbb{R}$ and $\ell(\cdot) : H \rightarrow \mathbb{R}$ defined by

$$a(\mathbf{u}, \mathbf{v}) = \int_\Omega \mathbb{C}\nabla\mathbf{u} \cdot \nabla\mathbf{v}, \tag{92}$$

$$\ell(\mathbf{v}) = \int_\Omega \mathbf{b} \cdot \mathbf{v}. \tag{93}$$

Note that Eq. (91) coincides with the Virtual Work Equation (49).

By Cauchy-Schwarz inequality we have

$$|\ell(\mathbf{v})| \leq \int_\Omega |\mathbf{v} \cdot \mathbf{b}| \leq \|\mathbf{b}\|_{L^2(\Omega)} \|\mathbf{v}\|_{H^1(\Omega)}, \tag{94}$$

so that $\ell(\cdot)$ is continuous.

The form $a(\cdot,\cdot)$ is obviously bilinear and it continuous since, by Cauchy-Schwarz inequality, we have

$$|a(\mathbf{v}_1, \mathbf{v}_2)| \quad \leq C\|\mathbf{v}_1\|_H \|\mathbf{v}_2\|_H, \tag{95}$$

with $C = \sup_{\mathbf{X}\in\Omega} \|\mathbb{C}(\mathbf{X})\|_\infty$.

To prove coercivity of $a(\cdot,\cdot)$ on $H \times H$ we need to show that there exists a constant $\alpha > 0$ such that

$$a(\mathbf{v}, \mathbf{v}) \geq \alpha\|\mathbf{v}\|_H^2 \qquad \text{for every } \mathbf{v} \in H.$$

By the strong convexity of $\mathbb{C}$ we have

$$a(\mathbf{v}, \mathbf{v}) \geq \xi_0 \int_\Omega \left|(\nabla\mathbf{v})^{\text{sym}}\right|^2. \tag{96}$$

In order to estimate from below the L^2-norm of the infinitesimal strain tensor $(\nabla\mathbf{v})^{\text{sym}}$ we shall need two important inequalities, see, for example, Kondrat'ev and Oleinik (1989).

Poincarè inequality

Let Ω be a bounded domain in $\mathbb{R}^n$ of Lipschitz class with constants r_0, M_0 and such that $|\Omega| \leq M_1 r_o^n$, $n \geq 2$. For every $\mathbf{v} \in H^1(\Omega, \mathbb{R}^n)$ such that

$$\mathbf{v} = \mathbf{0} \qquad \text{on } \partial\Omega \cap B_{r_0}(Q),$$

where Q is some point in $\partial\Omega$, we have

$$\int_\Omega |\mathbf{v}|^2 \leq C_P r_0^2 \int_\Omega |\nabla\mathbf{v}|^2, \tag{97}$$

where $C_P > 0$ is a constant only depending on M_0 and M_1.

Korn's inequality
Let Ω be a bounded domain in $\mathbb{R}^n$ of Lipschitz class with constant r_0, M_0, and such that $|\Omega| \leq M_1 r_0^n$, $n \geq 2$. For every $\mathbf{v} \in H^1(\Omega, \mathbb{R}^n)$ such that

$$\mathbf{v} = \mathbf{0} \qquad \text{on } \partial\Omega \cap B_{r_0}(Q),$$

where Q is some point in $\partial\Omega$, we have

$$\int_\Omega |\nabla \mathbf{v}|^2 \leq C_K \int_\Omega |(\nabla \mathbf{v})^{\mathrm{sym}}|^2, \tag{98}$$

where $C_K > 0$ is a constant only depending on M_0 and M_1.

Now, by Poincarè's and Korn's inequalities we have

$$\|\mathbf{v}\|^2_{H^1(\Omega)} \leq (1 + C_P r_0^2) C_K \int_\Omega |(\nabla \mathbf{v})^{\mathrm{sym}}|^2, \tag{99}$$

and therefore $a(\mathbf{v}, \mathbf{v})$ is coercive in H:

$$a(\mathbf{v}, \mathbf{v}) \geq \frac{\xi_0}{C_K(1 + C_P r_0^2)} \|\mathbf{v}\|^2_H \qquad \text{for every } \mathbf{v} \in H. \tag{100}$$

By Riesz representation theorem we can conclude that there exists a unique solution to (89). Finally, by choosing $\mathbf{v} = \mathbf{u}$ in (91) we have

$$\frac{\xi_0}{C_K(1 + C_P r_0^2)} \|\mathbf{u}\|^2_{H^1(\Omega)} \leq a(\mathbf{u}, \mathbf{u}) = \ell(\mathbf{u}) \leq \|\mathbf{b}\|_{L^2(\Omega)} \|\mathbf{u}\|_{H^1(\Omega)}$$

and the stability estimates follows:

$$\|\mathbf{u}\|_{H^1(\Omega)} \leq \frac{C_K(1 + C_P r_0^2)}{\xi_0} \|\mathbf{b}\|_{L^2(\Omega)}. \tag{101}$$

□

The existence, uniqueness and stability for other linearized elasticity problems can be proved similarly. In the sequel, we shall restrict the attention to the two cases in which $\partial_t \Omega = \emptyset$ or $\partial_d \Omega = \emptyset$.

Theorem 4.4. *(Existence and uniqueness for the Dirichlet problem with homogeneous data)*
Under the above assumptions, the problem

$$(LE)_{Dir}\left\{\begin{array}{ll} Div(\mathbb{C}\nabla \boldsymbol{u}) + \boldsymbol{b} = \boldsymbol{0} & in\ \Omega, \\ \boldsymbol{u} = \boldsymbol{0} & on\ \partial\Omega, \end{array}\right. \tag{102}$$

has a unique solution $\boldsymbol{u} \in H_0^1(\Omega) = \{\boldsymbol{f}:\Omega \to \mathbb{R}^n | \ \boldsymbol{f} \in H^1(\Omega),\ \boldsymbol{f} = \boldsymbol{0}\ on\ \partial\Omega\}$ *and*

$$\|\boldsymbol{u}\|_{H^1(\Omega)} \leq C\|\boldsymbol{b}\|_{L^2(\Omega)}, \tag{103}$$

where $C > 0$ *is a constant only depending on* r_0, M_0, M_1, ξ_0.

Proof. The proof follows the same lines of the proof of the $(\mathrm{LE})_{\mathrm{Mix}}$. Here, we shall need the following well-known version of Poincarè inequality in $H_0^1(\Omega)$. Let Ω be a bounded domain in $\mathbb{R}^n$, $n \geq 2$, with Lipschitz boundary $\partial\Omega$ with constants r_0, M_0 and such that $|\Omega| \leq M_1 r_0^n$. There exists $C_P > 0$, only depending on M_0 and M_1, such that

$$\int_\Omega |\mathbf{v}|^2 \leq r_0^2 C_P \int_\Omega |\nabla \mathbf{v}|^2 \qquad \text{for every } \mathbf{v} \in H_0^1(\Omega). \tag{104}$$

The Korn's inequality in $H_0^1(\Omega)$ reads as follows: under the above assumptions on Ω, we have

$$\int_\Omega |\nabla \mathbf{v}|^2 \leq 2\int_\Omega |(\nabla \mathbf{v})^{\mathrm{sym}}|^2 \qquad \text{for every } \mathbf{v} \in H_0^1(\Omega). \tag{105}$$

In this case, at least for regular functions, the above inequality follows from the identity

$$\int_\Omega |(\nabla \mathbf{v})^{\mathrm{sym}}|^2 = \frac{1}{2}\int_\Omega |\nabla \mathbf{v}|^2 + \frac{1}{2}\int_\Omega (\mathrm{div}\, \mathbf{v})^2 \qquad \text{for every } \mathbf{v} \in C_0^\infty(\Omega).$$

□

Theorem 4.5. *(Existence and uniqueness for the Neumann problem with homogeneous data)*
Let $\boldsymbol{b}$ *satisfying the global equilibrium conditions:* $\int_\Omega \boldsymbol{b} = \boldsymbol{0}$, $\int_\Omega \boldsymbol{X} \times \boldsymbol{b} = \boldsymbol{0}$. *Then, under the above assumptions, the problem*

$$(LE)_{Neu}\left\{\begin{array}{ll} Div(\mathbb{C}\nabla \boldsymbol{u}) + \boldsymbol{b} = \boldsymbol{0} & in\ \Omega, \\ (\mathbb{C}\nabla \boldsymbol{u})\, \boldsymbol{n} = \boldsymbol{0} & on\ \partial\Omega, \end{array}\right. \tag{106}$$

has a unique solution $\boldsymbol{u} \in H^1(\Omega)$ *satisfying*

$$\int_\Omega \boldsymbol{u} = \mathbf{0}, \qquad \int_\Omega \nabla \boldsymbol{u} - (\nabla \boldsymbol{u})^T = \mathbf{0}. \tag{107}$$

Moreover

$$\|\boldsymbol{u}\|_{H^1(\Omega)} \leq C \|\boldsymbol{b}\|_{L^2(\Omega)}, \tag{108}$$

where $C > 0$ *only depends on* r_0, M_0, M_1, ξ_0.

Proof. In this case it is convenient to work on the subspace H of $H^1(\Omega)$ defined by

$$H = \left\{ \mathbf{v} \in H^1(\Omega) \mid \int_\Omega \mathbf{v} = \mathbf{0}, \int_\Omega \nabla \mathbf{v} - (\nabla \mathbf{v})^T = \mathbf{0} \right\}.$$

The coerciveness of the bilinear form $a(\cdot,\cdot)$ in H^1 follows from the two inequalities below.

Poincarè's inequality
Under the above assumptions on Ω, we have

$$\int_\Omega \left| \mathbf{v} - \frac{1}{|\Omega|} \int_\Omega \mathbf{v} \right|^2 \leq r_0^2 C_P \int_\Omega |\nabla \mathbf{v}|^2 \quad \text{for every } \mathbf{v} \in H^1(\Omega), \tag{109}$$

where C_P only depends on M_0 and M_1.

Korn's inequality
Under the above assumptions on Ω, we have

$$\int_\Omega \left| \nabla \mathbf{v} - \frac{1}{|\Omega|} \int_\Omega \left(\nabla \mathbf{v} - (\nabla \mathbf{v})^{\mathrm{T}} \right) \right|^2 \leq C_K \int_\Omega |\nabla \mathbf{v}|^2 \quad \text{for every } \mathbf{v} \in H^1(\Omega), \tag{110}$$

where C_K only depends on M_0 and M_1.
By using these two inequalities, the proof of the Theorem follows the same lines of the cases $(\mathrm{LE})_{\mathrm{Mix}}$ and $(\mathrm{LE})_{\mathrm{Dir}}$. □

Remark 4.6. In all the previous problems, the bilinear form $a(\cdot,\cdot)$ is symmetric, that is $a(\mathbf{u},\mathbf{v}) = a(\mathbf{v},\mathbf{u})$, for every $\mathbf{v},\mathbf{u} \in H$. In this case the solution $\mathbf{u} \in H$ to the Linearized Elasticity problem can be found by solving the variational problem

$$I(\mathbf{u}) = \min_{\mathbf{v} \in H} I(\mathbf{v}),$$

where $I(\mathbf{v}) = \frac{1}{2} a(\mathbf{v},\mathbf{v}) - \ell(\mathbf{v})$ is the total energy of the problem.

Acknowledgments

I wish to thank Dr. Luca Della Longa for his assistance in typing the manuscript.

Bibliography

H. Brezis, *Analisi funzionale*, Liguori Editore, Napoli, 1986.

P.G. Ciarlet, *Mathematical Elasticity. Volume 1: Three Dimensional Elasticity*, Elsevier, Amsterdam, 2004.

G. Del Piero, *Lecture Notes in Continuum Mechanics*, UNESCO Ecole d'Ete 1986, Monastir, Tunisie, 1986.

G. Fichera, *Existence theorems in elasticity.* In *Handbuch der Physik*, VIa/2, Springer, Berlin–Heidelberg–New York, 1972.

M.E. Gurtin, *The linear theory of elasticity.* In *Handbuch der Physik* VIa/2, 1–295, Springer, Berlin–Heidelberg–New York, 1972.

V.A. Kondrat'ev and O.A. Oleinik, On the dependence of the constant in Korn's inequality on parameters characterizing the geometry of the region, *Russian Math. Surveys* **44**, 187–195, 1989.

R. Paroni, Constitutive equations and variational elasticity. In A. Morassi and R. Paroni, editors, *Classical and Advanced Theories of Thin Structures. Mechanical and Mathematical Aspects*, CISM Courses and Lectures. Springer, Wien–New York, 2008.

P. Podio-Guidugli, A Primer in Elasticity, *Journal of Elasticity* **58**, 1–104, 2000.

C. Truesdell, *The Elements of Continuum Mechanics*, Springer, Berlin–Heidelberg–New York, 1966.

T. Valent, *Boundary Value Problems of Finite Elasticity*, Springer, Berlin–Heidelberg–New York, 1988.

Constitutive Equations and Variational Elasticity

Roberto Paroni
Dipartimento di Architettura e Pianificazione,
Università degli Studi di Sassari,
Palazzo del Pou Salit, Piazza Duomo 6, 07041 Alghero (SS), Italy

Abstract A rather brief introduction to the non-linear theory of elasticity is given. In the presentation most of mechanical and mathematical concepts used are introduced from scratch.

1 Introduction

In these notes I briefly present the theory of constitutive equations, particularly focusing on hyperelastic materials, and I discuss the existence of solutions to the non-linear elastic problem.
In writing this manuscript I literally followed the notes of the lectures delivered at CISM. I did not try at all to be exhaustive on the argument, on the contrary I tried to give a simple and direct introduction to the subject. Since the audience had very different backgrounds I decided to introduce the concepts only after having shown that they were indeed necessary, following the belief that it is easier to accept and learn new material once is known what is needed for. As a result the manuscript looks a bit fragmented: a path which leads to the discussion of existence of solutions to the non-linear elastic problem is followed, but along the way several short detours are taken to introduce the mathematical "machinery" needed to go along. Following the path, I tried to be more intuitive than precise. Indeed all the technicalities have been swept under the carpet, so to speak, and only the major results and ideas are presented. Proofs are given only if simple, and along the way many results are left as exercise. The reader accustomed with Sobolev spaces and weak convergence can skip all the detours.
Needless to say the serious reader should consult the references quoted, since the present paper hopes to be only an informal introduction to the subject.
Finally, I wish to thank Dr. Giuseppe Zurlo for his assistance in the typing of the manuscript.

Notation. The notation used throughout is quite standard. With Lin^+, Sym^+ and SO(3) we denote the set of positive linear transformations, the set of positive definite symmetric transformations and the set of rotations, respectively. The Euclidean (Frobenius) inner product is denoted by $\cdot$ and the corresponding norm by $|\cdot|$. The gradient is denoted by ∇ and a bar over a set denotes the closure, for instance $\bar{\Omega}$ denotes the closure of Ω. Finally, the constants are denoted by the letter c: the value of c may change from line to line.

2 Elastic and Hyperelastic Materials

In general bodies made of different materials behave in different manners. This different behavior is specified, in continuum mechanics, with a constitutive equation, that is a relation between stress and strain. In what follows we summarize the principal concepts of the theory and we refer to the book by Truesdell (1991) for further details.
Let $\mathcal{B}$ be a body and $\boldsymbol{\kappa}$ be a placement of the body in the Euclidean space.

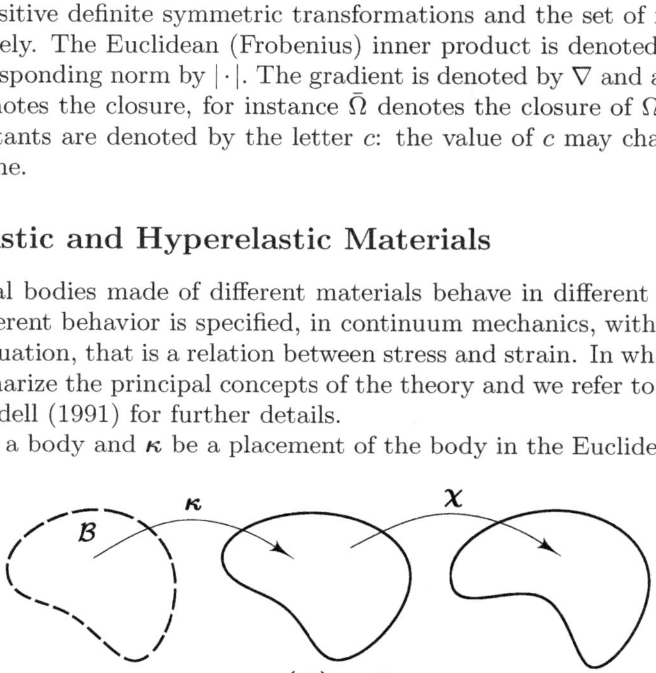

Figure 1.

Let us consider a second placement of the body and let χ be the mapping from the reference configuration $\boldsymbol{\kappa}$, see Figure 1, to the second placement considered. We shall denote by

$$\mathbf{F} := \nabla\chi$$

the deformation gradient.

Definition 2.1. A material is said to be **elastic** if the Cauchy stress $\boldsymbol{T}$ is determined by the deformation gradient. That is, there exists a mapping $\tilde{\mathcal{F}}_{\boldsymbol{\kappa}}$ such that

$$\boldsymbol{T}(\mathbf{x}) := \tilde{\mathcal{F}}_{\boldsymbol{\kappa}}(\mathbf{X}, \boldsymbol{F}(\mathbf{X})) \qquad \forall \mathbf{X} \in \boldsymbol{\kappa}(\mathcal{B}),$$

where $\mathbf{x} = \chi(\mathbf{X})$. The mapping $\tilde{\mathcal{F}}_{\boldsymbol{\kappa}}$ is called the response with respect to $\boldsymbol{\kappa}$.

Since the relation between the first Piola-Kirchhoff stress tensor $\mathbf{S}$ and the Cauchy stress tensor, see Morassi (2008), is given by

$$\mathbf{S} = (\det \mathbf{F})\mathbf{T}\mathbf{F}^{-T}$$

we have, for an elastic material

$$\mathbf{S} = (\det \mathbf{F})\tilde{\mathcal{F}}_{\boldsymbol{\kappa}}(\cdot, \mathbf{F})\mathbf{F}^{-T} =: \mathcal{F}_{\boldsymbol{\kappa}}(\cdot, \mathbf{F}).$$

Definition 2.2. An elastic material is **hyperelastic** if the Piola-Kirchhoff stress $\mathbf{S}$ has a potential, that is if there exists a scalar function

$$W_{\boldsymbol{\kappa}} : \boldsymbol{\kappa}(\mathcal{B}) \times \mathrm{Lin}^{+} \mapsto \mathbb{R}$$

such that

$$\boldsymbol{S} = \mathcal{F}_{\boldsymbol{\kappa}}(\cdot, \boldsymbol{F}) = DW_{\boldsymbol{\kappa}}(\cdot, \boldsymbol{F}).$$

The scalar function $W_{\boldsymbol{\kappa}}$ is called the stored-energy function.

The next theorem (whose proof can be found in the book by Gurtin (1981), pg. 186) justifies and motivates the definition of hyperelasticity.

Theorem 2.3. *An elastic body is hyperelastic if and only if, given any part* $\mathcal{P} \subseteq \boldsymbol{\kappa}(\mathcal{B})$ *and any time interval* $[t_0, t_1]$*, the work*

$$\int_{t_0}^{t_1} \int_{\mathcal{P}} \mathbf{S} \cdot \dot{\mathbf{F}} \, d\boldsymbol{X} \, dt$$

is nonnegative in closed processes.

3 Change of Reference Configuration

Let us consider two reference configurations $\boldsymbol{\kappa}_1$ and $\boldsymbol{\kappa}_2$ of the body $\mathcal{B}$ and let

$$\boldsymbol{p} := \boldsymbol{\kappa}_2 \circ \boldsymbol{\kappa}_1^{-1} = \boldsymbol{\chi}_2^{-1} \circ \boldsymbol{\chi}_1.$$

Let, see Figure 2, $\boldsymbol{\chi}_1$ and $\boldsymbol{\chi}_2$ be two mappings from $\boldsymbol{\kappa}_1$ and $\boldsymbol{\kappa}_2$, respectively, to the same placement, and let

$$\mathbf{F}_1 := \nabla\boldsymbol{\chi}_1, \qquad \mathbf{F}_2 := \nabla\boldsymbol{\chi}_2.$$

Since $\boldsymbol{\chi}_2 \circ \boldsymbol{p} = \boldsymbol{\chi}_1$ we let furthermore

$$\mathbf{P} := \mathbf{F}_2^{-1}\mathbf{F}_1.$$

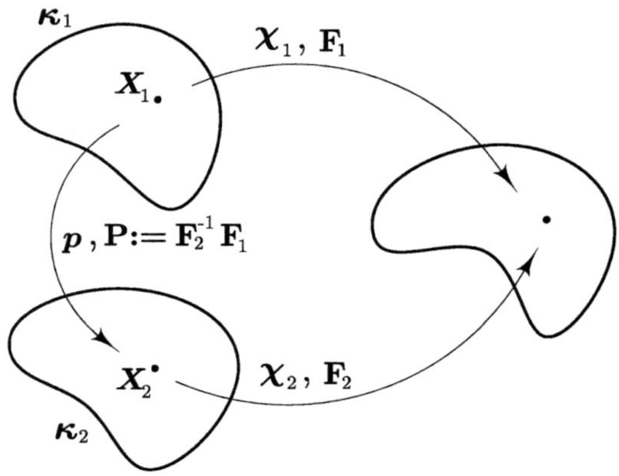

Figure 2.

Let $\mathbf{S}_1$ and $\mathbf{S}_2$ be the first Piola-Kirchhoff stress tensors with respect to the reference configurations $\boldsymbol{\kappa}_1$ and $\boldsymbol{\kappa}_2$, respectively. If $\mathbf{T}$ denotes the Cauchy stress tensor we have

$$\mathbf{S}_1 = (\det \mathbf{F}_1)\mathbf{T}\mathbf{F}_1^{-T}, \qquad \mathbf{S}_2 = (\det \mathbf{F}_2)\mathbf{T}\mathbf{F}_2^{-T}.$$

Thus

$$\mathbf{T} = (\det \mathbf{F}_2)^{-1}\mathbf{S}_2\mathbf{F}_2^T$$

and hence

$$\begin{aligned} \mathbf{S}_1 &= (\det \mathbf{F}_1)(\det \mathbf{F}_2)^{-1}\mathbf{S}_2\mathbf{F}_2^T\mathbf{F}_1^{-T} \\ &= (\det \mathbf{F}_2^{-1}\mathbf{F}_1)\mathbf{S}_2(\mathbf{F}_2^{-1}\mathbf{F}_1)^{-T} \\ &= (\det \mathbf{P})\mathbf{S}_2\mathbf{P}^{-T}. \end{aligned} \tag{1}$$

Let W_2 be the stored energy density with respect to the reference configuration $\boldsymbol{\kappa}_2$. Let us set

$$W_1(\boldsymbol{X}_1, \mathbf{F}_1) := (\det \mathbf{P})W_2(\boldsymbol{X}_2, \mathbf{F}_1\mathbf{P}^{-1}), \tag{2}$$

with

$$\boldsymbol{X}_2 = \boldsymbol{p}(\boldsymbol{X}_1).$$

Using the chain rule and equation (1) we can easily solve the following

Exercise 1. *Show that*

$$\mathbf{S}_1 = DW_1(\cdot, \mathbf{F}_1).$$

Equation (2) allows to calculate the stored-energy density with respect to a reference configuration once is known the stored-energy density with respect to another reference configuration. Also equation (2) shows that the definition of a hyperelastic body does not depend on the reference configuration.

4 Material Symmetry

Let $\boldsymbol{\kappa}$ be a reference configuration and $\boldsymbol{X}_o \in \boldsymbol{\kappa}(\mathcal{B})$, see Figure 3. Let us consider a particular case of the mapping $\boldsymbol{p}$ considered in the previous section

$$\boldsymbol{p}(\boldsymbol{X}) = \mathbf{Q}(\boldsymbol{X} - \boldsymbol{X}_o) + \boldsymbol{X}_o$$

with

$$\mathbf{Q} \in \mathrm{SO}(3),$$

where SO(3) denotes the set of rotations. The mapping $\boldsymbol{p}$ simply rotates the body around the point $\boldsymbol{X}_o$. Let us denote by $\mathbf{Q}\boldsymbol{\kappa}$ the placement $\boldsymbol{p} \circ \boldsymbol{\kappa}$. Let $W_{\boldsymbol{\kappa}}$ and $W_{\mathbf{Q}\boldsymbol{\kappa}}$ be the stored-energy densities with respect to the references $\boldsymbol{\kappa}$ and $\mathbf{Q}\boldsymbol{\kappa}$, respectively.

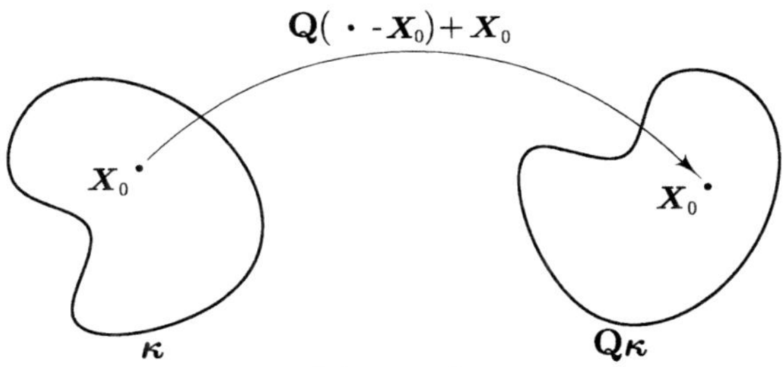

Figure 3.

If we can find a $\mathbf{Q} \in \mathrm{SO}(3)$ such that

$$W_{\boldsymbol{\kappa}}(\boldsymbol{X}_o, \mathbf{F}) = W_{\mathbf{Q}\boldsymbol{\kappa}}(\boldsymbol{X}_o, \mathbf{F}) \tag{3}$$

holds for every $\mathbf{F} \in \mathrm{Lin}^+$ we shall have shown that the response of the point $\boldsymbol{X}_o$ is just the same with respect to the two reference configurations considered, $\boldsymbol{\kappa}$ and $\mathbf{Q}\boldsymbol{\kappa}$. That is, the stored energy density can not distinguish the two reference configurations. The rotations $\mathbf{Q} \in \mathrm{SO}(3)$ for which (3) holds denote symmetries of the material occupying the placement $\boldsymbol{X}_o$.

We denote by

$$\mathcal{G}_{\boldsymbol{\kappa}}(\boldsymbol{X}_o) := \{\mathbf{Q} \in \mathrm{SO}(3) \ : \ \text{equation (3) holds}\} .$$

The set $\mathcal{G}_{\boldsymbol{\kappa}}(\boldsymbol{X}_o)$ is called the **symmetry group** of the point $\boldsymbol{X}_o$ in the placement $\boldsymbol{\kappa}$.

Exercise 2. *Show that* $\mathcal{G}_{\boldsymbol{\kappa}}(\mathbf{X}_o)$ *is a group.*

From (2) with $\boldsymbol{\kappa}_1 = \boldsymbol{\kappa}$ and $\boldsymbol{\kappa}_2 = \mathbf{Q}\boldsymbol{\kappa}$ we find

$$W_{\boldsymbol{\kappa}}(\boldsymbol{X}_o, \mathbf{F}) = W_{\mathbf{Q}\boldsymbol{\kappa}}(\boldsymbol{X}_o, \mathbf{F}\mathbf{Q}^T),$$

which is

$$W_{\boldsymbol{\kappa}}(\boldsymbol{X}_o, \mathbf{Q}\mathbf{F}) = W_{\mathbf{Q}\boldsymbol{\kappa}}(\boldsymbol{X}_o, \mathbf{F}) \qquad \forall \mathbf{F} \in \mathrm{Lin}^+. \tag{4}$$

Thus from (3) and (4) we can easily find that

$$\mathcal{G}_{\boldsymbol{\kappa}}(\boldsymbol{X}_o) = \left\{ \mathbf{Q} \in \mathrm{SO}(3) \,:\, W_{\boldsymbol{\kappa}}(\boldsymbol{X}_o, \mathbf{F}\mathbf{Q}) = W_{\boldsymbol{\kappa}}(\boldsymbol{X}_o, \mathbf{F}), \quad \forall \mathbf{F} \in \mathrm{Lin}^+ \right\}. \tag{5}$$

Definition 4.1. An elastic material is said to be **isotropic** at $\mathbf{X}_o$ if

$$\mathcal{G}_{\boldsymbol{\kappa}}(\mathbf{X}_o) = SO(3).$$

5 Material Frame Indifference

For brevity we shall state the principle of *Superimposed Rigid Motion* instead of the principle of *Material Frame Indifference.* These two principles are equivalent (see pg. 203 of Truesdell (1991)).
In the previous section we rotated the reference configuration, now we keep the reference configuration fixed and we superimpose a rigid motion to the actual placement.

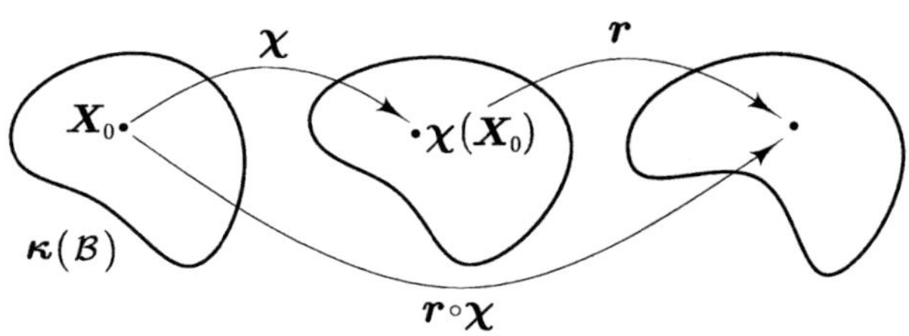

Figure 4.

Let $\boldsymbol{X}_o \in \boldsymbol{\kappa}(\mathcal{B})$ and $\boldsymbol{\chi}$ be a deformation, see Figure 4. Let us consider a rotation $\mathbf{Q} \in \mathrm{SO}(3)$ around the point $\boldsymbol{x}_o = \boldsymbol{\chi}(\boldsymbol{X}_o)$, that is

$$\boldsymbol{r}(\boldsymbol{x}) = \mathbf{Q}(\boldsymbol{x} - \boldsymbol{x}_o) + \boldsymbol{x}_o,$$

with $\boldsymbol{x} = \chi(\boldsymbol{X})$. We then have

$$\begin{aligned} \nabla\chi &= \mathbf{F}, \\ \nabla(\boldsymbol{r}\circ\chi) &= \nabla\left[\chi(\boldsymbol{X}_o) + \mathbf{Q}(\chi(\boldsymbol{X}) - \chi(\boldsymbol{X}_o)\right] = \mathbf{QF}. \end{aligned}$$

The *Principle of Material Frame Indifference* (more precisely the principle of *Superimposed Rigid Motions*) states that the stored energy density is invariant under superimposed rigid motions, that is

$$W_{\boldsymbol{\kappa}}(\boldsymbol{X}_o, \mathbf{F}) = W_{\boldsymbol{\kappa}}(\boldsymbol{X}_o, \mathbf{QF}) \tag{6}$$

for any $\mathbf{Q} \in \mathrm{SO}(3)$, $\mathbf{F} \in \mathrm{Lin}^+$ and $\boldsymbol{X}_o \in \boldsymbol{\kappa}(\mathcal{B})$. Let us briefly mention a couple of consequences. We know by the *Polar Decomposition Theorem*, see Morassi (2008) that

$$\mathbf{F} = \mathbf{RU},$$

with $\mathbf{R} \in \mathrm{SO}(3)$ and $\mathbf{U} \in \mathrm{Sym}^+$. Thus by Material Frame Indifference we have

$$W_{\boldsymbol{\kappa}}(\boldsymbol{X}_o, \mathbf{F}) = W_{\boldsymbol{\kappa}}(\boldsymbol{X}_o, \mathbf{RU}) = W_{\boldsymbol{\kappa}}(\boldsymbol{X}_o, \mathbf{U}),$$

which means that the stored energy density does not depend on the full deformation gradient but only on the stretch.

For an isotopic material $\mathcal{G}_{\boldsymbol{\kappa}} = \mathrm{SO}(3)$, thus we have that

$$W_{\boldsymbol{\kappa}}(\boldsymbol{X}_o, \mathbf{F}) \underset{\substack{\uparrow \\ \textit{symmetry}}}{=} W_{\boldsymbol{\kappa}}(\boldsymbol{X}_o, \mathbf{FQ}^T) \underset{\substack{\uparrow \\ \textit{M.F.I.}}}{=} W_{\boldsymbol{\kappa}}(\boldsymbol{X}_o, \mathbf{QFQ}^T)$$

or equivalently

$$W_{\boldsymbol{\kappa}}(\boldsymbol{X}_o, \mathbf{U}) = W_{\boldsymbol{\kappa}}(\boldsymbol{X}_o, \mathbf{QUQ}^T) \qquad \forall \mathbf{Q} \in \mathrm{SO}(3).$$

It then follows, from the spectral theorem, that $W_{\boldsymbol{\kappa}}$ depends only on $\mathbf{U}$ trough the principal stretches.

6 The Equilibrium Problem

Let $\Omega = \boldsymbol{\kappa}(\mathcal{B})$ be an open bounded smooth domain with boundary $\partial\Omega$. Let Γ_d, Γ_t be two subsets of $\partial\Omega$ such that

$$\Gamma_d \cap \Gamma_t = \emptyset, \qquad \Gamma_d \cup \Gamma_t = \partial\Omega.$$

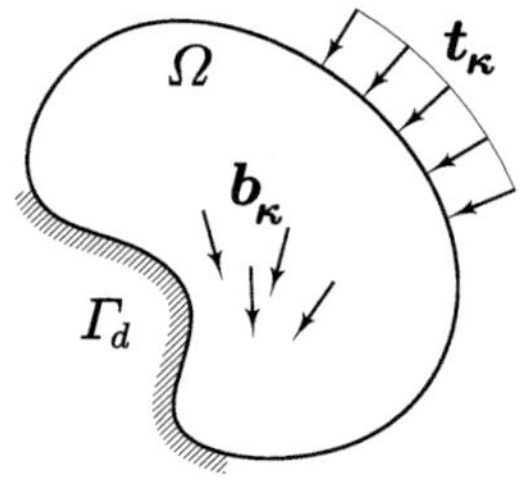

Figure 5.

We assume the body to be clamped on Γ_d, so that $\chi(\boldsymbol{X}) = \boldsymbol{X}$ on Γ_d, while on Γ_t we assume that the body is subject to surface forces $\boldsymbol{t}_\kappa$, see Figure 5.

We denote by $\boldsymbol{b}_\kappa$ the body forces, and throughout the section we assume that all fields are smooth. If $\boldsymbol{n}_\kappa$ denotes the outward normal to $\partial\Omega$ the equilibrium equations read

$$\begin{cases} \operatorname{Div}\mathbf{S} + \boldsymbol{b}_\kappa = \boldsymbol{0} & \text{in } \Omega, \\ \mathbf{S}\boldsymbol{n}_\kappa = \boldsymbol{t}_\kappa & \text{on } \Gamma_t, \\ \mathbf{S} = D_{\mathbf{F}} W_\kappa(\cdot, \mathbf{F}) & \text{in } \Omega, \\ \mathbf{S}\mathbf{F}^T \in \mathrm{Sym} & \text{in } \Omega, \\ \mathbf{F} = \nabla\chi & \text{in } \Omega, \\ \chi(\boldsymbol{X}) = \boldsymbol{X} & \text{on } \Gamma_d. \end{cases} \tag{7}$$

Let $\mathcal{A}$ be the set of kinematically admissible functions, that is

$$\mathcal{A} := \left\{ \boldsymbol{g} : \bar{\Omega} \mapsto \mathbb{R}^3 \ : \ \boldsymbol{g} \text{ is smooth and } \boldsymbol{g}(\boldsymbol{X}) = \boldsymbol{X} \text{ on } \Gamma_d \right\}.$$

Let us consider the functional

$$\mathcal{I} : \mathcal{A} \mapsto \mathbb{R}$$

defined by

$$\mathcal{I}(\boldsymbol{g}) = \int_\Omega W_\kappa(\boldsymbol{X}, \nabla \boldsymbol{g}(\boldsymbol{X}))\, d\boldsymbol{X} - \int_\Omega \boldsymbol{b}_\kappa \cdot \boldsymbol{g}\, d\boldsymbol{X} - \int_{\Gamma_t} \boldsymbol{t}_\kappa \cdot \boldsymbol{g}\, d\boldsymbol{X}.$$

We have the following

Exercise 3. *Show that the stationary points of the functional $\mathcal{I}$ are, at least formally, the solutions of the problem (7).*

Hint. *Let $\varphi : \bar{\Omega} \mapsto \mathbb{R}^3$ such that $\varphi(\mathbf{X}) = \mathbf{0}$ on Γ_d, so that if $\mathbf{f}$ is a stationary point of $\mathcal{I}$ we have that $\mathbf{f} + t\varphi \in \mathcal{A}$ for every $t \in \mathbb{R}$. Then study the identity*

$$0 = \lim_{t \to 0} \frac{\mathcal{I}(\mathbf{f} + t\varphi) - \mathcal{I}(\mathbf{f})}{t}.$$

The quantity $\int_\Omega W \, d\boldsymbol{X}$ is called the stored energy, while $\mathcal{I}(\boldsymbol{g})$ is called the total energy.

7 Minimum Energy Configurations

In this section we shall study the problem

$$\inf_{g \in \mathcal{A}} \mathcal{I}(\boldsymbol{g}).$$

This problem will give us the opportunity to introduce several technical tools needed in the course. The aim of this section is to discuss the existence of the minimum problem above. To do so we need to be quite precise on the function spaces we use. Let us for simplicity , take $\boldsymbol{t}_{\boldsymbol{\kappa}} = \boldsymbol{0}$ and

$$W_{\boldsymbol{\kappa}}(\boldsymbol{X}, \mathbf{F}) = W(\mathbf{F}),$$

so that

$$\mathcal{I}(\boldsymbol{g}) = \int_\Omega W(\nabla \boldsymbol{g}) \, d\boldsymbol{X} - \int_\Omega \boldsymbol{b} \cdot \boldsymbol{g} \, d\boldsymbol{X}. \tag{8}$$

We can take then

$$\mathcal{A} := \left\{ \boldsymbol{g} \in C^1(\bar{\Omega}) \, : \, \boldsymbol{g} = \boldsymbol{X} \;\; \text{on} \;\; \Gamma_d \right\}. \tag{9}$$

Also it natural to assume that the stored energy function will be large when the deformations are large, and in particular

$$W \nearrow +\infty \qquad \text{if} \qquad |\mathbf{F}| \to +\infty.$$

We augment this property by assuming that

$$W(\mathbf{F}) \geq c \, |\mathbf{F}|^p, \tag{10}$$

for some $c > 0$ and $1 < p < +\infty$.

7.1 A first detour: a paradigmatic scalar problem.

Before trying to minimize the energy (8) let us first consider a simpler scalar problem which, despite its simplicity, will highlight the main steps to be taken in order to deduce the existence of solutions.

Let $b \in \mathbb{R}$ and

$$f : \mathbb{R} \mapsto [0, +\infty),$$

be such that

$$f(x) \geq c|x|^p \qquad \forall x \in \mathbb{R}.$$

Let us see under which assumptions the problem

$$\inf_{x \in \mathbb{R}} \mathcal{I}(x) \qquad \text{with} \qquad \mathcal{I}(x) := f(x) - bx$$

has a minimum, that is, there exists $x_o \in \mathbb{R}$ such that

$$\inf_{x \in \mathbb{R}} \mathcal{I}(x) = \mathcal{I}(x_o).$$

Let $\{x_k\}$ be a minimizing sequence such that for every $k \in \mathbb{N}$ we have

$$\mathcal{I}(x_k) \leq \inf_{x \in \mathbb{R}} \mathcal{I}(x) + \frac{1}{k}.$$

Assuming that

$$m := \inf_{x \in \mathbb{R}} \mathcal{I}(x) < +\infty$$

we find

$$m + 1 \geq \mathcal{I}(x_k) \geq c|x_k|^p - bx_k \tag{11}$$

To continue we need the following

Exercise 4. *Let $a, b \geq 0$, $p, q > 1$ and $\frac{1}{p} + \frac{1}{q} = 1$. Prove that*

$$ab \leq \frac{a^p}{p} + \frac{b^q}{q}.$$

Hint. *Look for the minimum of*

$$h(x) = \frac{x^p}{p} + \frac{b^q}{q} - xb.$$

Hereafter for a given $p > 1$, p' will denote the number

$$p' = \frac{p}{p-1} \qquad \left(\text{i.e. } \frac{1}{p} + \frac{1}{p'} = 1\right).$$

Let $\varepsilon > 0$, then from Exercise 4 we have

$$-bx_k = -\frac{b}{\varepsilon}\varepsilon x_k \geq -\frac{\varepsilon^p |x_k|^p}{p} - \frac{|b|^{p'}}{\varepsilon^{p'} p'}$$

and hence, choosing ε small, from (11) we deduce

$$|x_k|^p \leq c\left(m + 1 + |b|^{p'}\right)$$

that is, $\{x_k\}$ is a bounded sequence. By the Bolzano-Weierstrass theorem we can deduce that there exists a subsequence $\{x_{k_j}\} \subset \{x_k\}$ and a point $x_o \in \mathbb{R}$ such that

$$x_{k_j} \to x_o.$$

Clearly x_o is the candidate minimizer. Indeed, assume that $f(x_{k_j})$ has a limit and that

$$\lim_{j\to+\infty} f(x_{k_j}) \geq f(x_o),$$

then

$$\mathcal{I}(x_o) \leq \lim_{j\to+\infty}\left[f(x_{k_j}) - bx_{k_j}\right] \leq \lim_{j\to+\infty}\left[\inf_{x\in\mathbb{R}} \mathcal{I}(x) + \frac{1}{k_j}\right],$$

and therefore

$$\mathcal{I}(x_o) \leq \inf_{x\in\mathbb{R}} \mathcal{I}(x),$$

which is

$$\mathcal{I}(x_o) = \inf_{x\in\mathbb{R}} \mathcal{I}(x).$$

The assumption that

$$\lim_{j\to+\infty} f(x_{k_j})$$

exists can be easily removed by considering the

$$\liminf_{j\to+\infty} f(x_{k_j})$$

which means that we consider the minimum limit of the convergent subsequences of $f(x_{k_j})$.

We can summarize the proof done in several steps:

A) show that the minimizing sequence is bounded;

B) extract a convergent subsequence from the minimizing sequence;

C) assumed f to be (sequentially) lower semicontinuous, that is, if $x_k \to x_o$ then

$$\liminf_{k\to+\infty} f(x_k) \geq f(x_o).$$

7.2 Back to the minimum energy configuration.

We now try to mimic, as much as possible, what we have done in Section 7.1. Let $\chi_k \in \mathcal{A}$ be a minimizing sequence such that

$$\mathcal{I}(\chi_k) \leq \inf_{g \in \mathcal{A}} \mathcal{I}(g) + \frac{1}{k}. \tag{12}$$

To start with, we need to check if χ_k is, in some sense, bounded, that is point A) of Section 7.1. We have

$$\mathcal{I}(\chi_k) \geq c \int_\Omega |\nabla \chi_k|^p \, d\boldsymbol{X} - \int_\Omega \boldsymbol{b} \cdot \chi_k \, d\boldsymbol{X}. \tag{13}$$

Let $f, g : \Omega \mapsto \mathbb{R}$ be integrable functions, then from Exercise 4 we have

$$\int_\Omega |f\, g| \, d\boldsymbol{X} \leq \frac{1}{p} \int_\Omega |f|^p \, d\boldsymbol{X} + \frac{1}{p'} \int_\Omega |g|^{p'} \, d\boldsymbol{X}. \tag{14}$$

It is convenient to introduce the following space, defined on the set of functions for which the integral make sense

$$L^p(\Omega) := \left\{ f : \Omega \mapsto \mathbb{R} \, : \, \int_\Omega |f|^p \, d\boldsymbol{X} < +\infty \right\}.$$

We shall denote by

$$||f||_{L^p(\Omega)} := \left(\int_\Omega |f|^p \, d\boldsymbol{X} \right)^{\frac{1}{p}}.$$

Remark. The space $L^p(\Omega)$ with the norm above is a Banach space.

We write $f_n \to f$ in $L^p(\Omega)$ if $||f_n - f||_{L^p(\Omega)} \to 0$. Equation (14) states therefore that if $f \in L^p(\Omega)$ and $g \in L^{p'}(\Omega)$ then $fg \in L^1(\Omega)$; this is also a consequence of Holder's inequality.

Exercise 5. (Holder's inequality) *Let* $f \in L^p(\Omega)$, $g \in L^{p'}(\Omega)$ *then* $fg \in L^1(\Omega)$ *and*

$$||fg||_{L^1(\Omega)} \leq ||f||_{L^p(\Omega)} ||g||_{L^{p'}(\Omega)}.$$

Hint. *Apply Exercise 4 with* $a = |f|/||f||_{L^p(\Omega)}$ *and* $b = |g|/||g||_{L^{p'}(\Omega)}$.

Let us go back to (13). As in the scalar case we have

$$\mathcal{I}(\chi_k) \geq c \int_\Omega |\nabla \chi_k|^p \, d\boldsymbol{X} - \frac{\varepsilon^p}{p} \int_\Omega |\chi_k|^p \, d\boldsymbol{X} - \frac{1}{\varepsilon^{p'} p'} \int_\Omega |\boldsymbol{b}|^{p'} \, d\boldsymbol{X}. \tag{15}$$

The term $\varepsilon^p \int_\Omega |\chi_k|^p \, d\boldsymbol{X}$ can be made small by choosing an ε small, but it cannot be set equal to zero. To handle it we use the following

Lemma 7.1. *Let $\mathcal{L}^3(\Omega) < +\infty$ (finite volume) and $\mathcal{H}^2(\Gamma_d) > 0$ (positive area). There exists a constant $\bar{c} > 0$ such that*

$$\int_\Omega |\boldsymbol{g}|^p \, d\mathbf{X} \le \bar{c} \left(\int_{\Gamma_d} |\boldsymbol{g}|^p \, d\mathbf{X} + \int_\Omega |\nabla \boldsymbol{g}|^p \, d\mathbf{X} \right) \qquad \forall \boldsymbol{g} \in C^1(\bar{\Omega}).$$

Proof. We give the proof for $\Omega = (0, \ell)^2 \subset \mathbb{R}^2$ and $\Gamma_d = \{0\} \times (0, \ell)$.

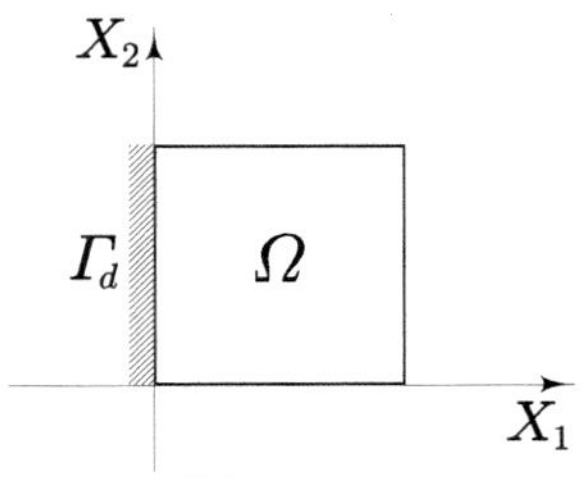

Figure 6.

Let $\boldsymbol{g} \in \mathcal{A}$, then $\boldsymbol{g}(0, X_2) = (0, X_2)$. We have

$$\boldsymbol{g}(X_1, X_2) - \boldsymbol{g}(0, X_2) = \int_0^{X_1} \frac{\partial \boldsymbol{g}(s, X_2)}{\partial X_1} \, ds,$$

thus

$$\begin{aligned} |\boldsymbol{g}(X_1, X_2)| - |\boldsymbol{g}(0, X_2)| \quad &\le \quad \int_0^\ell 1 \, |\nabla \boldsymbol{g}| \, dX_1 \le \\ \text{(by Holder's inequality)} \quad &\le \quad \left(\int_0^\ell 1^{p'} \right)^{\frac{1}{p'}} \left(\int_0^\ell |\nabla \boldsymbol{g}|^p \right)^{\frac{1}{p}} \end{aligned}$$

hence, since $(|a| + |b|)^p \le 2^p(|a|^p + |b|^p)$, we have

$$|\boldsymbol{g}(X_1, X_2)|^p \le 2^p |\boldsymbol{g}(0, X_2)|^p + 2^p \ell^{\frac{p}{p'}} \int_0^\ell |\nabla \boldsymbol{g}|^p.$$

Integrating over Ω we conclude the proof □

At this point from the Lemma above, from equation (15) and choosing an ε small we find, recalling that $\boldsymbol{g}(\boldsymbol{X}) = \boldsymbol{X}$ on Γ_d,

$$\mathcal{I}(\boldsymbol{\chi}_\kappa) + \int_{\Gamma_d} |\boldsymbol{X}|^p \, d\boldsymbol{X} \ge \frac{c}{2} \int_\Omega |\nabla \boldsymbol{\chi}_k|^p \, d\boldsymbol{X} + c_1 \int_\Omega |\boldsymbol{\chi}_k|^p \, d\boldsymbol{X} - c_2 \int_\Omega |\boldsymbol{b}|^{p'} \, d\boldsymbol{X},$$

that is

$$c_2 \int_\Omega |\boldsymbol{b}|^{p'}\, d\boldsymbol{X} + \int_{\Gamma_d} |\boldsymbol{X}|^p\, d\boldsymbol{X} + \mathcal{I}(\boldsymbol{\chi}_k) \geq c_3 \int_\Omega (|\nabla \boldsymbol{\chi}_k|^p + |\boldsymbol{\chi}_k|^p)\, d\boldsymbol{X}.$$

Assume now that

$$\inf_{g \in \mathcal{A}} \mathcal{I}(\boldsymbol{g}) < +\infty,$$

$$\boldsymbol{b} \in L^{p'}(\Omega) \qquad \text{so that} \qquad \int_\Omega |\boldsymbol{b}|^{p'}\, d\boldsymbol{X} < +\infty,$$

then, since $\int_{\Gamma_d} |\boldsymbol{X}|^p\, d\boldsymbol{X} < +\infty$, from equation (12) we deduce that

$$\int_\Omega |\,(\nabla \boldsymbol{\chi}_k|^p + |\boldsymbol{\chi}_k|^p)\, d\boldsymbol{X} \leq c_4 < +\infty. \tag{16}$$

Recall that $\boldsymbol{\chi}_k \in \mathcal{A}$ and hence $\boldsymbol{\chi}_k \in C^1(\bar{\Omega})$.

7.3 A brief detour: Sobolev spaces.

Let us define

$$||\boldsymbol{f}||_{W^{1,p}(\Omega)} := \left(\int_\Omega (|\nabla \boldsymbol{f}|^p + |\boldsymbol{f}|^p)\, d\boldsymbol{X} \right)^{1/p}.$$

It can be seen that $||\cdot||_{W^{1,p}(\Omega)}$ is a norm on $C^1(\bar{\Omega})$. We denote with $\tilde{W}^{1,p}(\Omega)$ the space $C^1(\bar{\Omega})$ equipped with the norm $||\cdot||_{W^{1,p}(\Omega)}$. It is easily shown that $\tilde{W}^{1,p}(\Omega)$ is not a complete space.

Exercise 6. *Let* $\Omega = (-1, 1)$. *Let*

$$f_n(x) = x^{\frac{2n}{2n-1}} \in C^1(\bar{\Omega})$$

and $f(x) = |x| \notin C^1(\bar{\Omega})$. *Show that*

$$||f_n - f||_{W^{1,2}(\Omega)} \to 0$$

as $n \to +\infty$ □

It is therefore useful to complete the space $\tilde{W}^{1,p}(\Omega)$; the problem is: how do we define the derivatives?

Definition 7.2. A function $\mathbf{F} \in L^p(\Omega)$ is said to be the **strong derivative** in $L^p(\Omega)$ of $f \in L^p(\Omega)$ if there exists a sequence $f_n \in C^1(\bar{\Omega})$ such that

$$f_n \to f \quad in \quad L^p(\Omega),$$

and

$$\nabla f_n \to \mathbf{F} \quad in \quad L^p(\Omega).$$

Clearly if $f \in C^1(\Omega)$ then $F = \nabla f$. Hereafter we shall denote by ∇f the strong derivative of f. We now define

$$W^{1,p}(\Omega) = \{f \in L^p(\Omega) : \nabla f \in L^p(\Omega)\}$$

that is $f \in W^{1,p}(\Omega)$ if $f \in L^p(\Omega)$ and there exists a function $\nabla f \in L^p(\Omega)$ and a sequence $\{f_n\} \subset C^1(\bar{\Omega})$ such that

$$f_n \to f \quad \text{in} \quad W^{1,p}(\Omega), \qquad \text{i.e.} \qquad ||f_n - f||_{W^{1,p}(\Omega)} \to 0,$$

that is

$$||f_n - f||_{L^p(\Omega)} \to 0 \quad \& \quad ||\nabla f_n - \nabla f||_{L^p(\Omega)} \to 0.$$

Let $\varphi \in C_0^\infty(\Omega)$ (i.e. $\varphi \in C^\infty(\Omega)$ and $\varphi = 0$ on a neighborhood of $\partial\Omega$); then, with f and f_n as above we have

$$\int_\Omega f_n \nabla\varphi = -\int_\Omega \varphi \nabla f_n.$$

Exercise 7. *Show that*

$$\int_\Omega f\,\nabla\varphi = -\int_\Omega \varphi\,\nabla f.$$

Definition 7.3. We say that $F \in L^p(\Omega)$ is the **weak derivative** of $f \in L^p(\Omega)$ if

$$\int_\Omega f\nabla\varphi = -\int_\Omega F\varphi, \qquad \forall\varphi \in C_0^\infty(\Omega).$$

We have just shown that strong derivatives are also weak derivatives. The converse is also true if Ω is a regular domain ($\partial\Omega$ is of class C^0). Therefore, in our case

$$\text{weak derivative} \Longleftrightarrow \text{strong derivative.}$$

Exercise 8. *Show that the weak derivatives (and therefore also the strong derivatives) are unique.*

7.4 Back again to the minimum energy configuration.

The minimizing sequence $\{\chi_k\} \subset \mathcal{A} \subset C^1(\bar{\Omega})$ satisfies, see equation (16),

$$\sup_k ||\chi_k||_{W^{1,p}(\Omega)} < +\infty.$$

We have thus concluded step A) mentioned in Section 7.1 We now start to worry about step B), that is: we need to extract a convergent subsequence from the minimizing sequence $\{\chi_k\}$.
We know that **if** there exists a subsequence $\{\chi_{k_j}\} \subset \{\chi_k\}$ converging to χ in the $W^{1,p}$ - norm, then $\chi \in W^{1,p}(\Omega)$. But the problem is: can we find a converging subsequence? The answer, as the following example shows, is no, we can not.

Example. Let $\Omega = (0, \pi)$, p=2 and

$$f_k(x) = \frac{\sin kx}{k}.$$

Then

$$\nabla f_k(x) = f_k'(x) = \cos kx,$$

and

$$||f_k||^2_{W^{1,2}(\Omega)} = \int_0^\pi (f_k^2 + f_k'^2)\, dx = \frac{\pi}{2k^2} + \frac{\pi}{2}.$$

Thus

$$\sup_k ||f_k||_{W^{1,2}(\Omega)} < +\infty.$$

Suppose there exists a $\{f_{k_j}\} \subset \{f_k\}$ converging in $W^{1,2}(\Omega)$ to f. Then, since

$$f_k \to f = 0 \quad \text{in} \quad L^p(\Omega)$$

we must have $f = 0$ and hence

$$\frac{\pi}{2} = ||f_{k_j}'||^2_{L^2(\Omega)} = ||f_{k_j}' - f'||^2_{L^2(\Omega)} \to 0,$$

which is a contradiction.

7.5 Another brief detour: weak convergence.

The previous example also shows that bounded sequences in L^p do not have, in general, a strongly convergent subsequence. Before giving the next definition it is convenient to recall that by Holder's inequality we have that if $f \in L^p(\Omega)$ and $g \in L^{p'}(\Omega)$, with $1 = 1/p + 1/p'$, then $fg \in L^1(\Omega)$.

Definition 7.4. Let $1 \le p < +\infty$, $\{f_n\} \subset L^p(\Omega)$ and $f \in L^p(\Omega)$. We say that f_n **weakly converges in** $L^p(\Omega)$ to f, and write

$$f_n \rightharpoonup f \quad \text{in} \quad L^p(\Omega),$$

if

$$\int_\Omega f_n g \to \int_\Omega fg, \qquad \forall g \in L^{p'}(\Omega).$$

The above definition is important because of the following

Theorem 7.5. *Let Ω be an open set, and $\{f_n\} \subset L^p(\Omega)$ with $1 < p < +\infty$. If*

$$\sup_n ||f_n||_{L^p(\Omega)} < +\infty$$

then there exist an $f \in L^p(\Omega)$ and a subsequence $\{f_{n_j}\}$ of $\{f_n\}$ such that

$$f_{n_j} \rightharpoonup f \quad \text{in} \quad L^p(\Omega).$$

We can summarize the theorem saying that bounded sequences in L^p have a weakly convergent subsequence. The next theorem instead clarifies the concept of weak convergence.

Theorem 7.6. *Let Ω be a bounded open set and $1 < p < +\infty$. Then*

$$f_n \rightharpoonup f \quad \text{in} \quad L^p(\Omega)$$

if and only if

$$\textit{1)} \quad \sup_n ||f_n||_{L^p(\Omega)} < +\infty,$$

$$\textit{2)} \quad \lim_{n\to+\infty} \int_D f_n(x)\,dx = \int_D f(x)\,dx \qquad \forall \text{ cube } D \subset \Omega.$$

In words, f_n weakly converges in L^p to f if and only if f_n is bounded in $L^p(\Omega)$ and the average, over every cube D, of f_n converges to the average of f, over the same cube D.

Exercise 9. (Riemann-Lebesgue Lemma) *Let $\Omega = (0, 2\pi)$. Show that*

$$\sin kx \rightharpoonup 0 \quad \text{in} \quad L^p(\Omega)$$

for every $1 < p < +\infty$.
Hint. *Use the theorem above.*

The next exercise deals with bounded sequences in $W^{1,p}$.

Exercise 10. *Let Ω be an open set, $1 < p < +\infty$. Let $\{f_n\} \subset W^{1,p}(\Omega)$ be such that*

$$\sup_n ||f_n||_{W^{1,p}(\Omega)} < +\infty.$$

Show that there exists $f \in L^p(\Omega)$, $F \in L^p(\Omega)$ and a subsequence $\{f_{n_j}\} \subset \{f_j\}$ such that

$$f_{n_j} \rightharpoonup f \quad \text{in} \quad L^p(\Omega),$$

$$\nabla f_{n_j} \rightharpoonup F \qquad in \qquad L^p(\Omega).$$

Moreover, show that

$$F = \nabla f.$$

We summarize the previous exercise as follows. If

$$\sup_n ||f_n||_{W^{1,p}(\Omega)} < +\infty,$$

then there exist $\{f_{n_j}\} \subset \{f_n\}$ and $f \in W^{1,p}(\Omega)$ such that

$$f_{n_j} \rightharpoonup f \qquad \text{in} \qquad W^{1,p}(\Omega),$$

that is

$$f_{n_j} \rightharpoonup f \quad \text{in} \quad L^p(\Omega) \qquad \& \qquad \nabla f_{n_j} \rightharpoonup \nabla f \quad \text{in} \quad L^p(\Omega).$$

Indeed it can be shown, by using Rellich's theorem, that

$$f_n \rightharpoonup f \qquad \text{in} \qquad W^{1,p}(\Omega)$$
$$\Downarrow$$
$$f_n \to f \quad \text{in} \quad L^p(\Omega) \quad \text{and} \quad \nabla f_n \rightharpoonup \nabla f \quad \text{in} \quad L^p(\Omega).$$

7.6 Re-back again to the minimum energy configuration.

Recall that the minimizing sequence $\{\chi_k\} \subset \mathcal{A}$ satisfies (see equation (16))

$$\sup_k ||\chi_k||_{W^{1,p}(\Omega)} < +\infty.$$

Hence there exist $\chi_0 \in W^{1,p}(\Omega)$ and $\{\chi_{k_j}\} \subset \{\chi_k\}$ such that

$$\chi_{k_j} \rightharpoonup \chi_0 \qquad \text{in} \qquad W^{1,p}(\Omega). \tag{17}$$

So we have finally achieved step B) mentioned at the end of Section 7.1. Now we tackle step C), that is: under which assumptions on W we have that

$$\liminf_{j\to+\infty} \mathcal{I}(\chi_{k_j}) \geq \mathcal{I}(\chi_0) \qquad \textbf{?}$$

Recall that

$$\mathcal{I}(\chi_{k_j}) = \int_\Omega W(\nabla\chi_{k_j})d\boldsymbol{X} - \int_\Omega \boldsymbol{b}\cdot\chi_{k_j}\, d\boldsymbol{X},$$

and that $\boldsymbol{b} \in L^{p'}(\Omega)$, thus

$$\lim_{j\to+\infty} \int_\Omega \boldsymbol{b}\cdot\chi_{k_j} = \int_\Omega \boldsymbol{b}\cdot\chi_0. \tag{18}$$

It thus suffices to show, under appropriate assumptions of W, that if

$$\chi_{k_j} \rightharpoonup \chi_0 \quad \text{in} \quad W^{1,p}(\Omega)$$

then

$$\liminf_{j \to +\infty} \int_\Omega W(\nabla \chi_{k_j})\, d\boldsymbol{X} \geq \int_\Omega W(\nabla \chi_0)\, d\boldsymbol{X},$$

that is, the functional

$$E(\chi) := \int_\Omega W(\nabla \chi)\, d\boldsymbol{X} \tag{19}$$

is sequentially-weakly-$W^{1,p}(\Omega)$-lower semicontinuous.

Before investigating under which assumption on W the functional E is sequentially-weakly-$W^{1,p}(\Omega)$-l.s.c., let us show that **if** it is, then there exists a minimizer. Indeed from (12) we have

$$\frac{1}{k_j} + \inf_{g \in \mathcal{A}} \mathcal{I}(\boldsymbol{g}) \geq \mathcal{I}(\chi_{k_j}) = \int_\Omega W(\nabla \chi_{k_j})\, d\boldsymbol{X} - \int_\Omega \boldsymbol{b} \cdot \chi_{k_j}\, d\boldsymbol{X},$$

thus, taking the lim inf and using (17) and (18) we obtain

$$\begin{aligned} \inf_{g \in \mathcal{A}} \mathcal{I}(\boldsymbol{g}) &\geq \liminf_{j \to +\infty} \int_\Omega W(\nabla \chi_{k_j})\, d\boldsymbol{X} - \lim_{j \to +\infty} \int_\Omega \boldsymbol{b} \cdot \chi_{k_j}, \\ &\geq \int_\Omega W(\nabla \chi_0)\, d\boldsymbol{X} - \int_\Omega \boldsymbol{b} \cdot \chi_0\, d\boldsymbol{X}, \\ &\geq \mathcal{I}(\chi_0). \end{aligned}$$

Remark. Indeed to conclude the proof we need to show that $\chi_0 \in \mathcal{A}$. This, as already known, is not true but

$$\chi_0 \in \bar{\mathcal{A}} := \left\{ \boldsymbol{g} \in W^{1,p}(\Omega) \; : \; \boldsymbol{g}(\boldsymbol{X}) = \boldsymbol{X} \quad \text{on} \quad \Gamma_d \right\}.$$

We refer to the book by Evans and R.F.Gariepy (1992) for the meaning of $\boldsymbol{g}(\boldsymbol{X}) = \boldsymbol{X}$ on Γ_d if $\boldsymbol{g}$ is simply a $W^{1,p}$ function. Once this is clear one can immediately work with the kinematically admissible set $\bar{\mathcal{A}}$ instead of $\mathcal{A}$ and achieve the end of the proof.

7.7 Convexity type of assumptions.

The only thing which remains to be investigated is: under which assumptions on W the functional (19) is sequentially-weakly-$W^{1,p}(\Omega)$-lower semicontinuous. The following theorem holds.

Theorem 7.7. *Let $p \in [1, +\infty)$, $f : \mathbb{R}^n \mapsto [0, +\infty]$ convex and lower semicontinuous, then*

$$F(u) := \int_\Omega f(\nabla u)\, d\mathbf{X}$$

is sequentially-weakly-$W^{1,p}(\Omega)$-l.s.c.

Thus, **if** the stored energy density W is convex and l.s.c. we deduce the existence of the minimizer.

Convexity of the stored energy density is, for many reasons, an unphysical assumption. Strict convexity has to be ruled out because would lead to the uniqueness of the minimizers, excluding therefore instabilities of the body.

Also consider the following two gradients of deformation

$$\mathbf{F}_1 = \nabla \boldsymbol{\chi}_1 = \begin{pmatrix} 1 & 0 & 0 \\ 2 & 1 & 0 \\ 2 & 0 & 1 \end{pmatrix}, \qquad \boldsymbol{\chi}_1 = (X_1, X_2 + 2X_1, X_3 + 2X_1),$$

and

$$\mathbf{F}_2 = \nabla \boldsymbol{\chi}_2 = \begin{pmatrix} 1 & 2 & 0 \\ 0 & 1 & 0 \\ 0 & 2 & 1 \end{pmatrix}, \qquad \boldsymbol{\chi}_2 = (X_1 + 2X_2, X_2, 2X_2 + X_3),$$

and note that

$$\frac{1}{2}\mathbf{F}_1 + \frac{1}{2}\mathbf{F}_2 = \begin{pmatrix} 1 & 1 & 0 \\ 1 & 1 & 0 \\ 1 & 1 & 1 \end{pmatrix}.$$

Thus

$$\det\left(\frac{1}{2}\mathbf{F}_1 + \frac{1}{2}\mathbf{F}_2\right) = 0,$$

and hence it is natural to require that

$$\lim_{t \to 1/2} W(t\mathbf{F}_1 + (1-t)\mathbf{F}_2) = +\infty.$$

But this can not happen if W is convex, since

$$W(t\mathbf{F}_1 + (1-t)\mathbf{F}_2) \le tW(\mathbf{F}_1) + (1-t)W(\mathbf{F}_2).$$

Finally the convexity assumption does not "get along" with the principle of Material Frame Indifference, as well. In fact

$$\begin{array}{ccccc} 0 & = & W(\mathbf{QF}) - W(\mathbf{F}) & \ge & D_F W(\mathbf{F}) \cdot (\mathbf{QF} - \mathbf{F}) \\ & \uparrow & & \uparrow & \\ & \text{M.F.I.} & & \text{convexity} & \\ & = & \mathbf{S}(\mathbf{F})\mathbf{F}^T \cdot (\mathbf{Q} - \mathbf{I}) & & \end{array}$$

thus

$$\mathbf{S}\mathbf{F}^T \cdot (\mathbf{Q} - \mathbf{I}) \leq 0 \Longleftrightarrow (\det \mathbf{F})^{-1}\mathbf{S}\mathbf{F}^T \cdot (\mathbf{Q} - \mathbf{I}),$$

which means

$$\mathbf{T} \cdot (\mathbf{Q} - \mathbf{I}) \leq 0.$$

Taking

$$\mathbf{Q} = \begin{pmatrix} 1 & 0 & 0 \\ 0 & -1 & 0 \\ 0 & 0 & -1 \end{pmatrix},$$

with respect to the principal axes of $\mathbf{T}$ we find

$$\sigma_2 + \sigma_3 \geq 0,$$

where σ_i are the eigenvalues of $\mathbf{T}$. Similarly one finds

$$\sigma_1 + \sigma_2 \geq 0 \qquad \& \qquad \sigma_1 + \sigma_3 \geq 0$$

which should hold for any deformation process. This is easily seen to be unphysical, for instance in an isotropic material a uniform compression would lead to a Cauchy stress of the form $\mathbf{T} = \sigma\mathbf{I}$ with $\sigma > 0$, i.e. tractions. Therefore W can **NOT** be convex.

Let us ask the opposite question: if E is sequentially-weakly-$W^{1,p}(\Omega)$-l.s.c. what can we say on W ?

Theorem 7.8. *Let $p \in (1, +\infty)$ and*

$$E(\boldsymbol{\chi}) = \int_\Omega W(\nabla\boldsymbol{\chi})\, d\boldsymbol{X}$$

be sequentially-weakly-$\mathrm{W}^{1,p}(\Omega)$*-l.s.c. Assume there exists $c > 0$ such that*

$$0 \leq W(\mathbf{F}) \leq c(1 + |\mathbf{F}|^p).$$

Then W is ***quasiconvex****, that is*

$$\int_\omega W(\mathbf{F})\, d\boldsymbol{X} \leq \int_\omega W(\mathbf{F} + \nabla\phi)\, d\boldsymbol{X}$$

for any $\phi \in C_0^\infty(\omega)$, $\mathbf{F} \in \mathbb{R}^{3\times 3}$ and for every open set ω.

Quasiconvexity 'says' that if we have a body subject to an homogenous boundary deformation $\mathbf{F}\boldsymbol{X} + \boldsymbol{c}$ and no body forces, than the minimizing configuration is the homogeneous one, i.e., $\mathbf{F}\boldsymbol{X} + \boldsymbol{c}$.

Also the converse of the previous theorem holds:

Theorem 7.9. *(Acerbi and Fusco (1984)) Let $p \in (1, +\infty)$, and*

$$E(\chi) = \int_\Omega W(\nabla\chi)\, d\boldsymbol{X}.$$

If W is quasiconvex and there exists $c > 0$ such that

$$0 \leq W(\mathbf{F}) \leq c(1 + |\mathbf{F}|^p) \tag{20}$$

then E is sequentially-weakly-$W^{1,p}(\Omega)$-l.s.c.

This theorem quite often is used in nonlinear elasticity even if the growth from above (20) is not physical. To see this, take

$$\mathbf{F}_n = \begin{pmatrix} 1 & 0 & 0 \\ 0 & 1 & 0 \\ 0 & 0 & 1/n \end{pmatrix}$$

then $\det \mathbf{F}_n = 1/n \to 0$, as $n \to +\infty$, so it is natural to have $W(\mathbf{F}_n) \nearrow +\infty$ as $n \to +\infty$. But by (20) we have

$$W(\mathbf{F}_n) \leq c(1 + 3^p),$$

for every n.

The growth from above (20) can be avoided if W is **policonvex**, that is: it exists a convex function

$$Z : \mathcal{M}^{3\times 3} \times \mathcal{M}^{3\times 3} \times \mathbb{R} \to \mathbb{R}$$

such that

$$W(\mathbf{F}) = Z(\mathbf{F}, \operatorname{cof}\mathbf{F}, \det \mathbf{F})$$

where

$$\operatorname{cof}\mathbf{F} = (\det \mathbf{F})\mathbf{F}^{-T}.$$

If W is policonvex one can work with the convex function Z and, essentially, apply the lower semicontinuity Theorem 7.7 for convex functions. We refer to the original work of Ball (1977) and also to the book by Ciarlet (1988).

In general we have

$$\text{convexity} \implies \text{policonvexity} \implies \text{quasiconvexity}.$$

That convexity implies policonvexity is obvious, while the proof that policonvexity implies quasiconvexity can be done with

Theorem 7.10. (Jensen's Inequality) *Let f be a convex function and $u \in L^1(\Omega)$. Then*

$$f\left(\fint_{\omega} u(\boldsymbol{X})\, d\boldsymbol{X}\right) \leq \fint_{\omega} f(u(\boldsymbol{X}))\, d\boldsymbol{X}$$

where

$$\fint_{\Omega} \cdot\, d\boldsymbol{X} := \frac{1}{|\Omega|}\int_{\Omega} \cdot\, d\boldsymbol{X}$$

and where $|\Omega|$ denotes the measure of Ω.

Theorem 7.11.

$$\text{policonvexity} \implies \text{quasiconvexity}.$$

Proof. Let there exists a convex function Z such that

$$W(\mathbf{F}) = Z(\mathbf{F}, \operatorname{cof}\mathbf{F}, \det\mathbf{F}),$$

and let us, for brevity, denote by

$$M(\mathbf{F}) = (\mathbf{F}, \operatorname{cof}\mathbf{F}, \det\mathbf{F}).$$

Let $\varphi \in C_0^\infty(\Omega)$, then by Jensen's inequality we have

$$\int_{\Omega} W(\mathbf{F}+\nabla\varphi)\, d\boldsymbol{X} = \int_{\Omega} Z(M(\mathbf{F}+\nabla\varphi))\, d\boldsymbol{X} \geq |\Omega| Z\left(\fint_{\Omega} M(\mathbf{F}+\nabla\varphi)\, d\boldsymbol{X}\right).$$

We note that all components of $M(\mathbf{F})$ are the determinant of a matrix. In fact, $\det\mathbf{F}$ is the determinant of $\mathbf{F}$, the components of $\operatorname{cof}\mathbf{F}$ are the components of the 2×2 sub-matrices of $\mathbf{F}$, and finally the components of $\mathbf{F}$ are the determinants of the 1×1 sub-matrices of $\mathbf{F}$ itself. So we can conclude the proof using the following

claim 7.12. Let $\boldsymbol{u} \in C^\infty(\Omega)$ then

$$\int_{\Omega} (\det \nabla\boldsymbol{u})\, d\boldsymbol{X}$$

depends only on the values of $\boldsymbol{u}$ and $\nabla\boldsymbol{u}$ on $\partial\Omega$.

Supposing the claim holds we find

$$\int_{\Omega} W(\mathbf{F}+\nabla\varphi)\, d\boldsymbol{X} \geq |\Omega| Z\left(\fint M(\mathbf{F}+\nabla\varphi)\right) =$$

$$= |\Omega| Z\left(\fint_{\Omega} M(\mathbf{F})\, d\boldsymbol{X}\right) = |\Omega| Z(M(\mathbf{F})) = |\Omega| W(\mathbf{F}) = \int_{\Omega} W(\mathbf{F})\, d\boldsymbol{X}.$$

Thus to conclude the proof it suffices to prove the claim.
PROOF OF THE CLAIM FOR $\boldsymbol{u} : \Omega \subset \mathbb{R}^2 \to \mathbb{R}^2$

$$
\begin{aligned}
(\det \nabla \boldsymbol{u}) &= \frac{\partial u_1}{\partial X_1}\frac{\partial u_2}{\partial X_2} - \frac{\partial u_1}{\partial X_2}\frac{\partial u_2}{\partial X_1} \\
&= \frac{\partial}{\partial X_1}\left(u_1 \frac{\partial u_2}{\partial X_2}\right) - u_1 \frac{\partial^2 u_2}{\partial X_1 \partial X_2} - \frac{\partial}{\partial X_2}\left(u_1 \frac{\partial u_2}{\partial X_1}\right) + u_1 \frac{\partial^2 u_2}{\partial X_1 \partial X_2}
\end{aligned}
$$

thus

$$
\int_\Omega (\det \nabla \boldsymbol{u})\, d\boldsymbol{X} = \int_{\partial\Omega} \left(u_1 \frac{\partial u_2}{\partial X_1} n_1 - u_1 \frac{\partial u_2}{\partial X_1} n_2\right) d\boldsymbol{X} \qquad \square
$$

Another very important type of convexity is rank-one convexity.

Definition 7.13. A function W is said to be **rank-one convex** if

$$
t \mapsto W(\boldsymbol{F} + t\mathbf{a} \otimes \mathbf{b})
$$

is convex for every $\boldsymbol{F} \in Lin$, $\mathbf{a}$,$\mathbf{b} \in \mathbb{R}^3$.

It can be proved that

$$
\text{quasiconvexity} \implies \text{rank-one convexity}
$$

and thus if W is not rank-one convex it is not quasiconvex. It has been proved that if W is a polynomial of degree two then quasiconvexity is equivalent to rank-one convexity.
We conclude by mentioning examples which show that in general the definitions of "convexity" given are not equivalent.
For $W : \mathbb{R}^{n\times m} \to \mathbb{R}$ we have

$$
\begin{array}{ccccccc}
\text{convexity} & \overset{\Rightarrow}{\not\Leftarrow} & \text{quasiconvexity} & \overset{\Rightarrow}{\not\Leftarrow} & \text{policonvexity} & \overset{\Rightarrow}{\not\Leftarrow} & \text{rank-one conv.} \\
& \uparrow & & & & \uparrow & \\
& W(\mathbf{F}) = \det \mathbf{F} & & & & \begin{array}{c} n \geq 3 \\ m \geq 2 \end{array} &
\end{array}
$$

Example. (Dacorogna and Marcellini (1988)) Let

$$
W_\delta : \mathbb{R}^{2\times 2} \to \mathbb{R}
$$

defined by

$$
W_\delta := |\xi|^4 - \delta |\xi|^2 \det \xi
$$

then

$$W_\delta \quad \text{is} \quad \begin{cases} \text{convex} & \Longleftrightarrow & |\delta| \leq \frac{4}{3}\sqrt{2} \approx 1.89 \\ \text{polyconvex} & \Longleftrightarrow & |\delta| \leq 2 \\ \text{quasiconvex} & \Longleftrightarrow & |\delta| \leq 2 + \eta \\ \text{rank-one convex} & \Longleftrightarrow & |\delta| \leq \frac{4}{3}\sqrt{3} \approx 2.31 \end{cases}$$

where $\eta > 0$. Whether or not $2 + \eta = \frac{4}{3}\sqrt{3}$ is still open.

Bibliography

E. Acerbi and N. Fusco. Semicontinuity problems in the calculus of variations. *Archive for Rational Mechanics and Analysis*, 86:125–145, 1984.

J. Ball. Convexity conditions and existence theorems in nonlinear elasticity. *Archive for Rational Mechanics and Analysis*, 63:337–403, 1977.

P.G. Ciarlet. *Mathematical Elasticity. Volume I: Three-Dimensional Elasticity.* North-Holland, 1988.

B. Dacorogna and P. Marcellini. A counterexample in the vectorial calculus of variations. In *Material instabilities in continuum mechanics*, pages 77–83. CRC Press, 1988.

L.C. Evans and R.F.Gariepy. *Measure Theory and Fine Properties of Functions.* CRC Press, 1992.

M.E. Gurtin. *An Introduction to Continuum Mechanics.* Academic Press, 1981.

A. Morassi. Strain, stress and linearized elasticity. In A. Morassi and R. Paroni, editors, *Classical and Advanced Theories of Thin Structures: Mechanical and Mathematical Aspects.* Springer, 2008.

C. Truesdell. *A First Course in Continuum Rational Mechanics.* Academic Press, 1991.

An introduction to Γ-convergence methods for thin structures

Danilo Percivale

Dipartimento di Ingegneria della Produzione,
Università di Genova,
Piazzale Kennedy, Fiera del Mare, Padiglione D, Genova, Italy

Abstract In these lectures an essential introduction to lower semicontinuity and Γ-convergence basic facts is given together with simple applications to thin structures in elasticity.

1 Introduction to lower semicontinuity, relaxation and Γ-convergence.

Let X be a metric space and $F : X \to \mathbf{R} \cup \{+\infty\}$.

Definition 1.1. We say that F is lower semicontinuous (briefly l.s.c.) if for every $v_h \to v$ in X,

$$\liminf_{h\to\infty} F(v_h) \geq F(v). \tag{1.0.1}$$

F is said upper semicontinuous (briefly u.s.c.) if $-F$ is l.s.c.

The following propositions states elementary properties of l.s.c functions.

Proposition 1.2. *i) If F and G are both l.s.c then $F + G$ is l.s.c.*
ii) If F and G are both l.s.c and $F + G$ is finite and continuous then both F and G are continuous.

Proof. i) is trivial so we prove ii). Since G is l.s.c. then $-G$ is upper s.c. then $F = F + G - G$ is u.s.c and then continuous.The same holds for G. □

Proposition 1.3. *Let (F_j) be a family of l.s.c functions. Then $\sup_j F_j$ is l.s.c.*

Proof. Let $v_h \to v$ in X. Then by lower semicontinuity of each F_j we get

$$\begin{aligned} F(v) = \sup_j F_j(v) \leq \sup_j(\liminf F(v_h)) \leq \\ \leq \liminf(\sup_j F_j(v_h)) = \liminf F(v_h) \end{aligned} \tag{1.0.2}$$

thus proving lower semicontinuity of F. □

Let now $V \subset X$. The problem of Calculus of Variations is to find

$$\min\{F(v) : v \in V\}.$$

To this aim are useful the following

Definition 1.4. Assume $\inf_V F > -\infty$. A sequence $\{v_h\} \subset V$ is said a minimizing sequence for F in V if $F(v_h) \to \inf_V F$.

Definition 1.5. Let $F : X \to \mathbf{R} \cup \{+\infty\}$ such that $\inf_V F > -\infty$. F is said coercive on V if every minimizing sequence admits a susequence which converges in V.

The following result clarifies connection between semicontinuity, coerciveness and existence of minimizers.

Theorem 1.6. *Let $F : X \to \boldsymbol{R} \cup \{+\infty\}$ be l.s.c., $V \subset X$, F coercive on V. Then there exists $v_0 \in V$ such that $F(v_0) = \min_V F$.*

Proof. Since F is coercive on V then there exists $v_h \in V$, $v_h \to v_0$ in V such that $F(v_h) \to \inf_V F$. Hence $\inf_V F \leq F(v_0) \leq \liminf F(v_h) = \inf_V F$ and then $F(v_0) = \min_V F$. □

In the application to the Calculus of Variations neither X nor its topology are a priori fixed, even if a natural choice exists in many situations. In particular we notice that we may assume that $V = X$ since the problem of finding $\min_V F$ is equivalent to find $\min_X \widetilde{F}$ where

$$\widetilde{F}(v) = \begin{cases} F(v) & \text{if } v \in V \\ \\ +\infty & \text{otherwise} \end{cases} . \tag{1.0.3}$$

It may happen that a natural topology (in which minimizing sequences are convergent) does not imply lower semicontinuity for F.

Example 1.7. Let $X = L^1(0,1)$, $V = \{u \in W^{1,1}(0.1) : u(0) = 0\} \subset X$ and

$$F(u) = \begin{cases} \displaystyle\int_0^1 (1 - |u'|^2)^2 \, dx \quad \text{if} u \in V \\ +\infty \quad \text{otherwise in} \quad L^1(0,1) \end{cases} \tag{1.0.4}$$

Let $u_h \in L^1(0,1)$ be a minimizing sequence : since $\inf F \leq F(0) = 0$ then $F(u_h) \leq 2$ for large h, $u_h \in V$ and recalling that $2(a+b)^2 \geq a^2 - 4b^2$

$$\int_0^1 |u_h'|^4 \, dx - 4 \leq 2 \int_0^1 (1 - |u_h'|^2)^2 \, dx \leq 4 \tag{1.0.5}$$

and then

$$|u_h(x)|^4 = \left(\int_0^x u_h'(t) \, dt \right)^4 \leq \int_0^1 |u_h'|^4 \leq 8. \tag{1.0.6}$$

By Rellich's theorem there exists a subsequence (not relabeled) such that $u_h \to u$ in L^1 and F is then coercive.
We show that F is not l.s.c.: let $v \equiv 0$ and $w(x) = 1 - 2|2x - 1|$. By denoting again with w the 1-periodic extension of w to the whole $[0, +\infty)$, we set $v_h(x) = h^{-1} w(hx)$. Then $|v_h'(x)| = |w'(hx)| = 1$ and $v_h \to 0$ in L^1, but $F(v_h) = 0 < F(v) = 1$ and F is not l.s.c. . (Nevertheless F has a minimizer: $v(x) = x$!!)

Example 1.8. Let $\Omega \subset \mathbf{R}^n$ an open bounded set , $\partial\Omega \in \mathrm{Lip}$

$$F(u) = \begin{cases} \displaystyle\int_\Omega |\nabla u|^2 \quad \text{if } u \in C^1(\bar{\Omega}) \\ +\infty \quad \text{otherwise in } L^2(\Omega) \end{cases} . \tag{1.0.7}$$

Let $u \in H^1(\Omega) \setminus C^1(\bar{\Omega})$ then there exists $u_h \in C^1(\bar{\Omega})$ such that $u_h \to u$ in $H^1(\Omega)$ (hence in L^2) and

$$F(u_h) = \int_\Omega |\nabla u_h|^2 \to \int_\Omega |\nabla u|^2 < F(u) = +\infty. \tag{1.0.8}$$

Then F is not lower semicontinuous.

Previous examples suggest the following

Definition 1.9. Let $F : X \to \mathbf{R} \cup \{+\infty\}$. The functional

$$\bar{F}(v) = \inf\{\liminf F(v_h) : v_h \to v\} \tag{1.0.9}$$

is called the relaxed functional of F.

It is possible to show that

$$\bar{F}(v) = \sup\{G(v) : G \text{ l.s.c.}, \ G \leq F\} \tag{1.0.10}$$

that is $\bar{F}$ is the greatest l.s.c. functional less or equal to F.
The following result exploits Definition1.9 and its proof is left to the reader.

Proposition 1.10. *$\bar{F}$ is the relaxation of F if and only if:*
i) for every $v_h \to v$ $\bar{F}(v) \leq \liminf F(v_h)$
ii) for every $v \in X$ there exists $\bar{v}_h \to v$ such that $F(\bar{v}_h) \to \bar{F}(v)$

We state now the most important property of the relaxed functional.

Theorem 1.11. *Let $F : X \to \boldsymbol{R} \cup \{+\infty\}$, $\inf F > -\infty$, F coercive. Then* $\inf F = \min \bar{F}$.

Proof. Let v_h be a minimizing sequence: then there exists a decreasing sequence $\varepsilon_h \to 0$ such that $\bar{F}(v_0) < \inf F + \varepsilon_h$ and by coerciveness we may assume that $v_h \to v_0$.By Proposition1.10 there exists $\bar{v}_h \to v_0$ such that $F(\bar{v}_h) \to \bar{F}(v_0)$ and

$$\bar{F}(v_0) \leq \liminf(F(v_h) - \varepsilon_h) \leq \inf F \leq \limsup F(\bar{v}_h) = \bar{F}(v_0). \tag{1.0.11}$$

Then $\bar{F}(v_0) = \inf F = \min F$ □

A natural generalization of relaxation is Γ-convergence.
Let $F_h, \ F : X \to \mathbf{R} \cup \{+\infty\}$, then

Definition 1.12. We set for every $u \in X$

$$F_+(u) =: \inf\{\limsup F_h(u_h) : \ u_h \to u\} =: \Gamma(X^-) \limsup F_h(u)$$

$$F_-(u) =: \inf\{\liminf F_h(u_h) : \ u_h \to u\} =: \Gamma(X^-) \liminf F_h(u) \tag{1.0.12}$$

If $F_+(u) = F_-(u) = F(u)$ for every $u \in X$ we set

$$F(u) =: \Gamma(X^-) \lim F_h(u) \tag{1.0.13}$$

and F is called the Γ- limit of the F_h.

Clearly $F_+(u) \geq F_-(u)$, and one can prove (see [17]) that both F_- and F_+ are l.s.c.;moreover the following characterization holds:

Proposition 1.13. *$F(u) = \Gamma(X^-)\lim F_h(u)$ if and only if*
i) $\forall u_h \to u, \quad \liminf F_h(u_h) \geq F(u).$
ii) $\forall u \in X \quad \exists \bar{u}_h \to u : F_h(\bar{u}_h) \to F(u)$
($\bar{u}_h$ is often called the "recovery sequence")

Proof. Assume that i) and ii) hold. From i) we get $F(u) \leq F_-(u)$ while ii) yields $F(u) \geq F_+(u)$ and since $F_+(u) \geq F_-(u)$ thesis follows.Conversely if $F_+(u) = F_-(u)$ for every $u \in X$, then $F_+(u) = F_-(u) = F(u) \leq \liminf F_h(u_h)$ for every $u_h \to u$ and i) is proven. Moreover for every $u \in X$ there exists $\bar{u}_h \to u$ such that

$$\limsup F_h(\bar{u}_h) = F_+(u) \leq \liminf F_h(\bar{u}_h)$$

thus proving ii). □

Remark 1.14. If $F_h \equiv F$ for every $h \in \mathbb{N}$ then

$$\Gamma(X^-)\lim F_h(u) = \bar{F}(u),$$

the relaxed functional of F.

As far as it concerns convergence of minimizing sequences, a new notion of coerciveness, independent of h, is needed. Namely

Definition 1.15. The family of functionals $(F_h)_{h\in\mathbb{N}}$ is called equicoercive if $\inf F_h > -\infty$ and for every $t \in \mathbf{R}$ there exists a sequentially compact set $E_t \subset X$ such that $\{u : \; F_h(u) \leq t\} \subset E_t$.

The following result is the natural extension of Theorem 1.11

Theorem 1.16. *Let $F_h, \; F : X \to \boldsymbol{R} \cup \{+\infty\}$ such that*
i) $\exists u_0 \in X : \; F(u_0) < +\infty$
ii) $\Gamma(X^-)\lim F_h(u) = F(u) \; \forall u \in X$
iii) $(F_h)_{h\in\mathbb{N}}$ is equicoercive.
Then $\inf F_h \to \min F$; moreover if $F_h(u_h) - \inf F_h \to 0$ then, up to subsequences, $u_h \to u_0$ and $F(u_0) = \min F$.

The proof of this theorem , very close to that of Theorem 1.11, is left to the reader (see again [17]).Another simple but useful result is the following

Theorem 1.17. *Let $\Gamma(X^-)\lim F_h(u) = F(u)$ and let $G : X \to \boldsymbol{R} \cup \{+\infty\}$ be continuous. Then $\Gamma(X^-)\lim(F_h(u) + G(u)) = F(u) + G(u)$.*

Example 1.18. Let $G_h : L^2(0,1) \to \mathbf{R} \cup \{+\infty\}$

$$G_h(u) = \begin{cases} \displaystyle\int_0^1 a_h |u'|^2\,dx - \int_0^1 fu\,dx & \text{if } u \in H_0^1(0,1) \\ +\infty & \text{otherwise} \end{cases} \tag{1.0.14}$$

where $f \in L^2(0,1)$, $0 < \lambda \le a_h \le \Lambda$, $a_h^{-1} \to a^{-1}$ in $w^* - L^\infty$. Then the Γ-limit of (G_h) is

$$G(u) = \begin{cases} \displaystyle\int_0^1 a|u'|^2\,dx - \int_0^1 fu\,dx & \text{if } u \in H_0^1(0,1) \\ +\infty & \text{otherwise} \end{cases} \tag{1.0.15}$$

Indeed let $u_h \to u$ in $L^2(0,1)$ and assume without restriction that $G_h(u_h) \le C$ (otherwise we may suppose $\liminf G_h(u_h) = +\infty$ and the inequality $\liminf G_h(u_h) \ge G(u)$ becomes obvious). Then a standard application of Poincarè inequality (try as an exercise!) show that $u_h \to u$ in $w - H^1(0,1)$. Let us consider the auxiliary problem

$$\min\left\{\int_0^1 a_h|v'|^2\,dx - \int_0^1 2au'v'\,dx\,:\; v \in H_0^1(0,1)\right\}; \tag{1.0.16}$$

if v_h is a minimizer then Euler equation yields $a_h v_h' - au' = c_h \in \mathbf{R}$ that is $v_h' = \frac{a}{a_h}u' + c_h a_h^{-1}$ and by using the boundary conditions we see that $c_h = (-\int_0^1 a_h^{-1})^{-1}\int_0^1 aa_h^{-1}u' \to 0$ which implies $v_h' \to u'$ weakly in L^2 and $a_h v_h' \to au'$ uniformly. Hence

$$\int_0^1 a_h|v_h'|^2\,dx \to \int_0^1 a|u'|^2\,dx \tag{1.0.17}$$

and

$$G_h(u_h) + \int_0^1 (fu_h - 2au'u_h')\,dx \ge \int_0^1 a_h|v_h'|^2\,dx - \int_0^1 2au'v_h'\,dx.$$

Then

$$\liminf G_h(u_h) \ge \int_0^1 a|u'|^2\,dx - \int_0^1 fu\,dx = G(u).$$

Let now $u \in H_0^1(0,1)$ and let v_h a minimizer of (1.0.16): then $v_h \to u$ in L^2 and by (1.0.17) we get $G_h(v_h) \to G(u)$ and the proof of Γ-convergence is complete.

2 Γ-convergence and thin structures

In this section we will show some applications of Γ-convergence to the theory of thin elastic and elasto-plastic structures: the main idea consists in considering a thin structure as a limit of a 3D body when one or two dimensions go to zero.We begin with the following simple

Example 2.1. Let $\Omega_h = \omega \times (-\frac{1}{h}, \frac{1}{h})$ where ω is a bounded open Lipschitz subset of $\mathbf{R}^2$ and let

$$F_h(u) = \begin{cases} \frac{1}{h}\int_{\Omega_h} |\nabla u|^2 \, dx_1 dx_2 \ \text{ if } \ u \in H_0^1(\Omega_h) \\ \\ +\infty \ \text{ otherwise in } L^2(\Omega_h). \end{cases} \tag{2.0.18}$$

It is worth noticing that the sequence of functionals is not defined on a fixed function space but a simple change of variables, namely

$$v(x_1, t) =: u(x_1, \frac{t}{h})$$

yields

$$F_h(u) = G_h(v) = \begin{cases} \int_{\Omega_1} (|v_{,1}|^2 + h^2 |v_{,t}|^2) \, dx_1 dt \ \text{ if } \ v \in H_0^1(\Omega_1) \\ \\ +\infty \ \text{ otherwise in } L^2(\Omega_1). \end{cases} \tag{2.0.19}$$

and we may compute $\Gamma \lim G_h(v)$ instead of $\Gamma \lim F_h(u)$. If we assume that $v_h \to v$ in $L^2(\Omega_1)$ and $G_h(v) \leq C$ then

$$\int_{\Omega_1} (|v_{h,1}|^2 + h^2 |v_{h,t}|^2) \, dx_1 dt \leq C$$

that is $v_{,t} = 0$ which means $v \in H^1(\Omega_1)$, v independent of t and then $v \in H^1(\omega)$. Obviously

$$G_h(v_h) \geq \int_{\Omega_1} |v_{h,1}|^2 \, dx_1 dt$$

hence

$$\liminf G_h(v_h) \geq \int_{\Omega_1} |v_{,1}|^2 \, dx_1 dt = 2 \int_{\omega} v_{,1}|^2 \, dx_1.$$

Let now any $v \in H_0^1(\Omega_1)$, v independent of t; then

$$G_h(v) = \int_{\Omega_1} |v_{,1}|^2 \, dx_1 dt = 2 \int_{\omega} v_{,1}|^2 \, dx_1$$

and this means that

$$\Gamma(L^2(\Omega_1)^-)\lim G_h(v) = G(v)$$

where

$$G(v) = \begin{cases} 2\int_\omega |v_{,1}|^2\,dx_1 \ \text{ if } \ v \in H_0^1(\Omega_1),\ v_{,t} = 0 \\ \\ +\infty \ \text{ otherwise in } L^2(\Omega_1). \end{cases} \tag{2.0.20}$$

A more delicate situation comes from 3D elasticity as we will show in our second example.

Example 2.2. Let again $\Omega_h = \omega \times (-\frac{1}{h}, \frac{1}{h})$ where ω is a bounded open Lipschitz subset of $\mathbf{R}^2$, let $U_h = \{\mathbf{u} \in (H^1(\Omega_h))^3 : \ \mathbf{u} = 0 \text{ in } \partial\omega \times (-\frac{1}{h}, \frac{1}{h})\}$ and

$$F_h(u) = \begin{cases} h^3\int_{\Omega_h}(\mu|\mathbb{E}(\mathbf{u})|^2 + \frac{\lambda}{2}|tr\mathbb{E}(\mathbf{u})|^2)\,dx_1dx_2 \ \text{ if } \ \mathbf{u} \in U_h \\ \\ +\infty \ \text{ otherwise in } (L^2(\Omega_h))^3 \end{cases} \tag{2.0.21}$$

where μ, λ are the Lamè constants and $\mathbb{E}(\mathbf{u})$ is the strain tensor.The change of variables

$$\mathbf{v}(\mathbf{x}, t) = \left(hu_1(\mathbf{x}, \frac{t}{h}),\ hu_2(\mathbf{x}, \frac{t}{h}),\ u_3(\mathbf{x}, \frac{t}{h})\right). \tag{2.0.22}$$

yields

$$\begin{cases} v_{\alpha,\beta} = hu_{\alpha,\beta} \\ v_{\alpha,t} = u_{\alpha,3},\ v_{3,\alpha} = u_{3,\alpha} \\ hv_{3,t} = u_{3,3} \end{cases} \tag{2.0.23}$$

and hence for every $\mathbf{u} \in U_h$ we have $\mathbf{v} \in V = \{\mathbf{v} \in H^1(\Omega_1) : \ \mathbf{v} = 0 \text{ in } \partial\omega \times (-1,1)\}$ and by writing briefly $\mathbb{E}_{ij}$ instead of $\mathbb{E}_{ij}(\mathbf{v})$ we get

$$F_h(\mathbf{u}) = G_h(\mathbf{v}) = \int_{\Omega_1} \mu(|\mathbb{E}_{\alpha\beta}|^2 + h^2|\mathbb{E}_{3\alpha}|^2 + h^4|\mathbb{E}_{33}|^2) + \frac{\lambda}{2}|\mathbb{E}_{11} + \mathbb{E}_{22} + h^2\mathbb{E}_{3:} \tag{2.0.24}$$

The same techniques of the previous example show that if $\mathbf{v}_h \to \mathbf{v}$ in L^2 then $\mathbf{v}_h \to \mathbf{v}$ in $w - H^1$ and $\mathbb{E}_{3\alpha}(\mathbf{v}) = \mathbb{E}_{33}(\mathbf{v}) = 0$ that is $v_3 \equiv v_3(x_1, x_2)$ and $v_\alpha = -x_3v_{3,\alpha}$. Since

$$\begin{aligned} &\mu(|\mathbb{E}_{\alpha\beta}|^2 + h^2|\mathbb{E}_{3\alpha}|^2 + h^4|\mathbb{E}_{33}|^2) + \tfrac{\lambda}{2}|\mathbb{E}_{11} + \mathbb{E}_{22} + h^2\mathbb{E}_{33}|^2 \geq \\ &\mu|\mathbb{E}_{\alpha\beta}|^2 + \tfrac{2\lambda\mu}{(\lambda+2\mu)^2}(\mathbb{E}_{11} + \mathbb{E}_{22})^2 \end{aligned} \tag{2.0.25}$$

then

$$\liminf G_h(\mathbf{v}_h) \geq \frac{2}{3}\mu \int_\omega |D^2 v_3|^2 + \frac{2\lambda}{(\lambda+2\mu)^2}|\Delta v_3|^2. \tag{2.0.26}$$

As far as it concerns the recovery sequence let $\mathbf{v} \in C^2(\bar{\Omega}_1)$, $\mathbf{v} = 0$ on $\partial\omega \times (-1,1)$ $v_3 = v_3(\mathbf{x})$, $v_\alpha = -x_3 v_{3,\alpha}$; then there exists a unique $\mathbf{p} \in C^1(\Omega_1)$ such that by setting $\mathbf{P}(\mathbf{x},t) = \int_0^t \mathbf{p}(\mathbf{x},\tau)\,d\tau$, $\sigma_h \in C_0^1(\omega)$, $\sigma_h(\mathbf{x}) \equiv 1$ if $d(\mathbf{x},\partial\omega) \geq \sqrt{h^{-1}}$ and

$$\mathbf{v}_h = (\mathbf{v}_\alpha + x_3^2 p_\alpha \sigma_h, u_3(\mathbf{x}) + h^{-2}\sigma_h(\mathbf{x})\mathbf{P}) \tag{2.0.27}$$

then

$$G_h(\mathbf{v}_h) = \frac{2}{3}\mu \int_\omega |D^2 v_3|^2 + \frac{2\lambda}{(\lambda+2\mu)^2}|\Delta v_3|^2\,dx_1dx_2 + o(1). \tag{2.0.28}$$

We have proven that

$$\Gamma \limsup G_h(\mathbf{v}) \leq G(\mathbf{v}),\ \forall \mathbf{v} \in V \cap C^2(\Omega_1) \tag{2.0.29}$$

where

$$G(v) = \begin{cases} \frac{2}{3}\mu \int_\omega |D^2 v_3|^2 + \frac{2\lambda}{(\lambda+2\mu)^2}|\Delta v_3|^2\,dx_1dx_2 \text{ if } \mathbf{v} \in V_\infty \\ \\ +\infty \quad \text{otherwise in} \quad L^2(\Omega_1) \end{cases} \tag{2.0.30}$$

and

$$V_\infty = \{\mathbf{v} \in V : v_3 \in H_0^2(\omega),\ v_\alpha = -x_3 v_{3,\alpha}\}.$$

By recalling now lower semicontinuity of the $\Gamma - \limsup$ we get

$$\Gamma - \limsup G_h(\mathbf{v}) \leq G(\mathbf{v})\ \forall \mathbf{v} \in V \tag{2.0.31}$$

hence

$$\Gamma - \lim G_h(\mathbf{v}) = G(\mathbf{v}). \tag{2.0.32}$$

3 The clamped elastic-plastic beam

A similar approach works also for the elastic-plastic plate or for the elastic-plastic beam : in the latter case a limit functional which takes into account formation of plastic hinges can be obtained, namely

$$\mathcal{F}(w) \ = \ \frac{1}{2}\int_{\mathbf{R}} |\ddot{w}|^2 - \int_{\mathbf{R}} w\,d\nu \ + \beta\,\sharp(S_{\dot{w}}) \ + \ \gamma \sum_{S_{\dot{w}}} |[\dot{w}]| \tag{3.0.33}$$

In the following $\beta > 0, \gamma > 0$ are given constants, $\sharp$ is the counting measure , $\nu = f\,dx + \nu_s$ is a Radon measure such that spt$f \subset [0, L]$ and spt$\nu_s \subset\subset (0, L)$. The total variation of ν will be denoted with $|\nu|_T$.
If $J \subset \mathbf{R}$ is an interval then $SBH(J)$ will denote the space of $W^{1,1}(J)$-functions whose second distributional derivative is a measure without Cantor part For every $v \in SBH(J)$, $S_{\dot v}$ will denote the set of jump points of $\dot v$ and $\ddot v$ will denote the absolutely continuous part of v''.

Lemma 3.1. *(Poincarè Inequality) Let $z \in SBH(\mathbf{R})$ with $z \equiv 0$ in $\mathbf{R}\backslash[0, L]$. Then*

$$\|z\|_\infty \le \frac{L}{8}|z''|_T$$

We want to summarize here some properties of minimizers of functional (3.0.33).
Now we can prove that a smallness condition (safe load condition) on $|\nu|_T$ entails the existence of minimizers.

Lemma 3.2. *(Safe load condition) Assume that*

$$|\nu|_T \; < \; \frac{8\,\gamma}{L} \tag{3.0.34}$$

then $\mathcal{F}$ achieves a finite minimum.

Proof. By Lemma 3.1 we get

$$\begin{aligned} \left|\int_{\mathbf{R}w\,d\nu}\right| \;&\le\; \frac{L}{8}|\nu|_T|w''|_T \le \frac{L}{8}|\nu|_T \sum_{S_{\dot w}} |[\dot w]| + \gamma\int_0^L |\ddot w| \le \\ &\le \frac{L}{8}|\nu|_T \sum_{S_{\dot w}} |[\dot w]| + \frac{1}{4}\int_0^L |\ddot w|^2 + \gamma^2 L \end{aligned} \tag{3.0.35}$$

for every $w \in SBH(0, L)$ such that spt$w \subset [0, L]$, hence, by Young inequality, we have for any admissible w

$$\mathcal{F}(w) \ge \beta\sharp(S_{\dot w}) + \left(\gamma - \frac{L}{8}|\nu|_T\right)\sum_{S_{\dot w}} |[\dot w]| - \gamma^2 L + \frac{1}{4}\int_0^L |\ddot w|^2$$

then if (3.2) holds existence of minimizers follows from a standard compactness and l.s.c argument ([PT2],[CLT2,3,4]). □

The first variation of the functional $\mathcal{F}$ yields, as usual,

Theorem 3.3. *(Euler equation) Let $w \in argmin\mathcal{F}$. Then $\ddot{w} \in C([0, L])$ and moreover*

$$(i) \qquad (\ddot{w})'' = \nu \qquad in\ (0, L)$$

$$(ii) \qquad \ddot{w}_- = \gamma sign([\dot{w}]) \qquad in\ S_{\dot{w}} \cap (0, L]$$

$$(iii) \qquad \ddot{w}_+ = \gamma sign([\dot{w}]) \qquad in\ S_{\dot{w}} \cap [0, L)$$

$$(iv) \qquad \dddot{w}_- = \dddot{w}_+ \qquad in\ (0, L) \setminus S_{\dddot{w}}.$$

In particular $\ddot{w} \in BH(0, L)$, hence $\dddot{w} = (\ddot{w})'$is continuous in $(0, L)$. We notice that if $\nu_s \equiv 0$ then $\dddot{w}_- = \dddot{w}_+$ on the whole $(0, L)$.

Remark 3.4. We notice that if $w \in \text{argmin}\mathcal{F}$ and $S_{\dot{w}} = \emptyset$ then $w(0) = w(L) = w'(0) = w'(L))$, $\ddot{w} = w''$ and condition i) of Theorem (3.3) entails $w'''' = \nu$ in $(0, L)$. Then $w \equiv u$, the solution of $u'''' = \nu;\ u \in H_0^2(0, L)$.

Another important consequence of Euler equation is the following

Lemma 3.5. *(Compliance identity)Assume that w satisfies conditions (i),(ii),(iii) of Theorem 3.3 . Then*

$$\mathcal{F}(w) = -\frac{1}{2}\int_0^L |\ddot{w}|^2\, dx + \beta \sharp (S_{\dot{w}}).$$

Proof. By i) we have $(\ddot{w})'' = \nu$ and taking into account that $\text{spt}\nu \subset [0, L]$ and $w'' = \ddot{w} + \sum_{S_{\dot{w}}} [\dot{w}]\, d\sharp \llcorner S_{\dot{w}}$ we get

$$\int_{\mathbf{R}} w\, d\nu = \int_0^L w\, d\nu = \int_0^L (\ddot{w})'' w = -\int_0^L (\ddot{w})' w' = \int_0^L |\ddot{w}|^2 + \sum_{S_{\dot{w}}} \ddot{w}[\dot{w}].$$

Recalling that $\ddot{w} = \gamma\text{sign}[\dot{w}]$

$$\int_0^L w\, d\nu = \int_0^L |\ddot{w}|^2 + \gamma \sum_{S_{\dot{w}}} |[\dot{w}]|$$

thesis follows by the definition of $\mathcal{F}$. □

4 The structure of non regular solutions

Theorem 4.1. *If $v \in argmin\mathcal{F}$ then $\sharp(S_{\dot{v}}) \leq 2$.*

Proof. Assume by contradiction that $w \in \text{argmin}\mathcal{F}$ and $x_1 < x_2 < x_3$ are three distinct points in $S_{\dot{w}}$. Then we can modify w by eliminating one of them and reducing at the same time the energy. Set

$$\lambda = \frac{x_2 - x_1}{x_3 - x_1}$$

$$0 < \varepsilon < \varepsilon_0 = \min\left\{ |[w](x_1)|\, , \frac{1-\lambda}{\lambda}\, |[w](x_3)| \right\}$$

and assume without restriction that $\varepsilon_0 =, |[w](x_1)|$. We define a function $W_\varepsilon \in SBH(\mathbf{R})$ such that $W_\varepsilon(0_-) = 0$ and

$$\dot{W}_\varepsilon \;=\; \dot{w} - \varepsilon\,\text{sign}([\dot{w}](x_1))\,\chi_{[x_1,x_2]} + \frac{\lambda\,\varepsilon\,\text{sign}([\dot{w}](x_1))}{1-\lambda}\,\chi_{[x_2,x_3]}, \qquad (4.0.36)$$

hence W_ε is different from w only outside $[x_1, x_3]$, $\int_0^L \dot{W}_\varepsilon = \int_0^L \dot{w} = 0$, spt$w \subset$ $[0, L]$ and

$$[\dot{W}_\varepsilon](x_2) \;=\; \frac{\varepsilon}{1-\lambda}\text{sign}[\dot{w}](x_1) + [\dot{w}](x_2).$$

Then either $[\dot{W}_{\varepsilon_0}](x_2)\,[\dot{w}](x_2) \;>\; 0$ or $[\dot{W}_{\varepsilon_0}](x_2)\,[\dot{w}](x_2) \;\leq\; 0$: in the first case $[\dot{W}_{\varepsilon_0}](x_1) = 0$, W fulfills all the Euler conditions in Theorem3.3 and $\ddot{W} = \ddot{w}$, hence the thesis by the compliance identity.
In the second case choose $\overline{\varepsilon} \in (0, \varepsilon_0]$ such that

$$[\dot{W}_{\overline{\varepsilon}}](x_2) \;=\; \frac{\overline{\varepsilon}}{1-\lambda}\text{sign}[\dot{w}](x_1) + [\dot{w}](x_2) = 0$$

then $W_{\overline{\varepsilon}}$ fulfills all conditions in Theorem 3.3 and the thesis follows by the compliance identity . Indeed, in any case for some $\overline{\varepsilon} \in (0, \varepsilon_0]$ either $[\dot{W}_{\overline{\varepsilon}}](x_1) = 0$ or $[\dot{W}_{\overline{\varepsilon}}](x_2) = 0$ and then $\sharp(S_{\dot{W}_{\overline{\varepsilon}}}) < \sharp(S_{\dot{v}})$ and $\tilde{v}(L) = 0$, $\ddot{W}_{\overline{\varepsilon}} = \ddot{v}$, $\dddot{v} = \dddot{W}_{\overline{\varepsilon}}$ and

$$\mathcal{F}(W_{\overline{\varepsilon}}) = -1/2\int_0^L |\ddot{W}_{\overline{\varepsilon}}|^2 + \beta\sharp(S_{\dot{W}}) \leq -1/2\int_0^L |\ddot{v}|^2 + \beta\sharp(S_{\dot{v}}) - \beta < \mathcal{F}(v)$$

□

We show now an example of a one parameter family of loads such that the corresponding regular solutions are not regular minimizers of $\mathcal{F}$.

Example 4.2. Let $\frac{27}{4} < k < 8$ and choose

$$\nu = \frac{k\,\gamma}{L}\,\delta_{2L/3} \tag{4.0.37}$$

so that

$$|\nu|_T = \frac{k\,\gamma}{L} < \frac{8\,\gamma}{L}$$

and the safe load condition is satisfied. Let u the solution of $u'''' = \nu;\ u \in H_0^2(0,L)$: we want to show that if

$$0 < \beta < \frac{L}{4}\,(\ddot{u}(L)-\gamma)^2$$

then there exists $w \in SBH(\mathbf{R})$, $w \equiv 0$ outside $(0,L)$ such that $\mathcal{F}(w) < \mathcal{F}(u)$.
We first observe that by (4.0.37) , we get

$$u''(L) = \frac{1}{L^2}\int_0^L \tau^2(L-\tau)\,d\nu(\tau) = \frac{4k\gamma}{27} > \gamma. \tag{4.0.38}$$

Let now v be solution of

$$\begin{cases} v'''' = 0 & \text{in } (0,L) \\ v(0) = v'(0) = v(L) = 0, & v''(L) = \gamma - u''(L), \end{cases}$$

explicitly

$$v(x) = \frac{\gamma - u''(L)}{4L}\,x^2(x-L)\,, \qquad v'(L) = \frac{L}{4}(\gamma - u''(L)).$$

Then $w = u + v$ is a solution of

$$\begin{cases} w'''' = f & (0,L) \\ w(0) = w'(0) = w(L) = 0; & \ddot{w}(L) = \gamma \end{cases}$$

If we define $w \equiv 0$ in $\mathbf{R}\setminus(0,L)$ then $S_{\dot{w}} = \{L\}$, $[\dot{w}](L) = \frac{L}{4}(u''(L)-\gamma) > 0$, $\ddot{w}(L) = \gamma = \gamma\text{sign}([w](L))$. Hence all conditions of Theorem 3.3 are fulfilled and taking into account that $u(0) = u(L) = \dot{u}(0) = \dot{u}(L) = 0 = v(0) = \dot{v}(0) = v(L)$ and $\dddot{v} \equiv$ const. , by compliance identity (Lemma 3.5) we get:

$$\mathcal{F}(w) = \beta - \frac{1}{2}\int_0^L |\ddot{w}|^2 = \beta - \frac{1}{2}\int_0^L |\ddot{u}|^2 - \int_0^L \ddot{u}\,\ddot{v} - \frac{1}{2}\int_0^L |\ddot{v}|^2 =$$

$$= \beta - \frac{1}{2}\int_0^L |\ddot u|^2 - (\dot u\,\ddot v)\big|_0^L + \int_0^L \dot u\,\dddot v - \frac{1}{2}(\dot v\,\ddot v)\big|_0^L + \frac{1}{2}\int_0^L \dot v\,\dddot v =$$

$$= \beta - \frac{1}{2}\int_0^L |\ddot u|^2 - \frac{1}{2}\dot v(L)\,\ddot v(L) =$$

$$= -\frac{1}{2}\int_0^L |\ddot u|^2 + \beta - \frac{L}{4}(\ddot u(L) - \gamma)^2 < -\frac{1}{2}\int_0^L |\ddot u|^2 = \mathcal{F}(u).$$

Unfortunately the analysis stops here since the problem of finding a double creased minimizer remains open as well as proving that there exists at most one crease.

Bibliography

[1] L. Ambrosio, N.Fusco, D.Pallara, *Functions of Bounded Variation and Free Discontinuity Problems,* Clarendon Press Oxford, (2000).

[2] L.Ambrosio, A.Coscia & G. Dal Maso, *Fine properties of functions with bounded deformation,* Arch.Rat. Mech.Anal.,139,3(1997), 201–238.

[3] L.Ambrosio & A.Braides, *Energies in SBV and variational models in fracture mechanics,* Homogeneization and Appl. to Material sciences, (Nice, 1995),9GAKUTO Int.Ser.Math.Sci Appl.

[4] G.Anzellotti, S.Baldo & D.Percivale, *Dimension reduction in variational problems, asymptotic development in Γ-convergence and thin structures in elasticity,* Asymptotic Anal., 9,(1994) 61–100.

[5] C.Baiocchi, G.Buttazzo, F.Gastaldi & F.Tomarelli, *General existence theorems for unilateral problems in continuum mechanics,* Arch.Rat.Mech. Anal.,100, 2(1988), 149-189.

[6] G.I.Barenblatt *The formation of equilibrium cracks during brittle fracture, general ideas and hypotheses. Axially symmetric cracks,* Appl.Math. Mech. (PMM) 23,(1959), 622-636.

[7] A.Braides, G.Dal Maso & A.Garroni, *Variational formulation of softening phenomena in fracture mechanics: the one–dimensional case,* Arch.Rat. Mech. Anal., 146,(1999) 23-58.

[8] A.Braides & I.Fonseca, *Brittle thin films,*Appl.Math.Opt.44(2001) 3, 299-323.

[9] A.Braides, I.Fonseca & G.Francfort, *3D-2D asymptotic analysis for inhomogeneous thin films,*Indiana Univ.Math.J 49(2000) 4, 1367-1404.

[10] M. Carriero, A. Leaci & F. Tomarelli, *Plastic free discontinuities and special bounded hessian,* C. R. Acad. Sci. Paris,314(1992), 595–600.

[11] M. Carriero, A. Leaci & F. Tomarelli, *Special Bounded Hessian and elastic-plastic plate,* Rend. Accad. Naz. delle Scienze (dei XL), (109)XV(1992), 223–258.

[12] M. Carriero, A. Leaci & F. Tomarelli, *Strong solution for an Elastic Plastic Plate,* Calc. Var.,2(1994), 219–240.

[13] M.Carriero, A.Leaci, F.Tomarelli, *Second Order Variational Problems with Free Discontinuity and Free Gradient Discontinuity, in:* Calculus of Variations: Topics from the Mathematical Heritage of Ennio De Giorgi, D. Pallara (Ed.), Quaderni di Matematica, 14, Series edited by Dipartimento di Matematica, Seconda Università di Napoli, (2004), 135–186.

[14] V.Casarino, D.Percivale, *A variational model for non linear elastic plates,* J. Convex Anal.,3(1996) 221-243.

[15] P.G.Ciarlet, *Mathematical Elasticity, vol II: Theory of Plates,* Studies in Math. and its Appl., North-Holland,(1997).

[16] F.Colombo, F.Tomarelli, *Boundary value problems and obstacle problem for elastic bodies with free cracks,* in: Calculus of Variations: Topics from the Mathematical Heritage of Ennio De Giorgi, D. Pallara (ed.), Quaderni di Matematica, Series edited by Dipartimento di Matematica, Seconda Universita' di Napoli, (2004), 221–243.

[17] G. Dal Maso *An Introduction to $\Gamma-convergence$.* Birkhauser (1993)

[18] E. De Giorgi *Free discontinuity problems in calculus of variations,* Frontiers in Pure & Applied Mathematics,R.Dautray ed., North-Holland, Amsterdam 1991, 55-61.

[19] E. De Giorgi & L. Ambrosio, *Un nuovo tipo di funzionale del Calcolo delle Variazioni,*Atti Accad. Naz. Lincei, Rend. Cl. Sci. Fis. Mat. Natur.,82(1988), 199-210.

[20] G.Del Piero,*One-dimensional ductile-brittle transition, yielding, and structured deformations* Variations of domain and free-boundary problems in solid mechanics (Paris 1997), 203–210, Solid Mech. Appl., 66, Kluwer Acad. Publ., Dordrecht (1999)

[21] G.Del Piero & D.R.Owen, *Structured deformations of continua,* Arch.Rat.Mech. Anal.,1241993, 99-155.

[22] G.Del Piero & L.Truskinovsky, *A one-dimensional model for localized and distributed failure,*J.Phys., IV France, pr. 8, (1998), 95-102.

[23] G.Francfort & J.J.Marigo, *Revisiting brittle fracture as an energy minimization problem,*J.Mech.Phys. Solids,46(1998), 1319-1342.

[24] A.A.Griffith, *The phenomenon of rupture and flow in solids,* Phyl.Trans. Roy.Soc. A,221, (1920) 163-198.

[25] G.Kirchhoff, *Über das Gleichgewicht und die Bewegung einer elastischen Scheibe,*J.Reine Angew.Math40(1850), 51-88

[26] D.Percivale, *Perfectly Plastic Plates: a variational definition* J.Reine Angew.Math.411(1990), 39-50.

[27] D.Percivale,*Thin elastic beams: St.Venant's problem,* Asympt. Anal.,20,(1999),39-60.

[28] D.Percivale & F.Tomarelli *Scaled Korn-Poincaré inequality in BD and a model of elastic plastic cantilever,*Asymptotic Analysis,23, (2000) 291-311.

[29] D.Percivale & F.Tomarelli*From SBD to SBH: the elastic plastic plate,* Interfaces and Free Boundaries 4(2002), 137-165.

[30] D.Percivale & F.Tomarelli*From Special Bonded Deformation to Special Bounded Hessian: the elastic plastic beam,* Quad.551/P, Dip.Mat. Politecnico Milano (2003), 1-41. to appear on M^3AS

[31] G.Savarè & F.Tomarelli, *Superposition and Chain Rule for Bounded Hessian Functions,*Advances in Math.140(1998), 237-281.

[32] M.A.Save & C.E.Massonet, *Plastic analysis and design of plates, shells and disks,*North–Holland Ser. Appl.Math Mech.(1972).

[33] R.Temam, *Problèmes Mathematiques en Plasticité,*Gauthier-Vllars,(1983), Paris.

[34] R.Temam & G.Strang *Functions of bounded deformation,*Arch.Rat.Mech. Anal.,75 (1980), 7–21.

[35] L.Truskinowsky *Fracture as a phase transition,*in R.C. Batra, M.F. Beatty, (eds), Contemporary Research in the Mechanics and Mathematics of Materials (ded.to J.L.Ericksen), C.I.M.N.E., Barcelona, (1996), 322-332

[36] P. Villaggio: *Qualitative Methods in Elasticity,*Nordhoff,(1977).

Concepts in the Mechanics of Thin Structures

Paolo Podio-Guidugli [*‡]

[*] Dipartimento di Ingegneria Civile, Università di Roma TorVergata, Roma, Italy

Abstract This writing is meant to restitute – a year later, to the best of my recollection – the contents of a series of six lectures I gave in Udine, at the Centro Internazionale di Scienze Meccaniche, within the framework of the Course on *Classical and Advanced Theories of Thin Structures: Mechanical and Mathematical Aspects* (June 5-9 2006). For this reason, I have made an effort to keep my presentation style informal and colloquial, even when, here and there, I have added some complementing material.

There are three parts. In the Premiss, I try and explain why and how, to my taste, the mechanics of thin structures should be presented; in particular, I introduce the direct and deductive approaches and I collect a few bits of history of the latter, in its two main variants, the asymptotic method and the method of internal constraints. Section 2 is devoted to an exposition of the method of formal scaling, a unified approach to classic rod and plate theories that is an outgrowth of the method of internal constraints. Finally, in Section 3, I discuss how to best approximate the stress field in a linearly elastic structure-like body, by exploiting the freedom in the choice of reactive stress fields that maintain the constraints.

1 Premiss

Why and how, in June 2006, should we be talking about thin structures for one week? These questions may well come to mind first. Let me try and offer some of the possible answers.

1.1 Why?

In technical mechanics, all structures are thin, in one way or another: some sort of thinness seems to be, more than expedient, indispensable to put together theories capable of giving explicit solutions to simple but important practical problems, such as the end deflection of a cantilever, say, or the buckling load of a pillar, the thrust of an arch, the natural vibration frequencies of a plate, et cetera. In this connection, it seems to me appropriate to recall – at times, verbatim – a long passage from (Podio-Guidugli, 1998).

In general, explicit solutions not only find a quantitative use, in that they allow for computing specific unknowns, but also a qualitative use, in that they help to interpret certain experimental findings in the light of the theory within which they were derived. Often, however, there is not a clear-cut distinction between quantitative and qualitative uses: for example, the benchmarking of algorithms and codes entails a not purely quantitative use of explicit solutions. This example prompts the observation that much current work in structure theory consists of computational developments, pertaining more to numerical analysis than to mechanics: in fact, a new discipline, *computational mechanics*, has acquired identity and dignity. In many cases of practical import, computational mechanics complements the program implicit in a Hadamard-like well-posedness cliché, by constructing solutions pointwise, with arbitrary precision, for each pointwise assignment of boundary data. Should we fear that " ... a day will come when rod and shell theories will loose their distinctive identities within computational mechanics and be subsumed under a general theory for the numerical treatment of three–dimensional problems, endowed with useful error estimates"? (S.S. Antman, on p. 600 of (Antman, 1995)).

In short, should we fear that computational mechanics phagocytize structural mechanics? Are we going to invest one week in the essentially inane effort to master a subject that is quickly becoming obsolete?

My answer to the former question parallels Antman's in (Antman, 1995), and is definitely negative; such should be the answer to the latter, then (but you may wish to postpone your own answer by a few days).

To put it flatly, rather than furnishing approximations useful in applications – something that computational mechanics does better – structural mechanics should pose and study *classes* of meaningful problems amenable to qualitative analysis – something that is out of the scope of computational mechanics. Computational mechanics cannot treat at one time classes of shapes, loads, and materials: for example, in the case of Saint-Venant Problem for a rod-like cylindrical body loaded at its ends, computational mechanics can compute with great precision the solution for *one* cylinder, *one* system of end loads, *one* material; but it cannot treat at one time a class of shapes (e.g., strictly speaking, it cannot predict what happens when a cylinder of given cross section becomes slender and slender, i.e., thinner and thinner), let alone treating classes of loads and classes of materials. Neither computational mechanics can establish the position of a given structure theory with respect to a parent, more general mechanical theory.

All in all, it seems to me that there are still good reasons to investigate the mechanical and mathematical foundations of structure theories, especially so in the light of the many interesting and recent contributions we are going to hear about from the other lecturers during this week.

1.2 How?

There are three main approaches to structure theories, one *direct* and two *deductive*. I shall spend most of my time in considering a type of structures having modest thickness, simple geometry and simple material response, that is to say, *linearly elastic plates*. The approach I shall take is one of the two deductive approaches; to put it into perspective, I shall now quickly sketch all three in the case of plates, taking freely from a paper written in collaboration with Antonio DiCarlo and Bill Williams (DiCarlo, P. Podio-Guidugli, and W.O. Williams, 2001).

The direct approach regards a plate as a two-dimensional structured continuum: kinematic descriptors are prescribed, balance equations are obtained from first principles, and constitutive prescriptions are posited *ab initio*. The two deductive approaches derive the balance equations and the constitutive assumptions for plates from those of a parent three-dimensional theory. In one of these approaches, the dependent theory is found as the limit of an approximation in which one body dimension, the thickness, is made to tend to zero. The other approach regards plates as three-dimensional bodies with special shape and partitionability, whose deformations are restricted to be, in some precise sense, 'plate-like'. I propose to refer to these approaches as, respectively, the *asymptotic scaling method* and the *method of internal constraints*; the *formal scaling method*, to be expounded later on, is an outgrowth of the method of internal constraints.

A comparative discussion of the approaches by asymptotic scaling and by introducing internal constraints is found in (DiCarlo, P. Podio-Guidugli, and W.O. Williams, 2001). Here I shall content myself with mentioning a couple of features of the second of these deductive approaches that I find especially attractive: constitutive prescriptions need not be developed afresh, as is inevitable with the direct approach; and, the generalized forces and stresses arising in the theory receive an intuitive interpretation.[1] The price to pay

[1] In truth, such an interpretation is intuitive mostly for those already accustomed to the standard notion of stress, that is, virtually all students of continuum mechanics after Cauchy. Nonetheless, some notion of "generalized stresses" foreran what was to become the standard concept of stress; when introducing the latter, Cauchy himself wrote (Cauchy, 1827): "The geometers who have investigated the equations of equilibrium or motion of thin plates or of surfaces ... have distinguished two kinds of forces, the

for these and other advantages is that, since the deformation class is special and hence the theory constrained, the posing of boundary-value problems is made subtler by the occurrence of reactions to the constraints. This drawback, however, can be turned into a benefit: in the third and last part of my course, I shall indicate how to choose the three-dimensional reactive stress field that best supplements a given two-dimensional shell solution.

1.3 A few bits of history of the deductive approaches to structure theories

The lines to follow have no pretenses of completeness, apart perhaps for those referring to my own work.

The methods of asymptotic expansion and of internal constraints. To derive from three-dimensional elasticity the equations of structures is a long-standing conceptual problem in continuum mechanics, whose substance is to describe the mechanical behavior of one or another class of thin continuous bodies with sufficient detail for applications, and yet without solving a full-blown three-dimensional initial/boundary-value problem. In the problem's long history, two main lines of reasoning emerge. One line, which can be traced back to Poisson (1828) and Cauchy (1829), is based on a judicious use of asymptotic expansion techniques in a thickness parameter. Another line, whose roots are in a famous paper by Kirchhoff (1850), sees structures as continuous bodies that are constitutively capable only of deformations of a specificied type. In a sense, the *method of asymptotic scaling* introduced by Ciarlet and Destuynder (1979) is a thickness expansion technique, made rigorous in the light of modern functional analysis; whereas the *method of internal constraints* I first proposed in (Podio-Guidugli, 1989), as well as the approach to rod and shell theories described by Antman and Marlow (1991), both rely on implementations of the idea of a restricted kinematics.

The asymptotic scaling method has been applied to a number of linear and nonlinear models of plates and shells; a major achievement of it is Ciarlet's 'justification' of von Kármán plate equations (Ciarlet, 1980, 1987); the reader is referred to (Ciarlet, 1997) for a detailed exposition and an inclusive list of references in the case of plates.

The method of internal constraints is based on the following sound and

one produced by dilatation or contraction, the other by the bending of these surfaces ... It has seemed to me that these two kinds of forces could be reduced to a single one, which ought to be called always *tension* or *pression*, a force which acts upon each element of a section chosen at will, not only in a flexible surface but also in a solid ... " (English translation by C. Truesdell, on p. 150 of (Truesdell, 1991)).

simple mechanical idea: any kinematical *Ansatz* on the solution of a three-dimensional problem for a plate-like or rod-like body has to be accompanied both by restrictions on the admissible data and by a system of reactive stresses that help maintaining the internal constraints implied by the *Ansatz* itself. It has been first applied to derive the classical equation of Germain-Lagrange for the equilibrium of an elastic plate of maximal response symmetry, acted upon by transverse loadings (Podio-Guidugli, 1989). Slightly more general versions of the method have been shown to work in the case of in-plane loadings (Lembo, 1989); for single or multilayered plates made of a monoclinic material and subject to arbitrary loadings, in both statical and dynamical situations (Lembo and Podio-Guidugli, 1991, 2001); and for shearable plates (Nardinocchi and Podio-Guidugli, 1994). The foundational aspects of a linear theory of thin elastic plates and shells obtained by means of the constraint method are the subject of (Podio-Guidugli, 1991) and (Podio-Guidugli, 1995) (see also (Podio-Guidugli, 1990)); a derivation of von Kármán equations in the spirit of the constraint method is given in (Podio-Guidugli, 2003).[2]

A structure theory is regarded as *consistent* with three-dimensional elasticity whenever both a rational deduction of the former from the latter is furnished and the error inherent in the approximations on which the deduction relies is estimated.[3] This is the conceptual standpoint of the asymptotic scaling method, according to which the classical plate equation, say, is obtained as a limit problem whose solution, the Kirchhoff-Love displacement field, consistently approximates the three-dimensional displacement when thickness tends to zero. The constraint method has a different standpoint: briefly, rather than justifying the Kirchhoff-Love's representation as an *approximation*, with the constraint method one seeks circumstances when Kirchhoff-Love's is an *exact* three-dimensional equilibrium solution in itself, and then simply integrate over the thickness to get the corresponding two-dimensional picture; neither approximations are made nor limits for

[2] The idea of an asymptotic scaling works also for *rod* problems and, with some care, for modelling *junctions* between elastic bodies of different "dimensions", such as blocks, plates, rods, *etc.* (Bourquin and Ciarlet, 1989; Ciarlet, H. Le Dret and R. Nzengwa, 1989; Ciarlet, 1990). Conceivably, the method of internal constraints works as well in the same circumstances: see, e.g., (Davì, 1989) for an application of the constraint method to derive rod equations, and (Nardinocchi and Podio-Guidugli, 2001) for an application to derive the equations of angle plates.

[3] The consistency issue is often discussed in the technical literature on plates and shells; suffice it to mention here the penetrating syntheses of Naghdi (1972) and of Koiter and Simmonds (1972).

the thickness tending to zero are taken; consistency with three-dimensional linear elasticity is automatic. In my view, *the two methods should be seen as complementary*: the asymptotic scaling method offers a mathematical justification for the assumptions on the form of the solution that are the starting point of the method of internal constraints; the latter method, on its part, provides a mechanical interpretation of the limit problem generated by the scaling procedures.[4]

In (Podio-Guidugli, 1995), the complementary role of the two methods is illustrated for the case when three-dimensional elastic equilibrium is approached through minimization of the energy functional. Other well-known variational characterizations of equilibrium are available; in particular, mixed variational principles have been used to derive lower-dimensional theories by way of scalings. Each variational statement employs a peculiar concept of solution, and hence leads to a peculiar scaling of the elastic state over the body: for example, the scaling of the stress field may not be the same when the minimum energy principle is used, and the scaling of stress follows from the scalings of displacement and constitutive law, and when a mixed stationarity principle is used, so that the scalings of displacement and stress are in principle independent. Consequently, the outcomes of scalings procedures based on different variational statements may differ. What variational statement to choose is at bottom a matter of taste but - remarkably, I believe - the complementary role of the scaling and constraint method would emerge clear in all cases.

Also, in (Podio-Guidugli, 1995) attention was confined to three-dimensional bodies having the form of a right cylinder of arbitrary, but constant, cross section, and hence to plates and rods with a straight axis. This was done for simplicity: both the asymptotic scaling method and the constraint method can treat shells and curved rods, provided one is willing to cope with the accompanying geometrical complications, some of which are far from trivial (Ciarlet, 2000, 2005).

The methods of $\Gamma-$convergence and of formal scaling. All classical theories of thin elastic structures admit a variational formulation, consisting of the minimization of a peculiar energy functional on a peculiar function space. An attractive manner to capture structure thinness is to establish the position of such a 'lower-dimensional' energy functional with respect to

[4]I first put forward these ideas during a visit I paid in May 1988 at the University of Paris VI on the invitation of Professor P.G. Ciarlet. Later, they were presented in a number of colloquium talks, workshops, and conferences, both in Italy and abroad; the manuscript (Podio-Guidugli, 1995) was circulated but never published, except for the sketch in (Podio-Guidugli, 1997).

the energy functional of some parent three-dimensional theory of elasticity. Basically, there are two successful ways to do so, the one mechanical, the other analytical. The internal constraint method we just reviewed regards structure thinness as the distinguishing common feature of a class of material bodies that are special for shape, partitionability, admissible motions, and stress response. Exploiting these specialties in a consistent fashion, a suitable three-dimensional energy functional is found, such as to yield the desired rod or plate functional by mere thickness integration, *whatever the geometric thinness.* This mechanical approach is at distinct variance with the analytical approach of the asymptotic expansion method and of the $\Gamma-$*convergence method*, which both achieve the same goal by taking a suitable limit of a standard three-dimensional energy functional when the geometric thinness tends to zero.

Along the week, we are going to hear about the recently proposed techniques of variational convergence by experts that have had a central role in developing them; this exonerates me from a detailed description I am poorly qualified for. Roughly speaking, given a parametric family of minimum problems $\min\limits_{u\in X_\varepsilon} F_\varepsilon(u)$, it may happen that a minimum problem $\min\limits_{u\in X} F(u)$ is found, such that minimizers and minima associated with the problems ruled by the functionals F_ε converge to minimizers and minima of the problem ruled by the functional F; if this is the case, the family of functionals F_ε is said to $\Gamma-$converge to the target functional F. To establish the relationship with three-dimensional elasticity of a given structure model through the notion of $\Gamma-$convergence, one has to find a family of three-dimensional functionals that, as the thickness parameter $\varepsilon \to 0+$, $\Gamma-$converges to a three-dimensional target functional in tight kinship with the lower-dimensional structure model at hand. Thus, $\Gamma-$convergence is basically a method to justify and validate a given structure model, not to deduce it, and even less to propose a new one.

As we shall see, the *method of formal scaling* bridges between the mechanical and analytical methods, in that it helps setting up a formal limit process that yields a complete taxonomy of plate and rod energy functionals, each of which suggests a potential target for a carefully devised process of variational convergence.

2 A Unified Approach to Classic Rod and Plate Theories

Traditionally, plate and rod theories are derived separately. I shall now present a unified method of deduction – firstly proposed in (Podio-Guidugli,

1995) as an evolution of the method of internal constraints, and based on a formal scaling of both data and unknown in terms of powers of the thickness – that yields a collection of energy functionals governing the statics of linearly elastic structures. This collection can be shown to be, in a precise sense, exhaustive; in particular, it includes the 'one-dimensional' functional whose Euler-Lagrange equations are the *Timoshenko's equations for shearable rods* and the 'two-dimensional' functional associated with the *Reissner-Mindlin's equations for shearable plates.* Validations of the same classic equations by way of variational convergence methods are, to my knowledge, not yet available. I shall suggest reasons for this, as well as a likely remedy.[5] Most of the developments to follow are taken from two joint papers with Bernadette Miara (Miara and Podio-Guidugli, 2006, 2007).

2.1 Timoshenko rods, Reissner-Mindlin plates, and three-dimensional linear elasticity

Let Ref be a given copy of a three-dimensional Euclidean point space $\mathcal{E}$. Consider a ε-family of regions in Ref having the form, for each fixed $\varepsilon \in]0,1]$, of a right cylinder $\mathcal{C}(\varepsilon) = \mathcal{P}(\varepsilon) \times \mathcal{R}(\varepsilon)$, of cross-section $\mathcal{P}(\varepsilon)$ and axis $\mathcal{R}(\varepsilon)$ of length $2h(\varepsilon)$, with $\mathcal{P}(1) = \mathcal{P}$, $\mathcal{R}(1) = \mathcal{R}$, and $h(1) = h$ (Fig. 1).

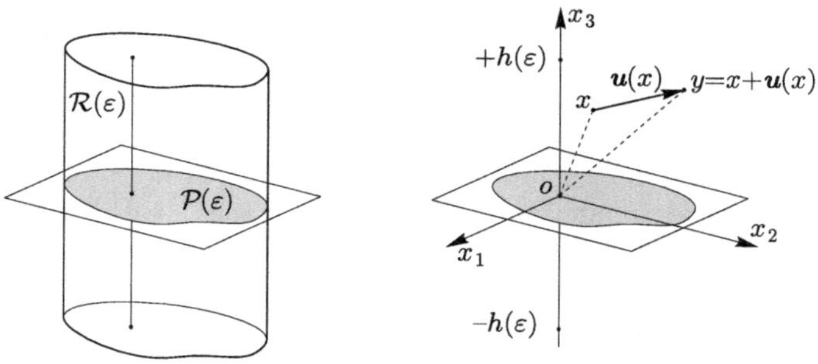

Figure 1. A right cylinder $\mathcal{C}(\varepsilon)$ seen as the cartesian product of a *plate-like* region $\mathcal{P}(\varepsilon)$ and a *rod-like* region $\mathcal{R}(\varepsilon)$.

Choose an origin point $o \in \mathcal{C}(\varepsilon)$ and an orthogonal triplet of unit vectors $\boldsymbol{e}_i$, with $\boldsymbol{e}_3$ parallel to $\mathcal{R}(\varepsilon)$. Denote by x_i the cartesian coordinates of a

[5]That the suggested remedy works was first announced and then proved, just a few months after the end of this course, in two papers written in collaboration with Roberto Paroni and Giuseppe Tomassetti (Paroni, P. Podio-Guidugli and G. Tomassetti, 2006, 2007), where a variational validation of the Reissner-Mindlin plate theory is achieved.

point $x \in \mathcal{C}(\varepsilon)$, and by y_i the cartesian coordinates of the point $y = x + \boldsymbol{u}(x)$ occupied by x in the current shape of $\mathcal{C}(\varepsilon)$, a region in another copy of $\mathcal{E}$ denoted by Cur. Finally, for any given ε, refer to $\mathcal{C}(\varepsilon)$ and $\boldsymbol{u}$ as, respectively, the *domain* and the *displacement.*

In addition to the domain $\mathcal{C}(\varepsilon)$, the *data* comprise:

- an *elasticity tensor* $\mathbb{C}$, a fourth-order tensor endowed with the standard index symmetries ($\mathbb{C}_{ijhk} = \mathbb{C}_{jihk} = \mathbb{C}_{ijkh} = \mathbb{C}_{hkij}$) and positive definite;
- a field of *distance loads* $\boldsymbol{d}$, defined over the interior part of $\mathcal{C}(\varepsilon)$;
- a field of *contact loads* $\boldsymbol{c}$ over $\partial\mathcal{C}(\varepsilon)$, with $\boldsymbol{c} = \boldsymbol{c}^{\pm}$ on top/bottom ends $\mathcal{P}^{\pm}$ of $\mathcal{C}(\varepsilon)$ and with $\boldsymbol{c} \equiv \boldsymbol{0}$ over $\partial\mathcal{P}(\varepsilon) \times \mathcal{R}(\varepsilon)$.

We formulate the following equilibrium problem of three-dimensional linear elasticity:

find an *elastic state* $(\boldsymbol{u}, \boldsymbol{E}, \boldsymbol{S})$ over the closure of $\mathcal{C}(\varepsilon)$, with
- $\boldsymbol{u}$ the *displacement* field,
- $\boldsymbol{E}(\boldsymbol{u}) = \operatorname{sym} \nabla \boldsymbol{u}$ the *strain* field,
- $\boldsymbol{S}(\boldsymbol{u}) = \mathbb{C}[\boldsymbol{E}(\boldsymbol{u})]$ the *stress* field,

such that

$$\Pi(\boldsymbol{u}) := \Sigma(\boldsymbol{u}) - \Delta(\boldsymbol{u}) = \min, \tag{1}$$

where

$$\Sigma(\boldsymbol{u}) := \int_{\mathcal{C}(\varepsilon)} \frac{1}{2} \boldsymbol{S}(\boldsymbol{u}) \cdot \boldsymbol{E}(\boldsymbol{u}) \quad \text{and} \quad \Delta(\boldsymbol{u}) := \int_{\mathcal{C}(\varepsilon)} \boldsymbol{d} \cdot \boldsymbol{u} + \int_{\mathcal{P}(\varepsilon)} \boldsymbol{c}^{\pm} \cdot \boldsymbol{u}^{\pm} \tag{2}$$

are, respectively, the *stored-energy functional* and the *load functional.*

Our current goal is to explore, quite informally, the relationships among this minimization problem and two 'lower-dimensional' minimization problems from structure theory. Stripped-to-the-bone versions of these two problems will be enough for our purpose; greater generality and more details are found in the Appendices of (Miara and Podio-Guidugli, 2006).

The equations of Timoshenko rods. In the simplest case, that is to say, *in-plane flexure of a straight rod* of cross-section $\mathcal{P}(\varepsilon)$ and unit length, Timoshenko's is a system of two coupled ODEs for a pair of scalar fields (v, ψ) defined over $\mathcal{R} \equiv (0,1)$:

$$\begin{aligned} &\big(s(\varepsilon)(v' - \psi)\big)' + f_0 = 0, \\ &\big(b(\varepsilon)\psi'\big)' + s(\varepsilon)(v' - \psi) + c_0 = 0, \end{aligned} \tag{3}$$

where

$$s(\varepsilon) = O(\varepsilon^2)$$

is the *shearing stiffness* and

$$b(\varepsilon) = O(\varepsilon^4)$$

is the *bending stiffness*, two positive constants that depend on the rod's constitution and the cross-section's geometry (Figure 2). The thinness pa-

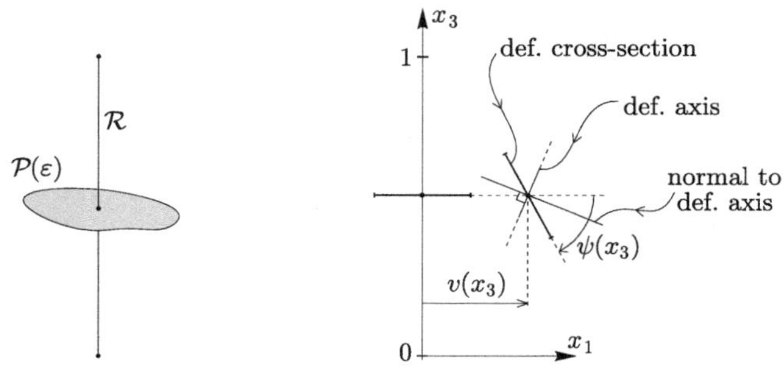

Figure 2.

rameter ε measures the ratio of the diameter of the rod's cross-section to the length of its axis. Equations (3) are the Euler-Lagrange equations associated with the 'one-dimensional' stored-energy functional:

$$\Sigma_T(v,\psi,\varepsilon) = \int_{\mathcal{R}} \frac{1}{2}\Big(s(\varepsilon)(v'-\psi)^2 + b(\varepsilon)\psi'^2\Big).^{6} \tag{4}$$

REMARK. A second look to system (3) reveals that *flexure* consists of a rather complex combination of *bending* and *shearing*. Flexure should not be confused with *pure bending*, and for good reasons.

Pure bending is in order when the kinematical constraint

$$v' - \psi = 0 \tag{5}$$

is enforced. Such a constraint is maintained by a *reactive shear force* $Q^{(R)}$, an unknown field that plays the same role as the pressure field in the mechanics of incompressible material bodies. $Q^{(R)}$ replaces for the active shear force $Q^{(A)} = s(\varepsilon)(v' - \psi)$ of the unconstrained case; system (3) becomes:

$$\begin{aligned} &\big(Q^{(R)}\big)' + f_0 = 0,\\ &\big(b(\varepsilon)v''\big)' + Q^{(R)} + c_0 = 0, \end{aligned} \tag{6}$$

[6]Needless to say, this functional is qualified as 'one-dimensional' because it has the form of an integral over a one-dimensional manifold.

whence the ‘pure’ (i.e., reaction-free) equation

$$\big(b(\varepsilon)v''\big)'' - f_0 + (c_0)' = 0$$

rules the choice of the basic unknown field, the transverse displacement v; when v has been determined, the relative rotation of the cross section ψ is computed from (5), the reactive shear force $Q^{(R)}$ from $(6)_2$.

Now, when a boundary-value problem ruled by system (3) is solved for (v, ψ), the transverse displacement v may be quite different from the corresponding field obtained by solving for $(v, Q^{(R)})$, with the same data, the problem ruled by system (6): the relationship between those two fields is of the same sort as the relationship between, say, the solutions of a motion problem, given the applied-force field and the initial conditions, for a material point being or not being constrained to move on a given surface.

Confusing the terms ‘flexure’ and ‘bending’ is pernicious, as is in general terminological confusion and imprecision, a mass disease that should be fought by all and everybody, but especially by those who advocate clear thinking and teaching.[7]

The Reissner-Mindlin plate equations. The Reissner-Mindlin equations are:

$$\begin{aligned}
&S(\varepsilon)(\Delta w + \operatorname{Div}\boldsymbol{\varphi}) + q_0 = 0,\\
&\frac{1}{2}B(\varepsilon)\big((1-\nu)\Delta\boldsymbol{\varphi} + (1+\nu)\nabla\operatorname{Div}\boldsymbol{\varphi}\big) - S(\varepsilon)(\nabla w + \boldsymbol{\varphi}) + \boldsymbol{m}_0 = \boldsymbol{0};
\end{aligned} \tag{7}$$

they govern the flexure of a plate occupying region $\mathcal{P} \equiv (0,1) \times (0,1)$, of thickness $2\,h(\varepsilon)$, *shearing stiffness* $S(\varepsilon) = O(\varepsilon)$, and *bending stiffness* $B(\varepsilon) = O(\varepsilon^3)$ (Figure 3); $S(\varepsilon), B(\varepsilon)$ are positive constants depending on the plate's constitution and its thickness, ν is a dimensionless material modulus, and the unknowns are a pair $(w, \boldsymbol{\varphi})$ of a scalar and a two-component vector field, defined over $\mathcal{P}$. The thinness parameter ε measures the ratio of the plate thickness to the plate diameter. The system (7) is composed by

---

[7] This reminds me of an instructive anecdote from Confucius' sayings, one out of many.

Once upon a time, a disciple asked Confucius: “Master! If a king entrusts you a province to rule according to your own ideas, what would you do first?” Answered Confucius: “First of all, I would rectify names.” And he went on and explained to his confused disciple: “If names are incorrect, if they do not reflect reality, then language has no object. If language has no object, action becomes impossible: all human affairs are disrupted, and to handle them becomes futile. Thus, the first task of a true statesman is to rectify names.”

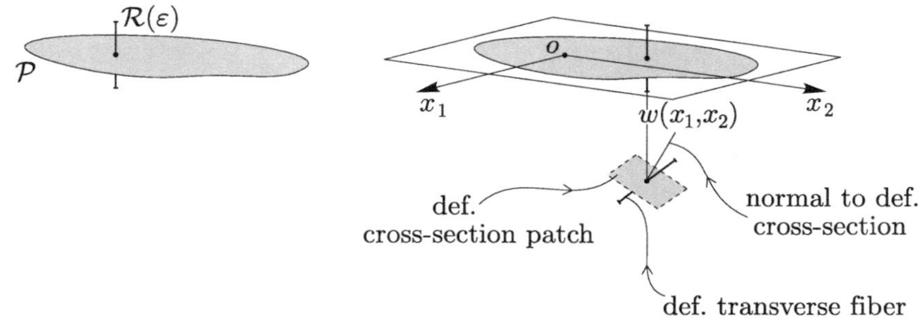

Figure 3.

the Euler-Lagrange equations associated with the 'two-dimensional' stored-energy functional

$$\Sigma_{RM}(w,\boldsymbol{\varphi},\varepsilon)=\int_{\mathcal{P}}\frac{1}{2}\Big(S(\varepsilon)|\nabla w+\boldsymbol{\varphi}|^2+B(\varepsilon)\big((1-\nu)|\mathrm{sym}\,\nabla\boldsymbol{\varphi}|^2+\nu(\mathrm{tr}\nabla\boldsymbol{\varphi})^2\big)\Big)\,. \tag{8}$$

Euristics. Can the three-dimensional minimization problem (1)-(2) be reduced to the lower-dimensional problems for Timoshenko rods and Reissner-Mindlin plates?

Let us give it for granted that the three-dimensional distance and contact loads have been chosen in such a way as to be compatible with the lower-dimensional load functionals. As to stored-energy functionals, ideally we would like to know whether there are choices of a type of material response and of two classes of displacement fields that would reduce $(2)_1$ to, respectively, (4), and (8):

$$\int_{\mathcal{C}(\varepsilon)}\frac{1}{2}\boldsymbol{S}(\boldsymbol{u})\cdot\boldsymbol{E}(\boldsymbol{u})=\begin{cases}\displaystyle\int_{\mathcal{R}}\Big(\int_{\mathcal{P}(\varepsilon)}(?)\Big)=\int_{\mathcal{R}}\frac{1}{2}\Big(s(\varepsilon)(v'-\psi)^2+b(\varepsilon)\psi'^2\Big),\\[2ex] \displaystyle\int_{\mathcal{P}}\Big(\int_{\mathcal{R}(\varepsilon)}(??)\Big)=\int_{\mathcal{P}}\frac{1}{2}\Big(S(\varepsilon)|\nabla w+\boldsymbol{\varphi}|^2+B(\varepsilon)(\dots)\Big).\end{cases} \tag{9}$$

As to material response, inspection of (4), and (8) reveals that at most five different material moduli are needed, namely, ν plus possibly one each for $s(\varepsilon), b(\varepsilon), S(\varepsilon)$, and $B(\varepsilon)$. Therefore, isotropic response – the naive choice based on the widespread belief that an engineer's world is isotropic by default – is out of question. A choice that embodies the maximal response symmetry compatible with the cartesian structure of the domain is *coherent transverse isotropy*, that is to say, transverse isotropy with respect to the direction $\boldsymbol{e}_3$ of the axis of the cylinder $\mathcal{C}(\varepsilon)$. The relative stress-strain law can be written in the following form:

$$\begin{aligned} S_{\alpha\beta} &= 2\mu E_{\alpha\beta} + (\lambda(E_{11}+E_{22}) + \tau_2 E_{33})\delta_{\alpha\beta}, \\ S_{3\alpha} &= 2\gamma E_{3\alpha}, \\ S_{33} &= \tau_1 E_{33} + \tau_2(E_{11}+E_{22}), \end{aligned} \tag{10}$$

where five independent material moduli appear; the constitutive inequalities

$$\mu > 0,\ \gamma > 0,\ \tau_1 > 0,\ \tau_1(\lambda+\mu) - \tau_2^2 > 0 \tag{11}$$

guarantee positivity of

$$\sigma = \frac{1}{2} S_{ij} E_{ij}\,, \tag{12}$$

the stored energy per unit volume of Ref.[8]

What remains for us to do is to make adroit selections of admissible three-dimensional displacement fields, 'rod-like' in Timoshenko's case and 'plate-like' in Reissner-Mindlin's; that is to say, two classes of admissible displacements parameterized, respectively, by the fields v, ψ over $\mathcal{R}$ and by the fields $w, \boldsymbol{\varphi}$ over $\mathcal{P}$.

To begin with, on formalizing the hints made explicit by Figure 2, we write a candidate Timoshenko displacement field in the cylinder $\mathcal{P}(\varepsilon) \times \mathcal{R}$ in the following form:

$$\boldsymbol{u}_T(x_1, x_2, x_3) = v(x_3)\boldsymbol{e}_1 - x_1\psi(x_3)\boldsymbol{e}_3\,.$$

Then, the strain field associated with $\boldsymbol{u}_T$ is:

$$\boldsymbol{E}(\boldsymbol{u}_T) = \frac{1}{2}(v' - \psi)(\boldsymbol{e}_1 \otimes \boldsymbol{e}_3 + \boldsymbol{e}_3 \otimes \boldsymbol{e}_1) - x_1\psi'\,\boldsymbol{e}_3 \otimes \boldsymbol{e}_3,$$

whence, with the use of (10) and (12), we have that

$$\sigma_T = S_{13}E_{13} + \frac{1}{2}S_{33}E_{33} = \frac{1}{2}\gamma(v'-\psi)^2 + \frac{1}{2}\tau_1\, x_1^2 \psi'^2\,.$$

On recalling (9), we see that

$$? = \int_{\mathcal{P}(\varepsilon)} \Big(\frac{1}{2}\gamma(v'-\psi)^2 + \frac{1}{2}\tau_1\, x_1^2\psi'^2\Big) = \frac{1}{2}\Big(\gamma A(\varepsilon)(v'-\psi)^2 + \tau_1\, I(\varepsilon)\psi'^2\Big),$$

---

[8] A transversely isotropic material is *isotropic* whenever

$$\gamma = \mu, \quad \tau_1 = \lambda + 2\mu, \quad \tau_2 = \lambda.$$

where

$$I(\varepsilon) = \int_{\mathcal{P}(\varepsilon)} x_1^2 .$$

Next, again building on the intuitive interpretation of the parameter fields w and $\boldsymbol{\varphi}$, we write a candidate Reissner-Mindlin displacement field in the cylinder $\mathcal{P} \times \mathcal{R}(\varepsilon)$ in the following form:

$$\boldsymbol{u}_{RM}(x_1, x_2) = x_3\, \boldsymbol{\varphi}(x_1, x_2) + u_3(x_1, x_2)\boldsymbol{e}_3, \quad \boldsymbol{\varphi}(x_1, x_2) = \varphi_\alpha(x_1, x_2)\boldsymbol{e}_\alpha,$$

and compute the associated strain field:

$$\boldsymbol{E}(\boldsymbol{u}_{RM}) = \mathrm{sym}\,\big((\nabla u_3 + \boldsymbol{\varphi}) \otimes \boldsymbol{e}_3\big) + x_3\, \mathrm{sym}\, \nabla\boldsymbol{\varphi}\,,$$

with

$$\mathrm{sym}\,\big((\nabla u_3 + \boldsymbol{\varphi}) \otimes \boldsymbol{e}_3\big) = \frac{1}{2}\big(u_{3,\alpha} + \varphi_\alpha\big)(\boldsymbol{e}_\alpha \otimes \boldsymbol{e}_3 + \boldsymbol{e}_3 \otimes \boldsymbol{e}_\alpha)$$

and

$$\mathrm{sym}\,\nabla\boldsymbol{\varphi} = \varphi_{1,1}\boldsymbol{e}_1 \otimes \boldsymbol{e}_1 + \frac{1}{2}\big(\varphi_{1,2} + \varphi_{2,1}\big)\big(\boldsymbol{e}_1 \otimes \boldsymbol{e}_2 + \boldsymbol{e}_2 \otimes \boldsymbol{e}_1\big) + \varphi_{2,2}\boldsymbol{e}_2 \otimes \boldsymbol{e}_2 \,.$$

Then, again with the use of (10) and (12), we have that

$$\begin{aligned}
\sigma_{RM} &= S_{\alpha 3}E_{\alpha 3} + S_{12}E_{12} + \frac{1}{2}(S_{11}E_{11} + S_{22}E_{22}) \\
&= \frac{1}{2}\gamma\, |\nabla u_3 + \boldsymbol{\varphi}|^2 + x_3^2\Big(\mu\, |\mathrm{sym}\,\nabla\boldsymbol{\varphi}|^2 + \frac{1}{2}\lambda(\mathrm{tr}\nabla\boldsymbol{\varphi})^2\Big)\,.
\end{aligned}$$

On recalling (9), we see that

$$\begin{aligned}
?? &= \int_{\mathcal{R}(\varepsilon)}\Big(\gamma|\nabla u_3 + \boldsymbol{\varphi}|^2 + x_3^2\Big(\mu\, |\mathrm{sym}\,\nabla\boldsymbol{\varphi}|^2 + \frac{1}{2}\lambda(\mathrm{tr}\nabla\boldsymbol{\varphi})^2\Big) \\
&= h(\varepsilon)\,\gamma\, |\nabla u_3 + \boldsymbol{\varphi}|^2 + \frac{2}{3}h(\varepsilon)^3\Big(\mu\, |\mathrm{sym}\,\nabla\boldsymbol{\varphi}|^2 + \frac{1}{2}\lambda(\mathrm{tr}\nabla\boldsymbol{\varphi})^2\Big).
\end{aligned}$$

For a complete reproduction of (9), we have only to set:

$$\begin{aligned}
s(\varepsilon) &= \gamma A(\varepsilon), \quad b(\varepsilon) = \tau_1 I(\varepsilon); \\
S(\varepsilon) &= 2\, h(\varepsilon)\gamma, \quad B(\varepsilon) = \frac{2}{3}h(\varepsilon)^3(\lambda + 2\mu), \quad \nu = \frac{\lambda}{\lambda + 2\mu}\,.
\end{aligned}$$

Interestingly, inequalities (11) imply that all stiffnesses $s(\varepsilon), \ldots, B(\varepsilon)$ are positive, as desired.

In conclusion, it seems to me that there are pretty good reasons to believe that both Timoshenko's and Reissner-Mindlin's models can be arrived starting from one and the same functional of three-dimensional transversely isotropic linear elasticity. What we are going to do next is to try and proceed in a systematic manner.

2.2 Scaling Procedure

The formal scaling method I propose involves first the (domain, solution) pair, then the strain measure, lastly the material moduli and the loads; the final product is a scaled three-dimensional functional. A key feature of this method is that none of the scaling exponents is chosen once and for all by *fiat*, although in a few instances – to be pointed out at occurrence – a choice is made so as to reflect certain reasonable prejudices; another key feature is that the exponent list must be such that the scaled energy functional stay bounded in the limit of vanishing thickness.

Scaling the domain and the solution. The common starting point of all scaling methods is to scale the cross-sectional coordinates x_α $(\alpha = 1,2)$ differently from the axial coordinate x_3, and to scale the in-plane displacement components u_α differently from the transverse component u_3. For each $\varepsilon > 0$ fixed, we scale the domain $\mathcal{C}(\varepsilon) = \mathcal{P}(\varepsilon) \times \mathcal{R}(\varepsilon)$ by mapping it one-to-one onto the region $\mathcal{C}(1) = \mathcal{P} \times \mathcal{R}$; our domain scaling is

$$x_\alpha = \varepsilon^p \bar{x}_\alpha \quad (\alpha = 1,2), \quad x_3 = \varepsilon^q \bar{x}_3; \tag{13}$$

likewise, our displacement scaling is

$$u_\alpha = \varepsilon^m \bar{u}_\alpha \quad (\alpha = 1,2), \quad u_3 = \varepsilon^n \bar{u}_3. \tag{14}$$

In relations (13) and (14), all entries in the exponent list (m, n, p, q) are *relative integers to be chosen later according to convenience.* Both cross-section coordinates of points are equally scaled, as well as both in-plane components of displacements, because neither we envisage any privileged in-plane direction in Ref nor we want to have any in Cur, as is the case with the classical structure theories we here aim to derive. This is the first instance where some a priori judgment mitigates our freedom in the choice of certain scaling exponents; there will be a few others.

REMARK. After Podio-Guidugli (1995, 1997), a (domain, displacement) scaling can be seen as a judicious choice of a continuous group of linear automorphisms $\mathcal{A}(\varepsilon)$ of $Ref \times Cur$, with thickness as the group parameter. In our present case, $\mathcal{A}(\varepsilon)$ has the following form, which generalizes slightly the standard choices:

$$\begin{bmatrix} \bar{x} \\ \bar{y} \end{bmatrix} = \mathcal{A}(\varepsilon) \begin{bmatrix} x \\ y \end{bmatrix}, \; \mathcal{A}(\varepsilon) = \begin{bmatrix} \boldsymbol{A}(\varepsilon) & \boldsymbol{0} \\ \boldsymbol{B}(\varepsilon) & \boldsymbol{C}(\varepsilon) \end{bmatrix}, \; \boldsymbol{B}(\varepsilon) = \boldsymbol{A}(\varepsilon) - \boldsymbol{C}(\varepsilon), \tag{15}$$

with

$$\boldsymbol{A}(\varepsilon) = diag < \varepsilon^{-p}, \varepsilon^{-p}, \varepsilon^{-q} >, \; \boldsymbol{C}(\varepsilon) = diag < \varepsilon^{-m}, \varepsilon^{-m}, \varepsilon^{-n} > .$$

One way to deduce (15) is as follows. For $\{\boldsymbol{c}_i \,|\, i=1,2,3\}$ an orthonormal basis, let $\bar{\boldsymbol{x}}=\bar{x}_i\boldsymbol{c}_i$, $\bar{\boldsymbol{u}}=\bar{u}_i\boldsymbol{c}_i$, $\boldsymbol{u}=u_i\boldsymbol{c}_i$, etc.. Then,

$$\bar{\boldsymbol{x}}=\boldsymbol{A}(\varepsilon)\boldsymbol{x}, \quad \bar{\boldsymbol{u}}=\boldsymbol{C}(\varepsilon)\boldsymbol{u};$$

moreover,

$$\boldsymbol{C}(\varepsilon)(\boldsymbol{y}-\boldsymbol{x})=\boldsymbol{C}(\varepsilon)\boldsymbol{u}(\boldsymbol{x})=\bar{\boldsymbol{u}}(\bar{\boldsymbol{x}})=\bar{\boldsymbol{y}}-\bar{\boldsymbol{x}}=\bar{\boldsymbol{y}}-\boldsymbol{A}(\varepsilon)\boldsymbol{x}.$$

Thus,

$$\bar{\boldsymbol{y}}=\big(\boldsymbol{A}(\varepsilon)-\boldsymbol{C}(\varepsilon)\big)\boldsymbol{x}+\boldsymbol{C}(\varepsilon)\boldsymbol{y}=\boldsymbol{B}(\varepsilon)\boldsymbol{x}+\boldsymbol{C}(\varepsilon)\boldsymbol{y}.$$

Note that $\boldsymbol{A}(1)=\boldsymbol{C}(1)=diag<1,1,1>$ if and only if $\mathcal{A}(1)$ is the identity of $Ref\times Cur$; and that $\mathcal{A}^{-1}(\varepsilon)$ has the same lower-triangular block structure as $\mathcal{A}(\varepsilon)$:

$$\mathcal{A}^{-1}(\varepsilon)=\begin{bmatrix}\boldsymbol{A}^{-1}(\varepsilon) & \boldsymbol{0}\\ \boldsymbol{D}(\varepsilon) & \boldsymbol{C}^{-1}(\varepsilon)\end{bmatrix}, \quad \boldsymbol{D}(\varepsilon)=\boldsymbol{A}^{-1}(\varepsilon)-\boldsymbol{C}^{-1}(\varepsilon).$$

Scaling the strain measure. The scaling of the deformation measure $\boldsymbol{E}$ follows from (13) and (14). One finds:

$$\begin{aligned}\boldsymbol{H} &= \varepsilon^{-p+m}\bar{H}_{\alpha\beta}\boldsymbol{c}_\alpha\otimes\boldsymbol{c}_\beta+\varepsilon^{-p+n}\bar{H}_{\alpha3}\boldsymbol{c}_\alpha\otimes\boldsymbol{c}_3\\ &+ +\varepsilon^{-q+m}\bar{H}_{3\alpha}\boldsymbol{c}_3\otimes\boldsymbol{c}_\alpha+\varepsilon^{-q+n}\bar{H}_{33}\boldsymbol{c}_3\otimes\boldsymbol{c}_3,\end{aligned}$$

where

$$\bar{H}_{ij}=\frac{\partial\bar{u}_i}{\partial\bar{x}_j}=\bar{u}_{i,j}\,.$$

Hence, in particular,

$$2(\mathrm{sym}\,\boldsymbol{H})_{\alpha3}=2E_{\alpha3}=\varepsilon^{-p+n}\bar{u}_{\alpha,3}+\varepsilon^{-q+m}\bar{u}_{3,\alpha}\,. \tag{16}$$

We see that the components $E_{\alpha3}$ of $\boldsymbol{E}$ consist of two addenda having generally different limit behavior. Instead, with a view toward capturing thinness via an ε-limit process to come, it would seem reasonable that the two addenda have the same weight whatever ε, so that $\boldsymbol{E}$ takes under scaling the desired structure:

$$\boldsymbol{E}=\varepsilon^{\alpha_1}\bar{E}_{\alpha\beta}\mathrm{sym}\,(\boldsymbol{c}_\alpha\otimes\boldsymbol{c}_\beta)+\varepsilon^{\alpha_2}\bar{E}_{\alpha3}\mathrm{sym}\,(\boldsymbol{c}_\alpha\otimes\boldsymbol{c}_3)+\varepsilon^{\alpha_3}\bar{E}_{33}\boldsymbol{c}_3\otimes\boldsymbol{c}_3, \tag{17}$$

and so preserves in full its role of linear strain measure. To allow for the displacement field $\boldsymbol{u}$, and hence $\bar{\boldsymbol{u}}$ as well, to have a generic dependence on all three space coordinates at the outset, we assume that

$$-p+n=-q+m; \tag{18}$$

it then follows from (16) that $\boldsymbol{E}$ has the structure displayed in (17).

REMARK. When plate equations are derived either by asymptotic analysis or by variational convergence, the standard (domain, displacement) scaling is

$$m = 1, \quad n = 0, \quad p = 0, \quad q = 1, \tag{19}$$

consistent with (18). In the linear case, the standard scaling is defined up to a multiplying factor; in the nonlinear case, that factor is "frozen" (Miara, 1994a,b), and the standard scaling becomes: $m = 2$, $n = 1$, $p = 0$, and $q = 1$.

Scaling the material moduli and the loads. Whatever the deduction method, the generality of the resulting plate and rod theories depends on the generality of the material response incorporated in the three-dimensional formulation one starts with. In this connection, our choice of the coherent transversely isotropic response (10) has already been motivated. We scale the material moduli in (10) as follows:

$$\bar{\lambda} = \varepsilon^{-r}\lambda, \ \bar{\mu} = \varepsilon^{-r}\mu, \ \bar{\gamma} = \varepsilon^{-u}\gamma, \ \bar{\tau}_1 = \varepsilon^{-v}\tau_1, \ \bar{\tau}_2 = \varepsilon^{-z}\tau_2 \,. \tag{20}$$

The moduli λ and μ are scaled by the same exponent because they appear additively in the last of (11); for that inequality to hold whatever $\varepsilon > 0$ and whatever $\tau_2 \neq 0$, we assume that

$$r + v - 2z = 0. \tag{21}$$

As to loads, we scale those per unit volume as follows:

$$\bar{d}_\alpha = \varepsilon^{-s} d_\alpha, \quad \bar{d}_3 = \varepsilon^{-t} d_3, \tag{22}$$

and those applied to the top and bottom ends of $\mathcal{C}(\varepsilon)$ as follows:

$$\bar{c}_\alpha^{\pm} = \varepsilon^{-y} c_\alpha^{\pm}, \quad \bar{c}_3^{\pm} = \varepsilon^{-w} c_3^{\pm}. \tag{23}$$

Once again, we have treated the in-plane components in the same way, both for volume loads and for surface loads. Moreover, for the work measures $b_\alpha u_\alpha d(vol)$ and $c_\alpha^{\pm} u_\alpha^{\pm} d(area)$ to scale the same, as well as the measures $b_3 u_3 d(vol)$ and $c_3^{\pm} u_3^{\pm} d(area)$, we assume that

$$s + q = y, \quad t + q = w. \tag{24}$$

The scaled stored-energy and load functionals. So far, we have stipulated the scalings detailed by (13), (14), (20), (22), and (23), and we have restricted them according to the prejudicial conditions (18), (21), and (24). We are now in a position to return to the definitions in (2) and construct:

- the *scaled stored-energy functional*:

$$\widehat{\Sigma}(\bar{\boldsymbol{u}},\varepsilon;\hat{\alpha},\hat{\beta}_1,\hat{\beta}_2,\hat{\gamma})=\varepsilon^{\hat{\alpha}}\mathrm{A}(\bar{\boldsymbol{u}})+\varepsilon^{\hat{\beta}_1}\mathrm{B}_1(\bar{\boldsymbol{u}})+\varepsilon^{\hat{\beta}_2}\mathrm{B}_2(\bar{\boldsymbol{u}})+\varepsilon^{\hat{\gamma}}\Gamma(\bar{\boldsymbol{u}}),\tag{25}$$

where

$$\begin{aligned}
\mathrm{A}(\bar{\boldsymbol{u}})&=\int_{\mathcal{C}(1)}\frac{1}{2}\bar{\tau}_1\bar{E}_{33}^2\,,\\
\mathrm{B}_1(\bar{\boldsymbol{u}})&=\int_{\mathcal{C}(1)}2\bar{\gamma}(\bar{E}_{13}^2+\bar{E}_{23}^2)\,,\quad \mathrm{B}_2(\bar{\boldsymbol{u}})=\int_{\mathcal{C}(1)}\big(\bar{\tau}_2\bar{E}_{33}(\bar{E}_{11}+\bar{E}_{22})\big)\,,\\
\Gamma(\bar{\boldsymbol{u}})&=\int_{\mathcal{C}(1)}\frac{1}{2}\Big((\bar{\lambda}+2\bar{\mu})(\bar{E}_{11}+\bar{E}_{22})^2-4\bar{\mu}(\bar{E}_{11}\bar{E}_{22}-\bar{E}_{12}^2)\Big),
\end{aligned}\tag{26}$$

with

$$\begin{aligned}
&\hat{\alpha}=-m+3n+p+v,\quad \hat{\beta}_1=m+n+p+u,\\
&\hat{\beta}_2=m+n+p+z,\quad \hat{\gamma}=3m-n+p+2z-v\,;
\end{aligned}\tag{27}$$

- the *scaled load functional*:

$$\widehat{\Delta}(\bar{\boldsymbol{u}},\varepsilon;\hat{\delta}_1,\hat{\delta}_2)=\varepsilon^{\hat{\delta}_1}\Delta_1(\bar{\boldsymbol{u}})+\varepsilon^{\hat{\delta}_2}\Delta_2(\bar{\boldsymbol{u}}),\tag{28}$$

where

$$\Delta_1(\bar{\boldsymbol{u}})=\int_{\mathcal{C}(1)}\bar{b}_\alpha\bar{u}_\alpha+\int_{\mathcal{P}(1)}\bar{c}_\alpha^{\pm}\bar{u}_\alpha^{\pm},\quad \Delta_2(\bar{\boldsymbol{u}})=\int_{\mathcal{C}(1)}\bar{b}_3\bar{u}_3+\int_{\mathcal{P}(1)}\bar{c}_3^{\pm}\bar{u}_3^{\pm}\,,\tag{29}$$

with

$$\hat{\delta}_1=2m-n+3p+s,\quad \hat{\delta}_2=m+3p+t\,.\tag{30}$$

We note that each list $\hat{\ell}=(\hat{\alpha},\hat{\beta}_1,\hat{\beta}_2,\hat{\gamma},\hat{\delta}_1,\hat{\delta}_2)$ of energy exponents consistent with the necessary condition

$$\hat{\alpha}-2\hat{\beta}_2+\hat{\gamma}=0\tag{31}$$

is compatible with infinitely many lists of scaling exponents $(m,n,p,u,v,z,s,t$ and we call

$$\widehat{\Pi}(\bar{\boldsymbol{u}},\varepsilon;\hat{\ell}):=\widehat{\Sigma}(\bar{\boldsymbol{u}},\varepsilon;\hat{\alpha},\hat{\beta}_1,\hat{\beta}_2,\hat{\gamma})-\widehat{\Delta}(\bar{\boldsymbol{u}},\varepsilon;\hat{\delta}_1,\hat{\delta}_2)$$

the *scaled energy functional.*

2.3 Taxonomy of Energy Functionals for Structure Mechanics

It seems to me reasonable to deem an energy-scaling list $\hat{\ell}$ admissible if the corresponding scaled energy does not blow up when the parameter ε becomes smaller and smaller. Consequently, I propose to call *an energy-scaling list admissible if the associated functional* $\hat{\Pi}$ *stays bounded above under the scaling group action*:

$$\lim_{\varepsilon \to 0+} \widehat{\Pi}(\bar{\boldsymbol{u}}, \varepsilon; \hat{\ell}) = \widehat{\Pi}(\bar{\boldsymbol{u}}, 0; \hat{\ell}) < +\infty, \tag{32}$$

at whatever field $\bar{\mathbf{u}}$ *keeps all the integrals in* (26) *and* (29) *finite.*

For the finiteness condition (32) to be satisfied,
(i) if anyone of the energy exponents $\hat{\alpha}, \hat{\beta}_1, \ldots, \hat{\delta}_2$ is positive [null], then the corresponding functional $\mathrm{A}, \mathrm{B}_1, \ldots, \hat{\Delta}_2$ drops off [is found in] the limit functional $\widehat{\Pi}(\bar{\boldsymbol{u}}, 0; \hat{\ell})$, and hence drops off [is found in] the energy functional $\Pi(\boldsymbol{u}, \varepsilon)$ arrived at by inverse scaling;[9]
(ii) if the energy exponent associated with anyone of the *constitutively positive* functionals $\mathrm{A}, \mathrm{B}_1, \Gamma$ is chosen negative, then that functional must be made to vanish identically; this can achieved only by restricting the class of functions on which the energy functional Π arrived at by inverse scaling is defined.

It is shown in (Miara and Podio-Guidugli, 2007) that condition (32) selects those lists of energy exponents which induce the three-dimensional energy functionals of interest in structure mechanics, including the energy functionals of the classic equilibrium theory of plates and rods. Precisely, three limit functionals are obtained: the first two describe (i) in-plane stretching and flexure of plates and (ii) axial stretching and flexure of straight rods; the third describes, according to the function space over which it is defined, (iii) antiplane shear of cylinders and (iv) torsion of either plate-like or rod-like cylinders. In this connection, we observe that Γ-convergence generally imports the functionals to be validated as Γ-limits from external sources: now, it seems to us that the limit functionals produced by our method of formal scaling may be regarded as natural candidate Γ-limits; and that, if this is done, Γ-convergence and formal scaling acquire a mutually complemental role, just as is the case for the asymptotic method and the constraint method (Podio-Guidugli, 1995).

The following abridged and partial account of the developments in (Miara and Podio-Guidugli, 2007) suffices for our current purposes. Nothing

[9] Needless to say, in the energy functional resulting from inverse scaling no trace is ever found of an addendum in the scaled energy functional that is proportional to a negative power of ε.

essential is lost, and lighter developments ensue, if we take

$$\hat{\delta}_1 = \hat{\delta}_2 = 0; \tag{33}$$

in addition, in order to concentrate on *shearable* structures, we assume that

$$\hat{\beta}_1 = 0.\text{[10]} \tag{34}$$

As to the energy exponents $\hat{\alpha}$ and $\hat{\gamma}$, we make an important preliminary observation, based on implication (ii) of the finiteness condition (32).

If $\hat{\gamma} < 0$, then the following *internal constraints* must prevail:

$$\bar{E}_{11} = \bar{E}_{22} = \bar{E}_{12} \equiv 0 \quad \text{in } \mathcal{C}(1) \tag{35}$$

must hold. Hence, the functionals B_2 and Γ are identically zero-valued and, moreover, the admissible displacement fields must be such that

$$u_1 = \hat{v}_1(x_3) - x_2\hat{\psi}_3(x_3), \quad u_2 = \hat{v}_2(x_3) + x_1\hat{\psi}_3(x_3) \tag{36}$$

(see Lembo and Podio-Guidugli (2001)). Likewise, if $\hat{\alpha} < 0$, then the internal constraint:

$$\bar{E}_{33} \equiv 0 \quad \text{in } \mathcal{C}(1). \tag{37}$$

must hold. Consequently, the functionals A and B_2 are identically zero-valued and, in addition, the admissible displacement fields must be such that

$$u_3 = \hat{u}_3(x_1, x_2).\text{[11]} \tag{38}$$

In the light of this observation, we examine the energetic consequences of the following two assignments of the pair $(\hat{\alpha}, \hat{\gamma})$:

- (*axial stretching and flexure of rod-like cylinders*) for

  $$\hat{\alpha} = 0, \quad \hat{\gamma} < 0, \tag{39}$$

  the scaled stored-energy functional takes the form:

  $$\widehat{\Sigma}_R(\bar{\boldsymbol{u}}, \varepsilon; \hat{\alpha}, \hat{\beta}_1) = \varepsilon^{\hat{\alpha}}\mathrm{A}(\bar{\boldsymbol{u}}) + \varepsilon^{\hat{\beta}_1}\mathrm{B}_1(\bar{\boldsymbol{u}}); \tag{40}$$

[10] It follows from these provisional assumptions that

$$s = -2m + n - 3p, \quad t = -m - 3p, \quad \text{and} \quad u = -(m + n + p).$$

Thus, the standard (domain,displacement) scaling (19) implies that $s = -2, t = -1$, the standard scaling of volume loads, and that $u = -1$.

[11] For either $\hat{\alpha}$ or $\hat{\gamma}$ negative, the functional B_2 (whose sign, we recall, is not definite) drops off the auxiliary functional $\widehat{\Pi}(\bar{\boldsymbol{u}}, \varepsilon; \hat{\ell})$, whatever $\hat{\beta}_2$. This is the reason why we need not consider that energy exponent in our discussion.

the associated energy functional is defined over the displacement class (36), and reads:

$$\Pi_R(\boldsymbol{u},\varepsilon)=\Sigma_R(\boldsymbol{u},\varepsilon)-\Delta(\boldsymbol{u},\varepsilon), \tag{41}$$

where

$$\Sigma_R(\boldsymbol{u},\varepsilon)=\int_{\mathcal{C}(\varepsilon)}\frac{1}{2}\tau_1 E_{33}^2(\boldsymbol{u})+\int_{\mathcal{C}(\varepsilon)}2\gamma\big(E_{13}^2(\boldsymbol{u})+E_{23}^2(\boldsymbol{u})\big)\,. \tag{42}$$

- *(in-plane stretching and flexure of plate-like cylinders)* for

$$\hat{\alpha}<0,\quad \hat{\gamma}=0, \tag{43}$$

the scaled stored-energy functional is

$$\widehat{\Sigma}_P(\bar{\boldsymbol{u}},\varepsilon;\hat{\beta}_1,\hat{\gamma})=\varepsilon^{\hat{\beta}_1}\mathrm{B}_1(\bar{\boldsymbol{u}})+\varepsilon^{\hat{\gamma}}\Gamma(\bar{\boldsymbol{u}}); \tag{44}$$

hence, the energy functional is

$$\Pi_P(\boldsymbol{u},\varepsilon)=\Sigma_P(\boldsymbol{u},\varepsilon)-\Delta(\boldsymbol{u},\varepsilon), \tag{45}$$

and is defined over the displacement class (38), where

$$\Sigma_P(\boldsymbol{u},\varepsilon)=\int_{\mathcal{C}(\varepsilon)}2\gamma\big(E_{13}^2(\boldsymbol{u})+E_{23}^2(\boldsymbol{u})\big)$$
$$+\int_{\mathcal{C}(\varepsilon)}\frac{1}{2}\Big((\lambda+2\mu)\big(E_{11}(\boldsymbol{u})+E_{22}(\boldsymbol{u})\big)^2-4\mu\big(E_{11}(\boldsymbol{u})E_{22}(\boldsymbol{u})-E_{12}^2(\boldsymbol{u})\big)\Big); \tag{46}$$

REMARK. Replacing assumption (34) with

$$\hat{\beta}_1<0 \tag{47}$$

opens the way to theories of *unsherable* plates (if (43) are also assumed to hold) and rods (if instead (39) hold). In fact, (47) implies that the following internal constraint must prevail:

$$E_{\alpha 3}=0\quad\Leftrightarrow\quad u_{\alpha,3}+u_{3,\alpha}=0\ \ (\alpha=1,2)\quad in\ \mathcal{C}(\varepsilon), \tag{48}$$

a system of PDEs that, when combined, for example, with $u_{3,3}=0$, the PDE equivalent to the constraint of transverse inextensibility (37), has for its general solution the well-known *Kirchhoff-Love* representation of the displacement field in a plate:

$$\boldsymbol{u}_{KL}=\boldsymbol{v}-x_3\nabla u_3+u_3\boldsymbol{e}_3,\quad \boldsymbol{v}=\hat{v}_\alpha(x_1,x_2)\boldsymbol{e}_\alpha,\quad u_3=\hat{u}_3(x_1,x_2)\,. \tag{49}$$

2.4 The Role of Second-Order Elastic Energy in the Derivation of Shearable Structure Models

The functional Π_P in (46) generalizes the Reissner-Mindlin plate functional, in that it is defined for all displacement fields having the form:

$$u_\alpha = \hat{u}_\alpha(x_1, x_2, x_3), \quad u_3 = \hat{u}_3(x_1, x_2), \tag{50}$$

whereas Reissner-Mindlin's is the restriction of Π_P to the following subcollection of the displacement class (50):

$$u_\alpha = \hat{v}_\alpha(x_1, x_2) + x_3\hat{\varphi}_\alpha(x_1, x_2), \quad u_3 = \hat{u}_3(x_1, x_2). \tag{51}$$

Likewise, the functional Π_R in (42) is a generalization of the Timoshenko rod functional, in that it is defined for all displacement fields having the form:

$$u_1 = \hat{v}_1(x_3) - x_2\hat{\psi}_3(x_3), \quad u_2 = \hat{v}_2(x_3) + x_1\hat{\psi}_3(x_3), \quad u_3 = \hat{u}_3(x_1, x_2, x_3),$$

a class larger than Timoshenko's class:

$$u_\alpha = \hat{v}_\alpha(x_3), \quad u_3 = \hat{w}(x_3) + x_2\hat{\psi}_1(x_3) - x_1\hat{\psi}_2(x_3).$$

To deduce from the finiteness condition (32) not only the desired limit functional but also the desired function class, the stored-energy functional must be appropriately modified. For brevity, I shall here confine myself to a quick treatment of the Reissner-Mindlin case, along the lines detailed in (Miara and Podio-Guidugli, 2007).

To begin with, note that relations (51) are the general solutions of the following system of PDEs:

$$u_{3,3} = 0, \quad u_{\alpha,33} = 0,$$

all of which can be interpreted as internal constraints, of the first and of the second order:

$$\begin{aligned} u_{3,3} = 0 \quad &\Leftrightarrow \quad \nabla \boldsymbol{u} \cdot \boldsymbol{e}_3 \otimes \boldsymbol{e}_3 = 0, \\ u_{\alpha,33} = 0 \quad &\Leftrightarrow \quad \nabla^{(2)} \boldsymbol{u} \cdot \big(\boldsymbol{e}_\alpha \otimes \boldsymbol{e}_3 \otimes \boldsymbol{e}_3\big) = 0\,. \end{aligned} \tag{52}$$

A straightforward measure to recover the second-order constraints from a finiteness requirement of type (32) consists in modifying the stored-energy functional so as to include the energy associated with the second-order displacement gradient $\nabla^{(2)}\boldsymbol{u}$. A minimally augmented form of the stored energy density (12) is:

$$\sigma_a = \sigma + \frac{1}{2}\,\tau_P \sum_\alpha (u_{\alpha,33})^2, \quad \tau_P > 0\,,$$

whence

$$\Sigma_a(\boldsymbol{u},\varepsilon)=\Sigma(\boldsymbol{u},\varepsilon)+\int_{\mathcal{C}(\varepsilon)}\frac{1}{2}\,\tau_P\sum_\alpha(u_{\alpha,33})^2\,.$$

On scaling the second-order material modulus:

$$\bar{\tau}_P=\varepsilon^{-i}\tau_P,$$

and on setting

$$\widehat{\Pi}_a(\bar{\boldsymbol{u}},\varepsilon;\hat{\ell},\iota):=\widehat{\Pi}(\bar{\boldsymbol{u}},\varepsilon;\hat{\ell})+\varepsilon^{\iota}\mathrm{I}(\bar{\boldsymbol{u}}),$$

where

$$\iota=2(m-p)+q+i,\quad \mathrm{I}(\bar{\boldsymbol{u}})=\int_{\mathcal{C}(1)}\frac{1}{2}\bar{\tau}_P\sum_\alpha(\bar{u}_{\alpha,33})^2,$$

the Reissner-Mindlin plate theory is arrived at by replacing the scaled energy functional $\widehat{\Pi}$ in (32) with its augmented version $\widehat{\Pi}_a$ and, moreover, by accompanying (43) with the assumption that

$$\iota<0\,.$$

REMARK. This example motivates the conjecture that it is because of overlooking the role of second-order constraints that so far, to my knowledge, neither variational convergence methods have been able to validate nor asymptotic methods have been able to produce lower-dimensional models of *shearable* structures (with the exception, in the latter case, of the recent work of I. Aganović, J. Tambaca and Z. Tutek (2006), where the additional complexity in the parent three-dimensional theory needed to capture flexure deformations is achieved through the use of *micropolar* linear elasticity).

3 The Best Approximate Stress Field in a Linearly Elastic Structure-like Body

In this third and last part of this series of lectures, I will show that, given an equilibrium problem for a plate-like body and a plate theory, the approximate evaluation of the stress field in the three-dimensional body can be substantially bettered by a well-tempered use of the reactive stress fields that maintain in the three-dimensional body the kinematical constraints implicit in the formulation of the given plate theory. To begin by an admittedly sibylline statement of the main take-home message to come, the basic <kinematic, dynamic> duality embodied in the internal working form 'explains' why second-order stored energy is important to the variational deduction of plate theories from three-dimensional elasticity and why second-order reactive stresses are expedient to sharpen the predictions of three-dimensional stress fields based on plate theories.

The material I shall present is taken from a paper recently written in collaboration with Marzio Lembo (Lembo and Podio-Guidugli, 2007).[12]

3.1 Setting the stage

Let us consider a three-dimensional equilibrium problem for a linearly elastic cylinder $\mathcal{C} = \mathcal{P} \times \mathcal{R}$, just as in Subsection 2.1 (here and henceforth, the parameter ε is dropped from the notation, because it is thought of as fixed once and for all). All the rest staying, let us contemplate the possibility that a consistent combination of Dirichlet and Neumann data $\boldsymbol{u}_D$ and $\boldsymbol{c}_N$ is specified over the mantle $\partial\mathcal{P} \times \mathcal{R}$ of $\mathcal{C}$. Finally, let us denote by $\widetilde{\mathcal{S}} := (\tilde{\boldsymbol{u}}, \widetilde{\boldsymbol{E}}, \widetilde{\boldsymbol{S}})$, with $\widetilde{\boldsymbol{E}} = \boldsymbol{E}(\tilde{\boldsymbol{u}}) := \operatorname{sym} \nabla \tilde{\boldsymbol{u}}$ and $\widetilde{\boldsymbol{S}} = \boldsymbol{S}(\tilde{\boldsymbol{u}}) := \mathbb{C}[\boldsymbol{E}(\tilde{\boldsymbol{u}})]$, the elastic state solving the problem at hand exactly.

Suppose that, with a view toward approximating a state $\widetilde{\mathcal{S}}$ too difficult or onerous to compute exactly, you decide to formulate and solve a corresponding two-dimensional equilibrium problem within some simple – but not too simple! – plate theory: say, a theory whose basic unknowns are two vector fields $\hat{\boldsymbol{u}}, \hat{\boldsymbol{\varphi}}$ over $\mathcal{P}$, the parameter fields to be used to construct an approximation of $\tilde{\boldsymbol{u}}$ in $\mathcal{C}$ having the following form:

$$\boldsymbol{u}(x_1, x_2, x_3) = \hat{\boldsymbol{u}}(x_1, x_2) + x_3 \hat{\boldsymbol{\varphi}}(x_1, x_2)\,. \tag{53}$$

A first step to take is to construct a set of two-dimensional data. A simple way to do so is to define the *distance force* and *couple fields* $\boldsymbol{q}$ and $\boldsymbol{t}$ per unit area of the domain $\mathcal{P}$ by a *fiber-wise equipollency rule*:

$$\boldsymbol{q} := \boldsymbol{c}^+ + \boldsymbol{c}^- + \int_{-h}^{+h} \boldsymbol{d}, \quad \boldsymbol{t} := h\, \boldsymbol{e}_3 \times (\boldsymbol{c}^+ - \boldsymbol{c}^-) + \int_{-h}^{+h} z\, \boldsymbol{e}_3 \times \boldsymbol{d}\, dz$$

at each interior point of $\mathcal{P}$; and to choose the Dirichlet and Neumann data $(\boldsymbol{w}, \boldsymbol{\phi})$ and $(\boldsymbol{f}, \boldsymbol{m})$ per unit length of the boundary curve $\partial\mathcal{P}$ according to a *fiber-wise equipotency rule*:

$$\boldsymbol{f} \cdot \boldsymbol{w} + \boldsymbol{m} \cdot \boldsymbol{\phi} := \int_{-h}^{+h} \boldsymbol{c}_N \cdot \boldsymbol{u}_D$$

at each point of $\partial\mathcal{P}$.[13] A second step is of course to determine the parameter fields so as to agree with these boundary data and with the balance

[12] The case of rod-like bodies and rod theories had been dealt with by us in (Lembo and Podio-Guidugli, 2001).

[13] Two systems of forces and couples are *equipollent* when they have the same resultant

equations of the chosen plate theory. The final step is to construct the approximate elastic state $\mathcal{S} = (\boldsymbol{u}, \boldsymbol{E} = \boldsymbol{E}(\boldsymbol{u}), \boldsymbol{S} = \mathbb{C}[\boldsymbol{E}(\boldsymbol{u})])$.

Now, such an elastic state is compatible with *one* set of three-dimensional data; this set belongs necessarily to the same equivalence class as the original three-dimensional set, and yet it can differ remarkably from it when a pointwise comparison is made. You do not have to believe me, just wait until you see the example I am going to give.

Consider a simply supported circular plate subject to uniform transverse load. On applying the Reissner-Mindlin theory, the fields $\hat{\boldsymbol{u}}$ and $\boldsymbol{\varphi}$ can be calculated explicitly and the corresponding elastic state in $\mathcal{C}$ constructed. Surprisingly enough, the equilibrium distance and contact loads $\boldsymbol{d}_{RM}$ and $\boldsymbol{c}_{RM}$ associated with the Reissner-Mindlin stress field $\boldsymbol{S}_{RM}$ turn out to be as in Fig. 4.

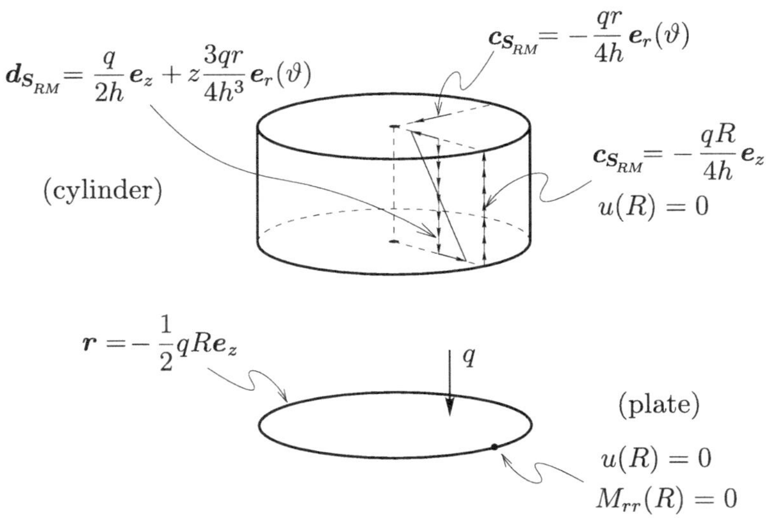

Figure 4. Simply supported plate subject to uniform transverse load ($\boldsymbol{r}$ denotes the uniform reaction offered by the support, along which the vertical displacement u is prohibited and the bending moment M_{rr} is null; from (Lembo and Podio-Guidugli, 2007)).

force and the same resultant moment, *equipotent* when they expend the same power in any given motion. The equipotency rule is a direct outcome of a virtual working approach to the deduction of structure theories (see (Lembo and Podio-Guidugli, 2001, 2007)).

This example demonstrates that, within the equivalence class of all cylinder problems whose data correspond to a given plate problem in the sense we have indicated, the one conceivably closest to that plate problem may well happen to have odd-looking pointwise data, whose distribution is far from being typical of the equivalence class at hand under almost all possible respects.

In general, nothing can be done to improve the approximation of the three-dimensional displacement and strain fields, and of the Dirichlet boundary conditions (unless, perhaps, one chooses a different plate theory; but this is another story). However, in the remain of my time, I will try and convince you that *the pointwise approximation of the three-dimensional stress field can be substantially bettered if the reaction stresses associated with the internal constraints implicit in the kinematical Ansatz are taken into account;* and that *the pointwise approximation of both applied loads and traction boundary conditions can be correspondingly bettered.*

The idea is not new: for unshearable plates, it has been advanced first in (Podio-Guidugli, 1989), where the role of the *reactive stress field* associated with the *simple internal constraint* implicit in the Kirchhoff-Love representation of the three-dimensional displacement has been pointed out. What is new is the application to plate theories based on a more general kinematics, in that such a kinematics, although not excessively detailed (as is appropriate for an approximation), calls for consideration of simple and *nonsimple internal constraints*, with accompanying reactive stress fields and *reactive hyperstress fields*, whose beneficial role can be demonstrated, both in general and *per exempla.*

3.2 The reactive stresses associated with the kinematical *Ansätze* of classic plate theories

Representation (53) requires that the deformations of material fibers parallel to the axis of $\mathcal{C}$ be homogeneous, in the sense that the directional derivative of the displacement field $\boldsymbol{u}$ along any given axial fiber turns out to be constant along that fiber. Indeed, as is straightforward to check, (53) is the general solution of the vectorial PDE

$$\boldsymbol{u}_{,33} = \boldsymbol{0}; \tag{54}$$

hence, for the fiber through the point of $\mathcal{P}$ of coordinates (x_1, x_2), we have from (53) that

$$\boldsymbol{u}_{,3} = \partial_{e_3}\boldsymbol{u} = (\nabla\boldsymbol{u})\boldsymbol{e}_3 = \hat{\boldsymbol{\varphi}}(x_1, x_2)\,.$$

Moreover, it follows from (53) that the strain components $E_{ij} = \boldsymbol{E} \cdot \boldsymbol{e}_i \otimes \boldsymbol{e}_j$ are:

$$\begin{aligned} E_{\alpha\beta} &= \frac{1}{2}(u_{\alpha,\beta} + u_{\beta,\alpha}) + x_3 \frac{1}{2}(\varphi_{\alpha,\beta} + \varphi_{\beta,\alpha}), \\ E_{\alpha 3} &= \frac{1}{2}(\varphi_\alpha + u_{3,\alpha}) + x_3 \frac{1}{2}\varphi_{3,\alpha}, \\ E_{33} &= \varphi_3. \end{aligned} \tag{55}$$

These relations show that a plate-like body compliant with prescription (53) is capable of a combination of flexural, shear, and extensional deformations, the latter both in the cross section and along the axis (DiCarlo, P. Podio-Guidugli, and W.O. Williams, 2001; Lembo and Podio-Guidugli, 2007); we may refer to the associated plate theory as to a theory of *transversely extensible plates.*

The kinematical *Ansatz* (53), while simple, is general enough to include two classical linear plate theories as special cases.

(i) To arrive at the Reissner-Mindlin theory, the axial fibers of cylinder $\mathcal{C}$ are thought of as subject, in addition to (54), to a prescription precluding their extension, namely,

$$\boldsymbol{u}_{,3} \cdot \boldsymbol{e}_3 = 0. \tag{56}$$

It follows from (54) and (56) that

$$\varphi_3 = \boldsymbol{\varphi} \cdot \boldsymbol{e}_3 = 0;$$

with this understanding, relation (53) retains its form (cf. (51)), while the components of $\boldsymbol{E}$ are given by $(55)_1$ and

$$E_{\alpha 3} = \frac{1}{2}(\varphi_\alpha + u_{3,\alpha}), \quad E_{33} = 0\,.$$

in the place of $(55)_2$ and $(55)_3$.

(ii) The Kirchhoff-Love theory is arrived at when, in addition to be thought of as inextensible, the axial fibers are required to remain orthogonal to the cross-sectional fibers:

$$E_{\alpha 3} = 0 \quad \Leftrightarrow \quad \varphi_\alpha + \widehat{u}_{3,\alpha} = 0\,. \tag{57}$$

Then, relation (53) takes the form (49), and the only nonnull components of $\boldsymbol{E}$ are those specified by $(55)_1$.

Equations (54), (56) and $(57)_1$ are all expressions of *internal constraints*, of diverse differential order. In fact, (56) and $(57)_1$ are of the first order:

$$\nabla \boldsymbol{u} \cdot \boldsymbol{e}_3 \otimes \boldsymbol{e}_3 = 0 \quad \text{and} \quad \nabla \boldsymbol{u} \cdot (\boldsymbol{e}_\alpha \otimes \boldsymbol{e}_3 + \boldsymbol{e}_3 \otimes \boldsymbol{e}_\alpha) = 0 \ (\alpha = 1, 2), \tag{58}$$

whereas (54) is of the second:

$$\nabla^{(2)}\boldsymbol{u}\cdot\boldsymbol{e}_i\otimes\boldsymbol{e}_3\otimes\boldsymbol{e}_3=0,\;\;(i=1,2,3)\,. \tag{59}$$

REMARK. Interestingly, a smooth first-order constraint has differential consequences that can legitimately be interpreted as higher-order constraints in themselves. For instance, differentiation of the inextensibility constraint $(58)_1$ yields the second-order constraint

$$\nabla^{(2)}\boldsymbol{u}\cdot\boldsymbol{e}_3\otimes(\boldsymbol{e}_i\otimes\boldsymbol{e}_3+\boldsymbol{e}_3\otimes\boldsymbol{e}_i)=0\,. \tag{60}$$

Combination of constraints (59) and (60) occurs in the Reissner-Mindlin plate theory; the additional second-order constraint induced by (59) is:

$$\nabla^{(2)}\boldsymbol{u}\cdot\boldsymbol{e}_3\otimes(\boldsymbol{e}_\alpha\otimes\boldsymbol{e}_3+\boldsymbol{e}_3\otimes\boldsymbol{e}_\alpha)=0. \tag{61}$$

An issue in the constitutive theory of continuous bodies is to decide *what reactive stress systems maintain a given internal constraint.* It is customary to assume that the reactive stress fields must be such as to expend no power in any admissible motion. In our present context, this characterization may be expressed in terms of the *internal working* pertaining to a body part $\mathcal{P}$:

$$\int_{\mathcal{P}}(\boldsymbol{S}\cdot\nabla\boldsymbol{u}+\mathsf{S}\cdot\nabla^{(2)}\boldsymbol{u}+\ldots)\,.^{14} \tag{62}$$

Precisely, each stress measure is split into an *active* and a *reactive* part:

$$\boldsymbol{S}=\boldsymbol{S}^A+\boldsymbol{S}^R,\quad \mathsf{S}=\mathsf{S}^A+\mathsf{S}^R,\;\ldots,$$

and it is stipulated that

$$\int_{\mathcal{P}}(\boldsymbol{S}^R\cdot\nabla\boldsymbol{u}+\mathsf{S}^R\cdot\nabla^{(2)}\boldsymbol{u}+\ldots)=0$$

for all body parts and all admissible motions. Thus, the first-order constraint $(58)_1$ is maintained by the reactive stress field

$$\boldsymbol{S}^R=\sigma\,\boldsymbol{e}_3\otimes\boldsymbol{e}_3, \tag{63}$$

[14]In cartesian components, the internal working per unit volume is given by

$$\boldsymbol{S}\cdot\nabla\boldsymbol{u}+\mathsf{S}\cdot\nabla^{(2)}\boldsymbol{u}=S_{ij}u_{i,j}+\mathsf{S}_{ijk}u_{i,jk},\quad \mathsf{S}_{ijk}=\mathsf{S}\cdot\boldsymbol{e}_i\otimes\boldsymbol{e}_j\otimes\boldsymbol{e}_k\,.$$

Since the second-gradient tensor $\nabla^{(2)}\boldsymbol{u}$ is symmetric with respect to the last two indices, it is convenient to assume that the *hyperstress* S has the same property: $\mathsf{S}_{ijk}=\mathsf{S}_{ikj}$.

where the scalar σ is arbitrary, in the sense that it is not constitutively determined; the reactive hyperstress maintaining the second-order constraints (59) that lead to the theory of transversely extensible plates turns out to be:

$$\mathbf{S}^R = \gamma_i \boldsymbol{e}_i \otimes \boldsymbol{e}_3 \otimes \boldsymbol{e}_3 \tag{64}$$

(three arbitrary scalars); and the reactive hyperstress corresponding to the combined constraints (59) and (61) is

$$\mathbf{S}^R = \gamma_i \boldsymbol{e}_i \otimes \boldsymbol{e}_3 \otimes \boldsymbol{e}_3 + \eta_\alpha \boldsymbol{e}_3 \otimes (\boldsymbol{e}_\alpha \otimes \boldsymbol{e}_3 + \boldsymbol{e}_3 \otimes \boldsymbol{e}_\alpha), \tag{65}$$

and features five arbitrary scalar parameters.

3.3 The simplest nonsimple materials for plate-like bodies

Let us now return to the internal working form (62): dropping all gradients higher than the first, we reduce it to the form appropriate for the so-called *simple* materials. We shall now focus on the simplest of all *non-simple* cases, which obtains when all gradients higher than the second are dropped. Then,

$$\int_{\mathcal{P}} (\boldsymbol{S} \cdot \nabla \boldsymbol{u} + \mathbf{S} \cdot \nabla^{(2)} \boldsymbol{u}) = \int_{\mathcal{P}} \boldsymbol{T} \cdot \nabla \boldsymbol{u} + \int_{\partial \mathcal{P}} (\boldsymbol{T}\boldsymbol{n} \cdot \boldsymbol{u} + \mathbf{S}\boldsymbol{n} \cdot \nabla \boldsymbol{u}),$$

with the *modified stress* $\boldsymbol{T}$ defined as follows:

$$\boldsymbol{T} := \boldsymbol{S} - \mathrm{Div}\mathbf{S}\,;$$

equilibrium is ruled by one field and two boundary operators, namely and respectively,

$$\boldsymbol{f} = -\mathrm{Div}\,\boldsymbol{T} \quad \text{and} \quad \boldsymbol{b}_1 = \boldsymbol{T}\boldsymbol{n} - {}^{s}\mathrm{Div}(\mathbf{S}\boldsymbol{n}) \;\;\&\;\; \boldsymbol{b}_2 = (\mathbf{S}\boldsymbol{n})\boldsymbol{n}\,.^{15} \tag{66}$$

Now, let

$$\boldsymbol{T} = \boldsymbol{T}^A + \boldsymbol{T}^R,$$

with

$$\boldsymbol{T}^A := \boldsymbol{S}^A - \mathrm{Div}\mathbf{S}^A \quad \text{and} \quad \boldsymbol{T}^R := \boldsymbol{S}^R - \mathrm{Div}\mathbf{S}^R$$

---

[15] In these equations, the divergence of the third-order tensor $\mathbf{S}$ is defined to be

$$\mathrm{Div}\mathbf{S} = \mathbf{S}_{,k}\boldsymbol{e}_k = \mathsf{S}_{ijk,k}\boldsymbol{e}_i \otimes \boldsymbol{e}_j;$$

for $\boldsymbol{n}$ the outer normal at regular ponts of the body's boundary, ${}^{s}\mathrm{Div}(\mathbf{S}\boldsymbol{n})$ denotes the surface divergence of the second-order tensor $\mathbf{S}\boldsymbol{n}$, defined to be

$${}^{s}\mathrm{Div}(\mathbf{S}\boldsymbol{n}) = \mathrm{Div}(\mathbf{S}\boldsymbol{n}) - \big(\partial_n(\mathbf{S}\boldsymbol{n})\big)\boldsymbol{n}.$$

the active and reactive parts of the modified stress $\boldsymbol{T}$. I shall now introduce two types of nonsimple material response for plate-like bodies, each of which is both sufficiently poor to leave the classic plate theories unchanged and yet sufficiently rich to better the approximation of three-dimensional stress fields substantially.

Firstly, I propose to consider the class of *transversely extensible* plate-like bodies, that is to say, those made of a second-grade material, whose *purely active stress* is delivered in terms of the strain tensor $\boldsymbol{E}$ by the constitutive mapping (10), and whose *purely reactive hyperstress* has the form (64); then,

$$\boldsymbol{T} = \boldsymbol{S}^A - \gamma_{i,3}\boldsymbol{e}_i \otimes \boldsymbol{e}_3, \tag{67}$$

with

$$[\boldsymbol{T}^A] = \begin{bmatrix} S^A_{11} & S^A_{12} & S^A_{13} \\ \cdot & S^A_{22} & S^A_{23} \\ \cdot & \cdot & S^A_{33} \end{bmatrix}, \quad [\boldsymbol{T}^R] = \begin{bmatrix} 0 & 0 & -\gamma_{1,3} \\ 0 & 0 & -\gamma_{2,3} \\ 0 & 0 & -\gamma_{3,3} \end{bmatrix}. \tag{68}$$

Next, I introduce the class of a *transversely inextensible* plate-like bodies, those made of such a second-grade material that *the stress has both an active part and a reactive part*, specified by

$$\begin{aligned} S^A_{\alpha\beta} &= 2\mu E_{\alpha\beta} + \lambda(E_{11} + E_{22})\delta_{\alpha\beta}, \\ S^A_{3\alpha} &= 2\gamma E_{3\alpha}, \end{aligned} \tag{69}$$

and by (63); and such that the *purely reactive hyperstress* has the form (65). In this case,

$$\boldsymbol{T} = \boldsymbol{S}^A - \gamma_{\alpha,3}\boldsymbol{e}_\alpha \otimes \boldsymbol{e}_3 + (\sigma - \gamma_{3,3} - \eta_{\alpha,\alpha})\boldsymbol{e}_3 \otimes \boldsymbol{e}_3 - \eta_{\alpha,3}\boldsymbol{e}_3 \otimes \boldsymbol{e}_\alpha, \tag{70}$$

with

$$\begin{aligned} [\boldsymbol{T}^A] &= \begin{bmatrix} S^A_{11} & S^A_{12} & S^A_{13} \\ \cdot & S^A_{22} & S^A_{23} \\ \cdot & \cdot & 0 \end{bmatrix}, \\ [\boldsymbol{T}^R] &= \begin{bmatrix} 0 & 0 & -\gamma_{1,3} \\ 0 & 0 & -\gamma_{2,3} \\ -\eta_{1,3} & -\eta_{2,3} & \sigma - \gamma_{3,3} - \eta_{\alpha,\alpha} \end{bmatrix}. \end{aligned} \tag{71}$$

It has been shown in (Lembo and Podio-Guidugli, 2007) that, when a plate-like body is thought of as made of either of the two constrained second-grade materials just considered, *the plate equations for the fields* $(\widehat{\boldsymbol{u}}, \boldsymbol{\varphi})$ *are the same* as those holding for a plate-like body made of a simple material whose active stress is delivered by, respectively, the constitutive mappings (10) and (69). Thus, as to approximating the three-dimensional displacement and strain fields, nothing is gained or lost. But, as to satisfying pointwise field equations and Neumann boundary conditions, much can be done, by inserting either of the representations (67)-(68) and (69)-(68) into (66): see the examples, analytical and numerical, given in (Lembo and Podio-Guidugli, 2007).

Comparison of equations (67)-(68) and (70)-(71) shows that addition of the simple constraint (56) both reduces to five the nonnull distinct components of the active stress and enriches the structure of the internal reactions, now depending on six scalar parameter fields instead of three. Consequently, while in both cases the admissible reactive stress fields may be expected to help medicate the equilibrium injuries generally inevitable when sticking to the outcome of a given kinematical *Ansatz*, the success of the cure might well be greater in the second case. We are in the presence of a sort of trade-off: a cruder kinematics prohibits consideration of certain deformations but allows for richer reaction fields, whence a better chance to approximate well the exact, and generally unknown, stress field and to match the given distance and contact loads pointwise. To return to computational mechanics in closing, although I know of no commercial code that does it, it seems to me plausible that algorithms accounting for reactions in a systematic way turn out to be computationally advantageous.

Bibliography

I. Aganović, J. Tambaca and Z. Tutek. Derivation and justification of the models of rods and plates from linearized three-dimensional micropolar elasticity. *J. Elasticity*, 84:131–152, 2006.

S.S. Antman. *Nonlinear Problems of Elasticity*. Springer-Verlag, 1995.

S.S. Antman and R.S. Marlow. Material constraints, Lagrange multipliers, and compatibility. Applications to rod and shell theories. *Arch. Rational Mech. Anal.*, 116:257–299, 1991.

F. Bourquin and P.G. Ciarlet. Modeling and justification of eigenvalue problems for junctions between elastic structures. *J. Funct. Anal.*, 87: 392–427, 1989.

A.-L. Cauchy. De la pression ou tension dans un corps solide. *Ex. de Math.*, 2:42–56, 1827.

A.-L. Cauchy. Sur l'équilibre et le mouvement d'une plaque solide. *Ex. de Math.*, 8:328–355, 1829.

P.G. Ciarlet. A justification of the von Kármán equations. *Arch. Rational Mech. Anal.*, 73:349–389, 1980.

P.G. Ciarlet. Recent progresses in the two-dimensional approximation of three-dimensional plate models in nonlinear elasticity. In *Numerical Ap-*

proximation of Partial Differential Equations (E.L. Ortiz Ed.), pages 3–19. North-Holland, 1987.

P.G. Ciarlet. *Plates and Junctions in Elastic Multi-Structures: An Asymptotic Analysis.* Springer-Verlag, 1990.

P.G. Ciarlet. *Mathematical Elasticity. Vol. II: Theory of Plates.* North-Holland, 1997.

P.G. Ciarlet. *Mathematical Elasticity. Vol. III: Theory of Shells.* North-Holland, 2000.

P.G. Ciarlet. *An Introduction to Differenial Geometry with Applications to Elasticity.* Springer, 2005.

P.G. Ciarlet and P. Destuynder. A justification of the two-dimensional linear plate model. *J. Mécanique*, 18:315–344, 1979.

P.G. Ciarlet, H. Le Dret and R. Nzengwa. Junctions between three-dimensional and two-dimensional linearly elastic structures. *J. Math. Pures Appl.*, 68:261–295, 1989.

F. Davì. Sul modello monodimensionale di un corpo elastico allungato. In *Atti IX Congr. AIMETA*, pages 137–140, 1989.

A. DiCarlo, P. Podio-Guidugli, and W.O. Williams. Shells with thickness distension. *Int. J. Solids Structures*, 38:1201–1225, 2001.

G. Kirchhoff. Uber das gleichgewicht und die bewegung einer elastischen scheibe. *J. reine angew. Math.*, 40:51–88, 1850.

W.T. Koiter and J.G. Simmonds. Foundations of shell theory. Dept. Mech. Engrg., Delft Uni. Tech. Report 473, 1972.

M. Lembo. The membranal and flexural equations of thin elastic plates deduced exactly from the three-dimensional theory. *Meccanica*, 24:93–97, 1989.

M. Lembo and P. Podio-Guidugli. Plate theory as an exact consequence of three-dimensional elasticity. *European J. Mech., A/Solids*, 10:1–32, 1991.

M. Lembo and P. Podio-Guidugli. Internal constraints, reactive stresses, and the Timoshenko beam theory. *J. Elasticity*, 65:131–148, 2001.

M. Lembo and P. Podio-Guidugli. How to use reactive stresses to improve plate-theory approximations of the stress field in a linearly elastic plate-like body. *Int. J. Solids Structures*, 44 (5):1337–1369, 2007.

B. Miara. Justification of the asymptotic analysis of elastic plates, I. The linear case. *Asymptotic Analysis*, 9:47–60, 1994a.

B. Miara. Justification of the asymptotic analysis of elastic plates, II. The non-linear case. *Asymptotic Analysis*, 9:119–134, 1994b.

B. Miara and P. Podio-Guidugli. Une approche formelle unifiée des theories classiques de plaques et poutres. *C.R. Acad. Sci. Paris, Ser. I*, 343: 675–678, 2006.

B. Miara and P. Podio-Guidugli. Deduction by scaling: a unified approach to classic plate and rod theories. *Asymptotic Analysis*, 51 (2):113–131, 2007.

P.M. Naghdi. The theory of shells and plates. In *Handbuch der Physik VIa/2*, pages 425–640. Springer Verlag, 1972.

P. Nardinocchi and P. Podio-Guidugli. The equations of Reissner-Mindlin plates obtained by the method of internal constraints. *Meccanica*, 29: 143–157, 1994.

P. Nardinocchi and P. Podio-Guidugli. Angle plates. *J. Elasticity*, 63:19–53, 2001.

R. Paroni, P. Podio-Guidugli and G. Tomassetti. The Reissner-Mindlin plate theory via γ-convergence. *C.R. Acad. Sci. Paris, Ser. I*, 343:437–440, 2006.

R. Paroni, P. Podio-Guidugli and G. Tomassetti. A justification of the Reissner-Mindlin plate theory through variational convergence. *Analysis and Applications*, 5:165–182, 2007.

P. Podio-Guidugli. An exact derivation of the thin plate equation. *J. Elasticity*, 22:121–133, 1989.

P. Podio-Guidugli. Constraint and scaling methods to derive shell theory from three-dimensional elasticity. *Riv. Mat. Univ. Parma*, 16:73–83, 1990.

P. Podio-Guidugli. *Lezioni sulla Teoria Lineare dei Gusci Elastici Sottili.* Masson, 1991.

P. Podio-Guidugli. Scaling and constraints in the mechanics of linearly elastic structures. Unpublished manuscript, 1995.

P. Podio-Guidugli. Old and new invariance methods in continuum mechanics. In *Proc. Modern Group Analysis VI (N.H. Ibragimov & F.M. Mahomed Ed.s)*, pages 41–52. New Age Int. Pub., 1997.

P. Podio-Guidugli. Some recent results on Saint-Venant problem. In *Atti dei Convegni Lincei N. 140, "Il Problema di de Saint-Venant: Aspetti Teorici e Applicativi"*, pages 35–45. Accademia Nazionale dei Lincei, 1998.

P. Podio-Guidugli. A new quasilinear model for plate buckling. *J. Elasticity*, 71:157–182, 2003.

S.-D. Poisson. Mémoire sur l'équilibre et le mouvement des corps élastiques. *Mém. Acad. Sci. Paris*, 3:357–570, 1828.

C. Truesdell. *A First Course in Rational Continuum Mechanics, Vol. 1, 2nd Ed.* Academic Press, 1991.

A Brief Introduction to Mathematical Shell Theory*

Philippe G. Ciarlet
City University of Hong Kong

Abstract In the first chapter, we study basic notions about surfaces, such as their two fundamental forms, the Gaussian curvature and covariant derivatives. We also state the fundamental theorem of surface theory, which asserts that the Gauß and Codazzi-Mainardi equations constitute sufficient conditions for two matrix fields defined in a simply-connected open subset of $\mathbb{R}^2$ to be the two fundamental forms of a surface in a three-dimensional Euclidean space. We also state the corresponding rigidity theorem.

The second chapter, which heavily relies on Chapter 1, begins by a detailed description of the nonlinear and linear equations proposed by W.T. Koiter for modeling thin elastic shells. These equations are "two-dimensional", in the sense that they are expressed in terms of two curvilinear coordinates used for defining the middle surface of the shell. The existence, uniqueness, and regularity of solutions to the linear Koiter equations is then established, thanks this time to a fundamental "Korn inequality on a surface" and to an "infinitesimal rigid displacement lemma on a surface".

*For the most part, this article is adapted from Chapters 2 and 4 of my book "An Introduction to Differential Geometry with Applications to Elasticity", Springer, Dordrecht, 2005, the writing of which was substantially supported by two grants from the Research Grants Council of Hong Kong Special Administrative Region, China [Project No. 9040869, CityU 100803 and Project No. 9040966, CityU 100604].

1 Differential Geometry of Surfaces

In this chapter, we study basic notions about surfaces, such as their two fundamental forms, covariant derivatives, etc., that are constantly used in mathematical shell theory. We also state the fundamental theorem of surface theory, which asserts that the Gauß and Codazzi-Mainardi equations constitute sufficient conditions for two matrix fields defined in a simply-connected open subset of $\mathbb{R}^2$ to be the two fundamental forms of a surface in a three-dimensional Euclidean space. We also state the corresponding rigidity theorem.

1.1 Surfaces Defined by Means of Curvilinear Coordinates

Latin indices and exponents range in the set $\{1, 2, 3\}$, *Greek* indices and exponents range in the set $\{1, 2\}$, and the summation convention be systematically used in conjunction with these rules. For instance, the relation (see Theorem 1.6.1)

$$\partial_\alpha(\eta_i \boldsymbol{a}^i) = (\eta_{\beta|\alpha} - b_{\alpha\beta}\eta_3)\boldsymbol{a}^\beta + (\eta_{3|\alpha} + b_\alpha^\beta \eta_\beta)\boldsymbol{a}^3$$

means that

$$\partial_\alpha\Big(\sum_{i=1}^{3} \eta_i \boldsymbol{a}^i\Big) = \sum_{\beta=1}^{2}(\eta_{\beta|\alpha} - b_{\alpha\beta}\eta_3)\boldsymbol{a}^\beta + \Big(\eta_{3|\alpha} + \sum_{\beta=1}^{2} b_\alpha^\beta \eta_\beta\Big)\boldsymbol{a}^3 \text{ for } \alpha = 1, 2.$$

Kronecker's symbols are designated by δ_α^β, $\delta_{\alpha\beta}$, or $\delta^{\alpha\beta}$ according to the context.

Let there be given a *three-dimensional Euclidean space* $\boldsymbol{E}^3$, equipped with an orthonormal basis consisting of three vectors $\widehat{\boldsymbol{e}}^i = \widehat{\boldsymbol{e}}_i$, and let $\boldsymbol{a} \cdot \boldsymbol{b}$, $|\boldsymbol{a}|$, and $\boldsymbol{a} \wedge \boldsymbol{b}$ denote the Euclidean inner product, the Euclidean norm, and the vector product of vectors $\boldsymbol{a}, \boldsymbol{b}$ in the space $\boldsymbol{E}^3$.

In addition, let there be given a *two-dimensional vector space*, in which two vectors $\boldsymbol{e}^\alpha = \boldsymbol{e}_\alpha$ form a basis. *This space will be identified with* $\mathbb{R}^2$. Let y_α denote the coordinates of a point $y \in \mathbb{R}^2$ and let $\partial_\alpha := \partial/\partial y_\alpha$ and $\partial_{\alpha\beta} := \partial^2/\partial y_\alpha \partial y_\beta$.

Finally, let there be given an *open* subset ω of $\mathbb{R}^2$ and a smooth enough mapping $\boldsymbol{\theta} : \omega \to \boldsymbol{E}^3$ (specific smoothness assumptions on $\boldsymbol{\theta}$ will be made later, according to each context). The set

$$\widehat{\omega} := \boldsymbol{\theta}(\omega)$$

is called a **surface** in $\boldsymbol{E}^3$.

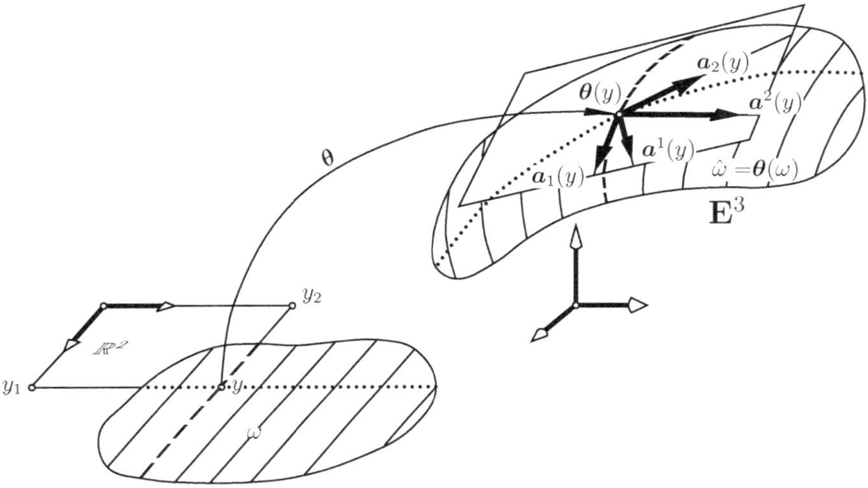

Figure 1.1.1. *Curvilinear coordinates on a surface and covariant and contravariant bases of the tangent plane.* Let $\widehat{\omega} = \boldsymbol{\theta}(\omega)$ be a surface in $\boldsymbol{E}^3$. The two coordinates y_1, y_2 of $y \in \omega$ are the curvilinear coordinates of $\widehat{y} = \boldsymbol{\theta}(y) \in \widehat{\omega}$. If the two vectors $\boldsymbol{a}_\alpha(y) = \partial_\alpha \boldsymbol{\theta}(y)$ are linearly independent, they are tangent to the coordinate lines passing through $\widehat{y}$ and they form the covariant basis of the tangent plane to $\widehat{\omega}$ at $\widehat{y} = \boldsymbol{\theta}(y)$. The two vectors $\boldsymbol{a}^\alpha(y)$ from this tangent plane defined by $\boldsymbol{a}^\alpha(y) \cdot \boldsymbol{a}_\beta(y) = \delta^\alpha_\beta$ form its contravariant basis.

If the mapping $\boldsymbol{\theta} : \omega \to \boldsymbol{E}^3$ is injective, each point $\widehat{y} \in \widehat{\omega}$ can be unambiguously written as

$$\widehat{y} = \boldsymbol{\theta}(y), \quad y \in \omega,$$

and the two coordinates y_α of y are called the **curvilinear coordinates** of $\widehat{y}$ (Figure 1.1.1). Well-known *examples* of surfaces and of curvilinear coordinates and their corresponding coordinate lines (defined in Section 1.2) are given in Figures 1.1.2 and 1.1.3.

Naturally, once a surface $\widehat{\omega}$ is defined as $\widehat{\omega} = \boldsymbol{\theta}(\omega)$, *there are infinitely many other ways of defining curvilinear coordinates on* $\widehat{\omega}$, depending on how the domain ω and the mapping $\boldsymbol{\theta}$ are chosen. For instance, a portion $\widehat{\omega}$ of a sphere may be represented by means of *Cartesian coordinates, spherical coordinates*, or *stereographic coordinates* (Figure 1.1.3). Incidentally, this example illustrates the variety of restrictions that have to be imposed on $\widehat{\omega}$ according to which kind of curvilinear coordinates it is equipped with!

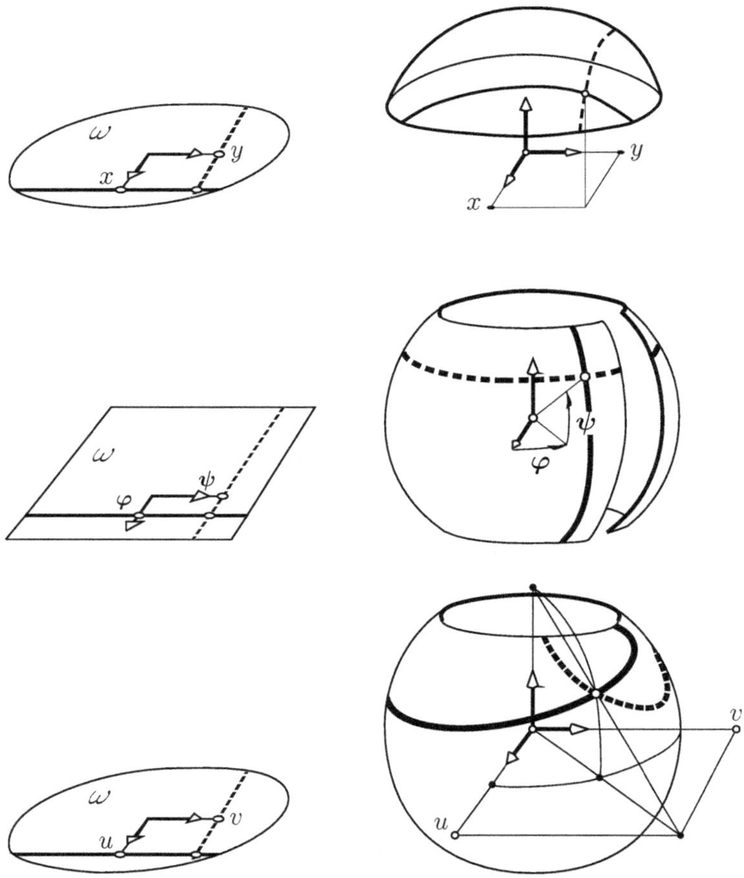

Figure 1.1.2. *Several systems of curvilinear coordinates on a sphere.* Let $\Sigma \subset \boldsymbol{E}^3$ be a sphere of radius R. A portion of Σ contained "in the northern hemisphere" can be represented by means of Cartesian coordinates, with a mapping $\boldsymbol{\theta}$ of the form:

$$\boldsymbol{\theta} : (x, y) \in \omega \to (x, y, \{R^2 - (x^2 + y^2)\}^{1/2}) \in \boldsymbol{E}^3.$$

A portion of Σ that excludes a neighborhood of both "poles" and of a "meridian" (to fix ideas) can be represented by means of spherical coordinates, with a mapping $\boldsymbol{\theta}$ of the form:

$$\boldsymbol{\theta} : (\varphi, \psi) \in \omega \to (R\cos\psi\cos\varphi, R\cos\psi\sin\varphi, R\sin\psi) \in \boldsymbol{E}^3.$$

A portion of Σ that excludes a neighborhood of the "North pole" can be represented by means of stereographic coordinates, with a mapping $\boldsymbol{\theta}$ of the form:

$$\boldsymbol{\theta} : (u, v) \in \omega \to \Big(\frac{2R^2 u}{u^2 + v^2 + R^2}, \frac{2R^2 v}{u^2 + v^2 + R^2}, R\frac{u^2 + v^2 - R^2}{u^2 + v^2 + R^2}\Big) \in \boldsymbol{E}^3.$$

The corresponding coordinate lines are represented in each case, with self-explanatory graphical conventions.

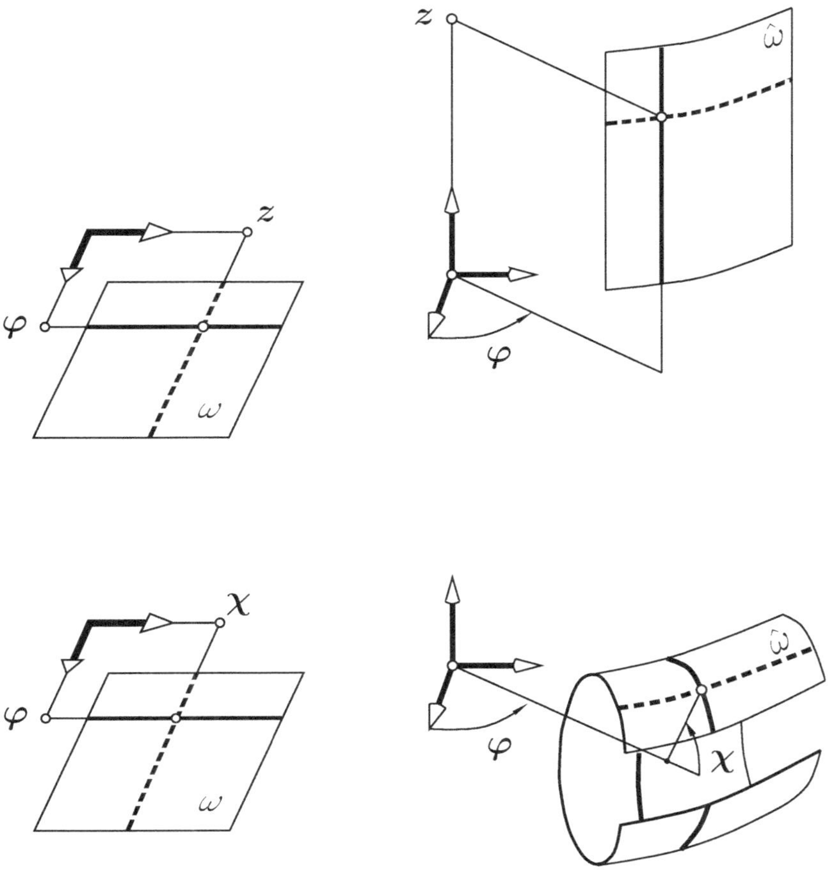

Figure 1.1.3. *Two familiar examples of surfaces and curvilinear coordinates.* A portion $\widehat{\omega}$ of a circular cylinder of radius R can be represented by a mapping $\boldsymbol{\theta}$ of the form

$$\boldsymbol{\theta} : (\varphi, z) \in \omega \to (R\cos\varphi, R\sin\varphi, z) \in \boldsymbol{E}^3.$$

A portion $\widehat{\omega}$ of a torus can be represented by a mapping $\boldsymbol{\theta}$ of the form

$$\boldsymbol{\theta} : (\varphi, \chi) \in \omega \to ((R + r\cos\chi)\cos\varphi, (R + r\cos\chi)\sin\varphi, r\sin\chi) \in \boldsymbol{E}^3,$$

with $R > r$.

The corresponding coordinate lines are represented in each case, with self-explanatory graphical conventions.

1.2 First Fundamental Form

Let ω be an open subset of $\mathbb{R}^2$ and let

$$\boldsymbol{\theta} = \theta_i \widehat{\boldsymbol{e}}^i : \omega \subset \mathbb{R}^2 \to \boldsymbol{\theta}(\omega) = \widehat{\omega} \subset \boldsymbol{E}^3$$

be a mapping that is *differentiable at a point* $y \in \omega$. If $\boldsymbol{\delta y}$ is such that $(y + \boldsymbol{\delta y}) \in \omega$, then

$$\boldsymbol{\theta}(y + \boldsymbol{\delta y}) = \boldsymbol{\theta}(y) + \boldsymbol{\nabla\theta}(y)\boldsymbol{\delta y} + o(\boldsymbol{\delta y}),$$

where the 3×2 matrix $\boldsymbol{\nabla\theta}(y)$ and the column vector $\boldsymbol{\delta y}$ are defined by

$$\boldsymbol{\nabla\theta}(y) := \begin{pmatrix} \partial_1\theta_1 & \partial_2\theta_1 \\ \partial_1\theta_2 & \partial_2\theta_2 \\ \partial_1\theta_3 & \partial_2\theta_3 \end{pmatrix}(y) \text{ and } \boldsymbol{\delta y} = \begin{pmatrix} \delta y_1 \\ \delta y_2 \end{pmatrix}.$$

Let the two vectors $\boldsymbol{a}_\alpha(y) \in \mathbb{R}^3$ be defined by

$$\boldsymbol{a}_\alpha(y) := \partial_\alpha\boldsymbol{\theta}(y) = \begin{pmatrix} \partial_\alpha\theta_1 \\ \partial_\alpha\theta_2 \\ \partial_\alpha\theta_3 \end{pmatrix}(y),$$

i.e., $\boldsymbol{a}_\alpha(y)$ *is the* α-th *column vector of the matrix* $\boldsymbol{\nabla\theta}(y)$. Then the expansion of $\boldsymbol{\theta}$ about y may be also written as

$$\boldsymbol{\theta}(y + \boldsymbol{\delta y}) = \boldsymbol{\theta}(y) + \delta y^\alpha \boldsymbol{a}_\alpha(y) + o(\boldsymbol{\delta y}).$$

If in particular $\boldsymbol{\delta y}$ is of the form $\boldsymbol{\delta y} = \delta t \boldsymbol{e}_\alpha$, where $\delta t \in \mathbb{R}$ and $\boldsymbol{e}_\alpha$ is one of the basis vectors in $\mathbb{R}^2$, this relation reduces to

$$\boldsymbol{\theta}(y + \delta t \boldsymbol{e}_\alpha) = \boldsymbol{\theta}(y) + \delta t \boldsymbol{a}_\alpha(y) + o(\delta t).$$

A mapping $\boldsymbol{\theta} : \omega \to \boldsymbol{E}^3$ is an **immersion at** $y \in \omega$ if it is differentiable at y and the 3×2 matrix $\boldsymbol{\nabla\theta}(y)$ is of rank two, or equivalently if the two vectors $\boldsymbol{a}_\alpha(y) = \partial_\alpha\boldsymbol{\theta}(y)$ are linearly independent.

Assume from now on in this section that *the mapping* $\boldsymbol{\theta}$ *is an immersion at* y. In this case, the last relation shows that *each vector* $\boldsymbol{a}_\alpha(y)$ *is tangent to the* α-th **coordinate line** *passing through* $\widehat{y} = \boldsymbol{\theta}(y)$, defined as the image by $\boldsymbol{\theta}$ of the points of ω that lie on a line parallel to $\boldsymbol{e}_\alpha$ passing through y (there exist t_0 and t_1 with $t_0 < 0 < t_1$ such that the α*-th* coordinate line is given by $t \in]t_0, t_1[\to \boldsymbol{f}_\alpha(t) := \boldsymbol{\theta}(y + t\boldsymbol{e}_\alpha)$ in a neighborhood of $\widehat{y}$; hence $\boldsymbol{f}'_\alpha(0) = \partial_\alpha\boldsymbol{\theta}(y) = \boldsymbol{a}_\alpha(y)$); see Figures 1.1.1, 1.1.2, and 1.1.3.

The vectors $\boldsymbol{a}_\alpha(y)$, which thus span the *tangent plane* to the surface $\widehat{\omega}$ at $\widehat{y} = \boldsymbol{\theta}(y)$, form the **covariant basis of the tangent plane** to $\widehat{\omega}$ at $\widehat{y}$; see Figure 1.1.1.

Returning to a general increment $\boldsymbol{\delta y} = \delta y^\alpha \boldsymbol{e}_\alpha$, we also infer from the expansion of $\boldsymbol{\theta}$ about y that ($\boldsymbol{\delta y}^T$ and $\boldsymbol{\nabla\theta}(y)^T$ respectively designate the transpose of the column vector $\boldsymbol{\delta y}$ and the transpose of the matrix $\boldsymbol{\nabla\theta}(y)$)

$$\begin{aligned}|\boldsymbol{\theta}(y+\boldsymbol{\delta y})-\boldsymbol{\theta}(y)|^2 &= \boldsymbol{\delta y}^T\boldsymbol{\nabla\theta}(y)^T\boldsymbol{\nabla\theta}(y)\boldsymbol{\delta y}+o(|\boldsymbol{\delta y}|^2)\\ &= \delta y^\alpha \boldsymbol{a}_\alpha(y)\cdot\boldsymbol{a}_\beta(y)\delta y^\beta + o(|\boldsymbol{\delta y}|^2).\end{aligned}$$

In other words, the principal part with respect to $\boldsymbol{\delta y}$ of the length between the points $\boldsymbol{\theta}(y+\boldsymbol{\delta y})$ and $\boldsymbol{\theta}(y)$ is $\{\delta y^\alpha \boldsymbol{a}_\alpha(y)\cdot\boldsymbol{a}_\beta(y)\delta y^\beta\}^{1/2}$. This observation suggests to define a matrix $(a_{\alpha\beta}(y))$ of order two by letting

$$a_{\alpha\beta}(y) := \boldsymbol{a}_\alpha(y)\cdot\boldsymbol{a}_\beta(y) = \left(\boldsymbol{\nabla\theta}(y)^T\boldsymbol{\nabla\theta}(y)\right)_{\alpha\beta}.$$

The elements $a_{\alpha\beta}(y)$ of this symmetric matrix are called the **covariant components of the first fundamental form**, also called the **metric tensor**, *of the surface* $\widehat{\omega}$ at $\widehat{y} = \boldsymbol{\theta}(y)$.

Note that *the matrix* $(a_{\alpha\beta}(y))$ *is positive definite* since the vectors $\boldsymbol{a}_\alpha(y)$ are assumed to be linearly independent.

The two vectors $\boldsymbol{a}_\alpha(y)$ being thus defined, *the four relations*

$$\boldsymbol{a}^\alpha(y)\cdot\boldsymbol{a}_\beta(y) = \delta^\alpha_\beta$$

unambiguously define two linearly independent vectors $\boldsymbol{a}^\alpha(y)$ *in the tangent plane.* To see this, let *a priori* $\boldsymbol{a}^\alpha(y) = Y^{\alpha\sigma}(y)\boldsymbol{a}_\sigma(y)$ in the relations $\boldsymbol{a}^\alpha(y)\cdot\boldsymbol{a}_\beta(y) = \delta^\alpha_\beta$. This gives $Y^{\alpha\sigma}(y)a_{\sigma\beta}(y) = \delta^\alpha_\beta$; hence $Y^{\alpha\sigma}(y) = a^{\alpha\sigma}(y)$, where

$$(a^{\alpha\beta}(y)) := (a_{\alpha\beta}(y))^{-1}.$$

Hence $\boldsymbol{a}^\alpha(y) = a^{\alpha\sigma}(y)\boldsymbol{a}_\sigma(y)$. These relations in turn imply that

$$\begin{aligned}\boldsymbol{a}^\alpha(y)\cdot\boldsymbol{a}^\beta(y) &= a^{\alpha\sigma}(y)a^{\beta\tau}(y)\boldsymbol{a}_\sigma(y)\cdot\boldsymbol{a}_\tau(y)\\ &= a^{\alpha\sigma}(y)a^{\beta\tau}(y)a_{\sigma\tau}(y) = a^{\alpha\sigma}(y)\delta^\beta_\sigma = a^{\alpha\beta}(y),\end{aligned}$$

and thus the vectors $\boldsymbol{a}^\alpha(y)$ are linearly independent since the matrix $(a^{\alpha\beta}(y))$ is positive definite. We would likewise establish that $\boldsymbol{a}_\alpha(y) = a_{\alpha\beta}(y)\boldsymbol{a}^\beta(y)$.

The two vectors $\boldsymbol{a}^\alpha(y)$ form the **contravariant basis of the tangent plane** to the surface $\widehat{\omega}$ at $\widehat{y} = \boldsymbol{\theta}(y)$ (Figure 1.1.1) and the elements $a^{\alpha\beta}(y)$ of the symmetric matrix $(a^{\alpha\beta}(y))$ are called the **contravariant components of the first fundamental form**, or **metric tensor**, of the surface $\widehat{\omega}$ at $\widehat{y} = \boldsymbol{\theta}(y)$.

Let us record for convenience the fundamental relations that exist between the vectors of the covariant and contravariant bases of the tangent plane and the covariant and contravariant components of the first fundamental tensor:

$$a_{\alpha\beta}(y) = \boldsymbol{a}_\alpha(y) \cdot \boldsymbol{a}_\beta(y) \quad \text{and} \quad a^{\alpha\beta}(y) = \boldsymbol{a}^\alpha(y) \cdot \boldsymbol{a}^\beta(y),$$
$$\boldsymbol{a}_\alpha(y) = a_{\alpha\beta}(y)\boldsymbol{a}^\beta(y) \quad \text{and} \quad \boldsymbol{a}^\alpha(y) = a^{\alpha\beta}(y)\boldsymbol{a}_\beta(y).$$

A mapping $\boldsymbol{\theta} : \omega \to \boldsymbol{E}^3$ is an **immersion** if it is an immersion at each point in ω, i.e., if $\boldsymbol{\theta}$ is differentiable in ω and the two vectors $\partial_\alpha \boldsymbol{\theta}(y)$ are linearly independent at each $y \in \omega$.

If $\boldsymbol{\theta} : \omega \to \boldsymbol{E}^3$ is an immersion, the vector fields $\boldsymbol{a}_\alpha : \omega \to \mathbb{R}^3$ and $\boldsymbol{a}^\alpha : \omega \to \mathbb{R}^3$ respectively form the **covariant**, and **contravariant, bases of the tangent planes**.

1.3 Areas and Lengths on a Surface

We now review fundamental formulas expressing *area* and *length elements* at a point $\widehat{y} = \boldsymbol{\theta}(y)$ of the surface $\widehat{\omega} = \boldsymbol{\theta}(\omega)$ in terms of the matrix $(a_{\alpha\beta}(y))$; see Figure 1.3.1.

These formulas highlight in particular the crucial rôle played by the matrix $(a_{\alpha\beta}(y))$ for computing "metric" notions at $\widehat{y} = \boldsymbol{\theta}(y)$.

Theorem 1.3.1. *Let ω be an open subset of $\mathbb{R}^2$, let $\boldsymbol{\theta} : \omega \to \boldsymbol{E}^3$ be an injective and smooth enough immersion, and let $\widehat{\omega} = \boldsymbol{\theta}(\omega)$.*

(a) *The area element $\mathrm{d}\widehat{a}(\widehat{y})$ at $\widehat{y} = \boldsymbol{\theta}(y) \in \widehat{\omega}$ is given in terms of the area element $\mathrm{d}y$ at $y \in \omega$ by*

$$\mathrm{d}\widehat{a}(\widehat{y}) = \sqrt{a(y)}\,\mathrm{d}y, \text{ where } a(y) := \det(a_{\alpha\beta}(y)).$$

(b) *The length element $\mathrm{d}\widehat{\ell}(\widehat{y})$ at $\widehat{y} = \boldsymbol{\theta}(y) \in \widehat{\omega}$ is given by*

$$\mathrm{d}\widehat{\ell}(\widehat{y}) = \left\{\delta y^\alpha a_{\alpha\beta}(y)\delta y^\beta\right\}^{1/2}.$$

Proof. The relation (a) between the area elements is well known.

The expression of the length element in (b) recalls that $\mathrm{d}\widehat{\ell}(\widehat{y})$ is by definition the principal part with respect to $\boldsymbol{\delta y} = \delta y^\alpha \boldsymbol{e}_\alpha$ of the length $|\boldsymbol{\theta}(y + \boldsymbol{\delta y}) - \boldsymbol{\theta}(y)|$, whose expression precisely led to the introduction of the matrix $(a_{\alpha\beta}(y))$. □

The relations found in Theorem 1.3.1 are used for computing *surface integrals* and *lengths* on the surface $\widehat{\omega}$ by means of integrals inside ω, i.e., in terms of the *curvilinear coordinates* used for defining the surface $\widehat{\omega}$ (see again Figure 1.3.1). In what follows, $\boldsymbol{\theta} : \omega \to \boldsymbol{E}^3$ is again an injective and smooth enough immersion.

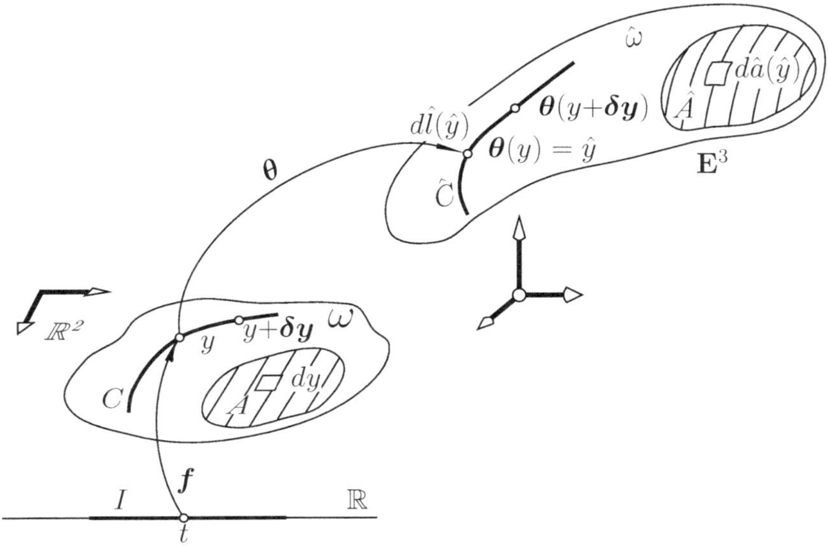

Figure 1.3.1. *Area and length elements on a surface.* The elements $\mathrm{d}\widehat{a}(\widehat{y})$ and $\mathrm{d}\widehat{\ell}(\widehat{y})$ at $\widehat{y} = \boldsymbol{\theta}(y) \in \widehat{\omega}$ are related to $\mathrm{d}y$ and $\boldsymbol{\delta y}$ by means of the covariant components of the metric tensor of the surface $\widehat{\omega}$; cf. Theorem 1.3.1. The corresponding relations are used for computing the area of a surface $\widehat{A} = \boldsymbol{\theta}(A) \subset \widehat{\omega}$ and the length of a curve $\widehat{C} = \boldsymbol{\theta}(C) \subset \widehat{\omega}$, where $C = \boldsymbol{f}(I)$ and I is a compact interval of $\mathbb{R}$.

Let A be a bounded, open, and connected subset of $\mathbb{R}^2$ with a Lipschitz-continuous boundary, that satisfies $\overline{A} \subset \omega$. Let $\widehat{A} := \boldsymbol{\theta}(A)$, and let $\widehat{f} \in L^1(\widehat{A})$ be given. Then

$$\int_{\widehat{A}} \widehat{f}(\widehat{y})\,\mathrm{d}\widehat{a}(\widehat{y}) = \int_A (\widehat{f} \circ \boldsymbol{\theta})(y)\sqrt{a(y)}\,\mathrm{d}y.$$

In particular, the *area* of $\widehat{A}$ is given by

$$\mathit{area}\,\widehat{A} := \int_{\widehat{A}} \mathrm{d}\widehat{a}(\widehat{y}) = \int_A \sqrt{a(y)}\,\mathrm{d}y.$$

Consider next a *curve* $C = \boldsymbol{f}(I)$ in ω, where I is a compact interval of $\mathbb{R}$ and $\boldsymbol{f} = f^\alpha \boldsymbol{e}_\alpha : I \to \omega$ is a smooth enough injective mapping. Then the *length* of the *curve* $\widehat{C} := \boldsymbol{\theta}(C) \subset \widehat{\omega}$ is given by

$$\textit{length}\,\widehat{C} := \int_I \left|\frac{\mathrm{d}}{\mathrm{d}t}(\boldsymbol{\theta} \circ \boldsymbol{f})(t)\right| \mathrm{d}t = \int_I \sqrt{a_{\alpha\beta}(\boldsymbol{f}(t))\frac{\mathrm{d}f^\alpha}{\mathrm{d}t}(t)\frac{\mathrm{d}f^\beta}{\mathrm{d}t}(t)}\,\mathrm{d}t.$$

The last relation shows in particular that the lengths of curves inside the surface $\boldsymbol{\theta}(\omega)$ are precisely those induced by the Euclidean metric of the space $\boldsymbol{E}^3$. For this reason, the surface $\boldsymbol{\theta}(\omega)$ is said to be **isometrically imbedded** in $\boldsymbol{E}^3$.

1.4 Second Fundamental Form; Curvature on a Surface

Consider a surface given as the image $\boldsymbol{\theta}(\omega) \subset \boldsymbol{E}^3$ of a *two-dimensional* open set $\omega \subset \mathbb{R}^2$ by a smooth enough immersion $\boldsymbol{\theta} : \omega \subset \mathbb{R}^2 \to \boldsymbol{E}^3$. Then it is intuitively clear that this surface *cannot* be defined by its "metric" alone, i.e., by its sole first fundamental form.

As suggested by Figure 1.4.1, the "missing information" is provided by the "curvature" of a surface. A natural way to give substance to this otherwise vague notion consists in specifying how the *curvature of a curve on a surface* can be computed. As shown in this section, solving this question relies on the knowledge of the *second fundamental form* of a surface, which naturally appears for this purpose through its covariant components (Theorem 1.4.3).

Consider as in Section 1.1 a *surface* $\widehat{\omega} = \boldsymbol{\theta}(\omega)$ in $\boldsymbol{E}^3$, where ω is an open subset of $\mathbb{R}^2$ and $\boldsymbol{\theta} : \omega \subset \mathbb{R}^2 \to \boldsymbol{E}^3$ is a smooth enough immersion. For each $y \in \omega$, the vector

$$\boldsymbol{a}_3(y) := \frac{\boldsymbol{a}_1(y) \wedge \boldsymbol{a}_2(y)}{|\boldsymbol{a}_1(y) \wedge \boldsymbol{a}_2(y)|}$$

is thus well defined, has Euclidean norm one, and is normal to the surface $\widehat{\omega}$ at the point $\widehat{y} = \boldsymbol{\theta}(y)$.

Remark 1.4.1. The denominator in the definition of $\boldsymbol{a}_3(y)$ may be also written as

$$|\boldsymbol{a}_1(y) \wedge \boldsymbol{a}_2(y)| = \sqrt{a(y)},$$

where $a(y) := \det(a_{\alpha\beta}(y))$. This relation, which holds in fact even if $a(y) = 0$, will be established in the course of the proof of Theorem 2.3.2. □

Fix $y \in \omega$ and consider a *plane* P *normal to* $\widehat{\omega}$ *at* $\widehat{y} = \boldsymbol{\theta}(y)$, i.e., a plane that contains the vector $\boldsymbol{a}_3(y)$. The intersection $\widehat{C} = P \cap \widehat{\omega}$ is thus a *planar curve* on $\widehat{\omega}$.

As shown in Theorem 1.4.3, it is remarkable that the *curvature* of $\widehat{C}$ at $\widehat{y}$ can be computed by means of the covariant components $a_{\alpha\beta}(y)$ of the

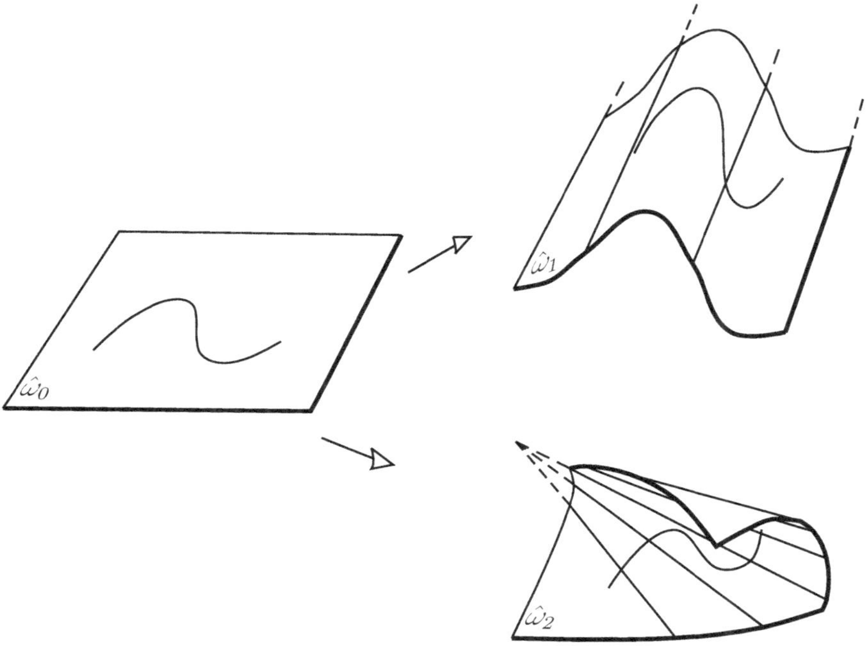

Figure 1.4.1. *A metric alone does not define a surface in* $\boldsymbol{E}^3$. A flat surface $\widehat{\omega}_0$ may be deformed into a portion $\widehat{\omega}_1$ of a cylinder or a portion $\widehat{\omega}_2$ of a cone without altering the length of any curve drawn on it (cylinders and cones are instances of "developable surfaces", i.e., that can be, at least locally, continuously "rolled out", or "developed", onto a plane, without changing the metric of the intermediary surfaces in the process; see Stoker [1969, Chapter 5, Sections 2 to 6] or Klingenberg [1973, Section 3.7]). Yet it should be clear that in general $\widehat{\omega}_0$ and $\widehat{\omega}_1$, or $\widehat{\omega}_0$ and $\widehat{\omega}_2$, or $\widehat{\omega}_1$ and $\widehat{\omega}_2$, are not identical surfaces *modulo* an "isometry" of $\boldsymbol{E}^3$, i.e., up to a translation and a rotation in $\boldsymbol{E}^3$.

first fundamental form of the surface $\widehat{\omega} = \boldsymbol{\theta}(\omega)$ introduced in Section 1.2, together with the covariant components $b_{\alpha\beta}(y)$ of the *"second" fundamental form* of $\widehat{\omega}$. The definition of the **curvature** *of a planar curve* is recalled in Figure 1.4.2.

If the algebraic curvature of $\widehat{C}$ at $\widehat{y}$ is $\neq 0$, it can be written as $\dfrac{1}{R}$, and R is then called the **algebraic radius of curvature** of the curve $\widehat{C}$ at $\widehat{y}$. This means that the **center of curvature** of the curve $\widehat{C}$ at $\widehat{y}$ is the point $(\widehat{y} + R\boldsymbol{a}_3(y))$; see Figure 1.4.3. While R is *not* intrinsically defined, as its sign changes in any system of curvilinear coordinates where the normal vector $\boldsymbol{a}_3(y)$ is replaced by its opposite, *the center of curvature is intrinsically defined.*

If the curvature of $\widehat{C}$ at $\widehat{y}$ is 0, the radius of curvature of the curve $\widehat{C}$ at $\widehat{y}$ is said to be *infinite*; for this reason, it is customary to still write the curvature as $\dfrac{1}{R}$ in this case.

Note that *the real number* $\dfrac{1}{R}$ *is always well defined* by the formula given in the next theorem, since the symmetric matrix $(a_{\alpha\beta}(y))$ is positive definite. This implies in particular that *the radius of curvature never vanishes along a curve on a surface* $\boldsymbol{\theta}(\omega)$ *defined by a mapping* $\boldsymbol{\theta}$ *satisfying the assumptions of the next theorem, hence in particular of class* $\mathcal{C}^2$ *on* ω.

Remark 1.4.2. It is intuitively clear that if $R = 0$, the mapping $\boldsymbol{\theta}$ "cannot be too smooth". Think of a surface made of two portions of planes intersecting along a segment, which thus constitutes a *fold* on the surface. Or think of a surface $\boldsymbol{\theta}(\omega)$ with $0 \in \omega$ and $\boldsymbol{\theta}(y_1, y_2) = |y_1|^{1+\alpha}$ for some $0 < \alpha < 1$ and (y_1, y_2) in a neighborhood of 0, so that $\boldsymbol{\theta} \in \mathcal{C}^1(\omega; \boldsymbol{E}^3)$ but $\boldsymbol{\theta} \notin \mathcal{C}^2(\omega; \boldsymbol{E}^3)$: The radius of curvature of a curve on such a surface corresponding to a constant y_2 vanishes at $y_1 = 0$. □

Theorem 1.4.3. *Let ω be an open subset of $\mathbb{R}^2$, let $\boldsymbol{\theta} \in \mathcal{C}^2(\omega; \boldsymbol{E}^3)$ be an injective immersion, and let $y \in \omega$ be fixed.*

Consider a plane P normal to $\widehat{\omega} = \boldsymbol{\theta}(\omega)$ at the point $\widehat{y} = \boldsymbol{\theta}(y)$. The intersection $P \cap \widehat{\omega}$ is a curve $\widehat{C}$ on $\widehat{\omega}$, which is the image $\widehat{C} = \boldsymbol{\theta}(C)$ of a curve C in the set $\overline{\omega}$. Assume that, in a sufficiently small neighborhood of y, the restriction of C to this neighborhood is the image $\boldsymbol{f}(I)$ of an open interval $I \subset \mathbb{R}$, where $\boldsymbol{f} = f^\alpha \boldsymbol{e}_\alpha : I \to \mathbb{R}$ is a smooth enough injective mapping that satisfies $\dfrac{\mathrm{d}f^\alpha}{\mathrm{d}t}(t)\, \boldsymbol{e}_\alpha \neq \boldsymbol{0}$, where $t \in I$ is such that $y = \boldsymbol{f}(t)$ (Figure 1.4.3).

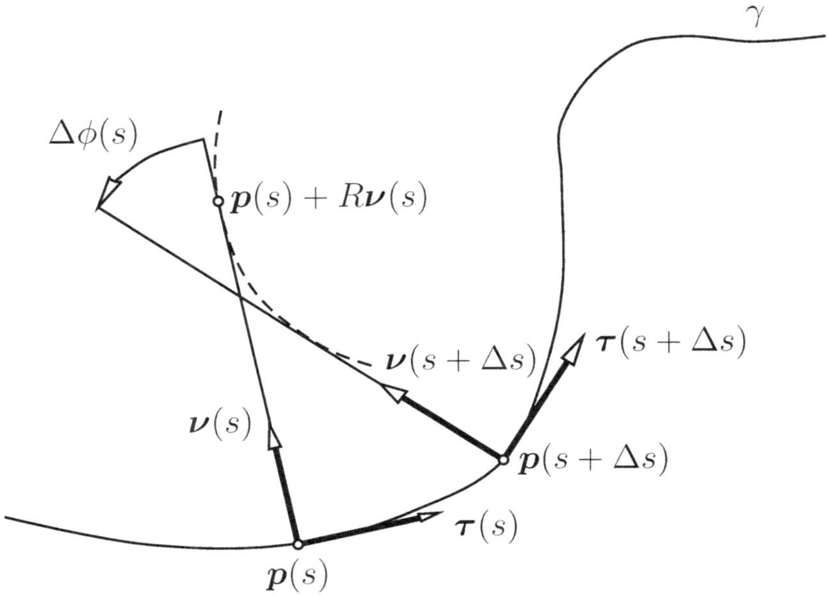

Figure 1.4.2. *Curvature of a planar curve.* Let γ be a smooth enough planar curve, parametrized by its curvilinear abscissa s. Consider two points $\boldsymbol{p}(s)$ and $\boldsymbol{p}(s+\Delta s)$ with curvilinear abscissae s and $s+\Delta s$ and let $\Delta\phi(s)$ be the algebraic angle between the two normals $\boldsymbol{\nu}(s)$ and $\boldsymbol{\nu}(s+\Delta s)$ (oriented in the usual way) to γ at those points. When $\Delta s \to 0$, the ratio $\dfrac{\Delta\phi(s)}{\Delta s}$ has a limit, called the "curvature" of γ at $\boldsymbol{p}(s)$. If this limit is non-zero, its inverse R is called the "algebraic radius of curvature" of γ at $\boldsymbol{p}(s)$ (the sign of R depends on the orientation chosen on γ).

The point $\boldsymbol{p}(s)+R\boldsymbol{\nu}(s)$, which is intrinsically defined, is called the "center of curvature" of γ at $\boldsymbol{p}(s)$: It is the center of the "osculating circle" at $\boldsymbol{p}(s)$, i.e., the limit as $\Delta s \to 0$ of the circle tangent to γ at $\boldsymbol{p}(s)$ that passes through the point $\boldsymbol{p}(s+\Delta s)$. The center of curvature is also the limit as $\Delta s \to 0$ of the intersection of the normals $\boldsymbol{\nu}(s)$ and $\boldsymbol{\nu}(s+\Delta s)$. Consequently, the centers of curvature of γ lie on a curve (dashed on the figure), called "la développée" in French, that is tangent to the normals to γ.

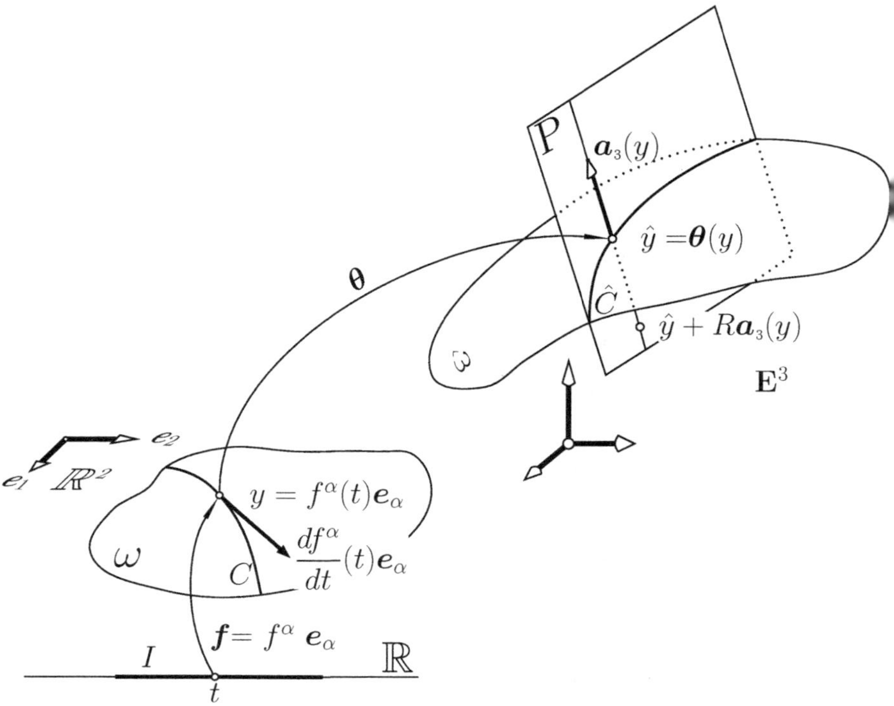

Figure 1.4.3. *Curvature on a surface.* Let P be a plane containing the vector $\boldsymbol{a}_3(y) = \dfrac{\boldsymbol{a}_1(y) \wedge \boldsymbol{a}_2(y)}{|\boldsymbol{a}_1(y) \wedge \boldsymbol{a}_2(y)|}$, which is normal to the surface $\widehat{\omega} = \boldsymbol{\theta}(\omega)$. The algebraic curvature $\dfrac{1}{R}$ of the planar curve $\widehat{C} = P \cap \widehat{\omega} = \boldsymbol{\theta}(C)$ at $\widehat{y} = \boldsymbol{\theta}(y)$ is given by the ratio

$$\frac{1}{R} = \frac{b_{\alpha\beta}(\boldsymbol{f}(t))\dfrac{\mathrm{d}f^\alpha}{\mathrm{d}t}(t)\,\dfrac{\mathrm{d}f^\beta}{\mathrm{d}t}(t)}{a_{\alpha\beta}(\boldsymbol{f}(t))\dfrac{\mathrm{d}f^\alpha}{\mathrm{d}t}(t)\,\dfrac{\mathrm{d}f^\beta}{\mathrm{d}t}(t)},$$

where $a_{\alpha\beta}(y)$ and $b_{\alpha\beta}(y)$ are the covariant components of the first and second fundamental forms of the surface $\widehat{\omega}$ at $\widehat{y}$ and $\dfrac{\mathrm{d}f^\alpha}{\mathrm{d}t}(t)$ are the components of the vector tangent to the curve $C = \boldsymbol{f}(I)$ at $y = \boldsymbol{f}(t) = f^\alpha(t)\boldsymbol{e}_\alpha$. If $\dfrac{1}{R} \neq 0$, the center of curvature of the curve $\widehat{C}$ at $\widehat{y}$ is the point $(\widehat{y} + R\boldsymbol{a}_3(y))$, which is intrinsically defined in the Euclidean space $\boldsymbol{E}^3$.

Then the curvature $\frac{1}{R}$ *of the planar curve* $\widehat{C}$ *at* $\widehat{y}$ *is given by the ratio*

$$\frac{1}{R} = \frac{b_{\alpha\beta}(\boldsymbol{f}(t))\frac{\mathrm{d}f^{\alpha}}{\mathrm{d}t}(t)\,\frac{\mathrm{d}f^{\beta}}{\mathrm{d}t}(t)}{a_{\alpha\beta}(\boldsymbol{f}(t))\frac{\mathrm{d}f^{\alpha}}{\mathrm{d}t}(t)\,\frac{\mathrm{d}f^{\beta}}{\mathrm{d}t}(t)},$$

where $a_{\alpha\beta}(y)$ *are the covariant components of the first fundamental form of* $\widehat{\omega}$ *at* y (Section 1.1) *and*

$$b_{\alpha\beta}(y) := \boldsymbol{a}_3(y) \cdot \partial_\alpha \boldsymbol{a}_\beta(y) = -\partial_\alpha \boldsymbol{a}_3(y) \cdot \boldsymbol{a}_\beta(y) = b_{\beta\alpha}(y).$$

Proof. (i) We first establish a well-known formula giving the curvature $\frac{1}{R}$ of a planar curve. Using the notations of Figure 1.4.2, we note that

$$\sin \Delta\phi(s) = \boldsymbol{\nu}(s) \cdot \boldsymbol{\tau}(s + \Delta s) = -\{\boldsymbol{\nu}(s + \Delta s) - \boldsymbol{\nu}(s)\} \cdot \boldsymbol{\tau}(s + \Delta s),$$

so that

$$\frac{1}{R} := \lim_{\Delta s \to 0} \frac{\Delta\phi(s)}{\Delta s} = \lim_{\Delta s \to 0} \frac{\sin \Delta\phi(s)}{\Delta s} = -\frac{\mathrm{d}\boldsymbol{\nu}}{\mathrm{d}s}(s) \cdot \boldsymbol{\tau}(s).$$

(ii) The curve $(\boldsymbol{\theta} \circ \boldsymbol{f})(I)$, which is *a priori* parametrized by $t \in I$, can be also parametrized by its curvilinear abscissa s in a neighborhood of the point $\widehat{y}$. There thus exist an interval $\tilde{I} \subset I$ and a mapping $\boldsymbol{p} : J \to P$, where $J \subset \mathbb{R}$ is an interval, such that

$$(\boldsymbol{\theta} \circ \boldsymbol{f})(t) = \boldsymbol{p}(s) \text{ and } (\boldsymbol{a}_3 \circ \boldsymbol{f})(t) = \boldsymbol{\nu}(s) \text{ for all } t \in \tilde{I},\ s \in J.$$

By (i), the curvature $\frac{1}{R}$ of $\widehat{C}$ is given by

$$\frac{1}{R} = -\frac{\mathrm{d}\boldsymbol{\nu}}{\mathrm{d}s}(s) \cdot \boldsymbol{\tau}(s),$$

where

$$\begin{aligned}
\frac{\mathrm{d}\boldsymbol{\nu}}{\mathrm{d}s}(s) &= \frac{\mathrm{d}(\boldsymbol{a}_3 \circ \boldsymbol{f})}{\mathrm{d}t}(t)\frac{\mathrm{d}t}{\mathrm{d}s} = \partial_\alpha \boldsymbol{a}_3(\boldsymbol{f}(t))\frac{\mathrm{d}f^{\alpha}}{\mathrm{d}t}(t)\,\frac{\mathrm{d}t}{\mathrm{d}s}, \\
\boldsymbol{\tau}(s) &= \frac{\mathrm{d}\boldsymbol{p}}{\mathrm{d}s}(s) = \frac{\mathrm{d}(\boldsymbol{\theta} \circ \boldsymbol{f})}{\mathrm{d}t}(t)\frac{\mathrm{d}t}{\mathrm{d}s} \\
&= \partial_\beta \boldsymbol{\theta}(\boldsymbol{f}(t))\frac{\mathrm{d}f^{\beta}}{\mathrm{d}t}(t)\,\frac{\mathrm{d}t}{\mathrm{d}s} = \boldsymbol{a}_\beta(\boldsymbol{f}(t))\frac{\mathrm{d}f^{\beta}}{\mathrm{d}t}(t)\,\frac{\mathrm{d}t}{\mathrm{d}s}.
\end{aligned}$$

Hence

$$\frac{1}{R} = -\partial_\alpha \boldsymbol{a}_3(\boldsymbol{f}(t)) \cdot \boldsymbol{a}_\beta(\boldsymbol{f}(t)) \frac{\mathrm{d}f^\alpha}{\mathrm{d}t}(t) \frac{\mathrm{d}f^\beta}{\mathrm{d}t}(t) \left(\frac{\mathrm{d}t}{\mathrm{d}s}\right)^2.$$

To obtain the announced expression for $\frac{1}{R}$, it suffices to note that

$$-\partial_\alpha \boldsymbol{a}_3(\boldsymbol{f}(t)) \cdot \boldsymbol{a}_\beta(\boldsymbol{f}(t)) = b_{\alpha\beta}(\boldsymbol{f}(t)),$$

by definition of the functions $b_{\alpha\beta}$ and that (Theorem 1.3.1 (b))

$$\mathrm{d}s = \left\{\delta y^\alpha a_{\alpha\beta}(y) \delta y^\beta\right\}^{1/2} = \left\{a_{\alpha\beta}(\boldsymbol{f}(t)) \frac{\mathrm{d}f^\alpha}{\mathrm{d}t}(t) \frac{\mathrm{d}f^\beta}{\mathrm{d}t}(t)\right\}^{1/2} \mathrm{d}t.$$

□

Remark 1.4.4. The knowledge of the curvatures of curves contained in planes *normal to* $\widehat{\omega}$ suffices for computing the curvature of *any* curve on $\widehat{\omega}$. More specifically, the radius of curvature $\widetilde{R}$ at $\widehat{y}$ of any smooth enough curve $\widetilde{C}$ (planar or not) on the surface $\widehat{\omega}$ is given by $\frac{\cos\varphi}{\widetilde{R}} = \frac{1}{R}$, where φ is the angle between the "principal normal" to $\widetilde{C}$ at $\widehat{y}$ and $\boldsymbol{a}_3(y)$ and $\frac{1}{R}$ is given in Theorem 1.4.3; see, e.g., Stoker [1969, Chapter 4, Section 12]. □

The elements $b_{\alpha\beta}(y)$ of the symmetric matrix $(b_{\alpha\beta}(y))$ defined in Theorem 1.4.3 are called the **covariant components of the second fundamental form** of the surface $\widehat{\omega} = \boldsymbol{\theta}(\omega)$ at $\widehat{y} = \boldsymbol{\theta}(y)$.

1.5 Principal Curvature; Gaussian Curvature

The analysis of the previous section suggests that precise information about the shape of a surface $\widehat{\omega} = \boldsymbol{\theta}(\omega)$ in a neighborhood of one of its points $\widehat{y} = \boldsymbol{\theta}(y)$ can be gathered by letting the plane P turn around the normal vector $\boldsymbol{a}_3(y)$ and by following in this process the variations of the curvatures at $\widehat{y}$ of the corresponding planar curves $P \cap \widehat{\omega}$, as given in Theorem 1.4.3.

As a first step in this direction, we show that *these curvatures span a compact interval of* $\mathbb{R}$. In particular then, they "stay away from infinity".

Note that this compact interval contains 0 if, and only if, the radius of curvature of the curve $P \cap \widehat{\omega}$ is infinite for at least one such plane P.

Theorem 1.5.1. (a) *Let the assumptions and notations be as in* Theorem 1.4.3. *For a fixed $y \in \omega$, consider the set $\mathcal{P}$ of all planes P normal to the surface $\widehat{\omega} = \boldsymbol{\theta}(\omega)$ at $\widehat{y} = \boldsymbol{\theta}(y)$. Then the set of curvatures of the associated planar curves $P \cap \widehat{\omega}$, $P \in \mathcal{P}$, is a compact interval of $\mathbb{R}$, denoted* $\left[\frac{1}{R_1(y)}, \frac{1}{R_2(y)}\right]$.

(b) *Let the matrix* $(b_\alpha^\beta(y))$, α *being the row index, be defined by*

$$b_\alpha^\beta(y) := a^{\beta\sigma}(y) b_{\alpha\sigma}(y),$$

where $(a^{\alpha\beta}(y)) = (a_{\alpha\beta}(y))^{-1}$ (Section 1.2) *and the matrix* $(b_{\alpha\beta}(y))$ *is defined as in* Theorem 1.4.3. *Then*

$$\frac{1}{R_1(y)} + \frac{1}{R_2(y)} = \operatorname{tr}(b_\alpha^\beta(y)),$$
$$\frac{1}{R_1(y)R_2(y)} = \det(b_\alpha^\beta(y)) = \frac{\det(b_{\alpha\beta}(y))}{\det(a_{\alpha\beta}(y))}.$$

(c) *If* $\frac{1}{R_1(y)} \neq \frac{1}{R_2(y)}$, *there is a unique pair of orthogonal planes* $P_1 \in \mathcal{P}$ *and* $P_2 \in \mathcal{P}$ *such that the curvatures of the associated planar curves* $P_1 \cap \widehat{\omega}$ *and* $P_2 \cap \widehat{\omega}$ *are precisely* $\frac{1}{R_1(y)}$ *and* $\frac{1}{R_2(y)}$.

Proof. (i) Let $\Delta(P)$ denote the intersection of $P \in \mathcal{P}$ with the tangent plane T to the surface $\widehat{\omega}$ at $\widehat{y}$, and let $\widehat{C}(P)$ denote the intersection of P with $\widehat{\omega}$. Hence $\Delta(P)$ is tangent to $\widehat{C}(P)$ at $\widehat{y} \in \widehat{\omega}$.

In a sufficiently small neighborhood of $\widehat{y}$ the restriction of the curve $\widehat{C}(P)$ to this neighborhood is given by $\widehat{C}(P) = (\boldsymbol{\theta} \circ \boldsymbol{f}(P))(I(P))$, where $I(P) \subset \mathbb{R}$ is an open interval and $\boldsymbol{f}(P) = f^\alpha(P)\boldsymbol{e}_\alpha : I(P) \to \mathbb{R}^2$ is a smooth enough injective mapping that satisfies $\frac{\mathrm{d}f^\alpha(P)}{\mathrm{d}t}(t)\boldsymbol{e}_\alpha \neq \mathbf{0}$, where $t \in I(P)$ is such that $y = \boldsymbol{f}(P)(t)$. Hence the line $\Delta(P)$ is given by

$$\Delta(P) = \left\{\widehat{y} + \lambda \frac{\mathrm{d}(\boldsymbol{\theta} \circ \boldsymbol{f}(P))}{\mathrm{d}t}(t); \lambda \in \mathbb{R}\right\} = \left\{\widehat{y} + \lambda \xi^\alpha \boldsymbol{a}_\alpha(y); \lambda \in \mathbb{R}\right\},$$

where $\xi^\alpha := \frac{\mathrm{d}f^\alpha(P)}{\mathrm{d}t}(t)$ and $\xi^\alpha \boldsymbol{e}_\alpha \neq \mathbf{0}$ by assumption.

Since the line $\{y + \mu\xi^\alpha \boldsymbol{e}_\alpha; \mu \in \mathbb{R}\}$ is tangent to the curve $C(P) := \boldsymbol{\theta}^{-1}(\widehat{C}(P))$ at $y \in \omega$ (the mapping $\boldsymbol{\theta} : \omega \to \mathbb{R}^3$ is injective by assumption) for each such parametrizing function $\boldsymbol{f}(P) : I(P) \to \mathbb{R}^2$ and since the vectors $\boldsymbol{a}_\alpha(y)$ are linearly independent, there exists a bijection between the set of all lines $\Delta(P) \subset T$, $P \in \mathcal{P}$, and the set of all lines supporting the nonzero tangent vectors to the curves $C(P)$.

Hence Theorem 1.4.3 shows that when P varies in $\mathcal{P}$, the curvature of the corresponding curves $\widehat{C} = \widehat{C}(P)$ at $\widehat{y}$ takes the same values as does the ratio $\frac{b_{\alpha\beta}(y)\xi^\alpha\xi^\beta}{a_{\alpha\beta}(y)\xi^\alpha\xi^\beta}$ when $\boldsymbol{\xi} := \begin{pmatrix}\xi_1 \\ \xi_2\end{pmatrix}$ varies in $\mathbb{R}^2 - \{\mathbf{0}\}$.

(ii) Let the symmetric matrices $\boldsymbol{A}$ and $\boldsymbol{B}$ of order two be defined by

$$\boldsymbol{A} := (a_{\alpha\beta}(y)) \text{ and } \boldsymbol{B} := (b_{\alpha\beta}(y)).$$

Since $\boldsymbol{A}$ is positive definite, it has a (unique) *square root* $\boldsymbol{C}$, i.e., a symmetric positive definite matrix $\boldsymbol{C}$ such that $\boldsymbol{A} = \boldsymbol{C}^2$. Hence the ratio

$$\frac{b_{\alpha\beta}(y)\xi^\alpha\xi^\beta}{a_{\alpha\beta}(y)\xi^\alpha\xi^\beta} = \frac{\boldsymbol{\xi}^T\boldsymbol{B}\boldsymbol{\xi}}{\boldsymbol{\xi}^T\boldsymbol{A}\boldsymbol{\xi}} = \frac{\boldsymbol{\eta}^T\boldsymbol{C}^{-1}\boldsymbol{B}\boldsymbol{C}^{-1}\boldsymbol{\eta}}{\boldsymbol{\eta}^T\boldsymbol{\eta}}, \text{ where } \boldsymbol{\eta} = \boldsymbol{C}\boldsymbol{\xi},$$

is nothing but the *Rayleigh quotient* associated with the symmetric matrix $\boldsymbol{C}^{-1}\boldsymbol{B}\boldsymbol{C}^{-1}$. When $\boldsymbol{\eta}$ varies in $\mathbb{R}^2 - \{\boldsymbol{0}\}$, this Rayleigh quotient thus spans the compact interval of $\mathbb{R}$ whose end-points are the smallest and largest eigenvalue, respectively denoted $\frac{1}{R_1(y)}$ and $\frac{1}{R_2(y)}$, of the matrix $\boldsymbol{C}^{-1}\boldsymbol{B}\boldsymbol{C}^{-1}$ (for a proof, see, e.g., Ciarlet [1982, Theorem 1.3-1]). This proves (a).

Furthermore, the relation

$$\boldsymbol{C}(\boldsymbol{C}^{-1}\boldsymbol{B}\boldsymbol{C}^{-1})\boldsymbol{C}^{-1} = \boldsymbol{B}\boldsymbol{C}^{-2} = \boldsymbol{B}\boldsymbol{A}^{-1}$$

shows that the eigenvalues of the symmetric matrix $\boldsymbol{C}^{-1}\boldsymbol{B}\boldsymbol{C}^{-1}$ coincide with those of the (in general non-symmetric) matrix $\boldsymbol{B}\boldsymbol{A}^{-1}$. Note that $\boldsymbol{B}\boldsymbol{A}^{-1} = (b_\alpha^\beta(y))$ with $b_\alpha^\beta(y) := a^{\beta\sigma}(y)b_{\alpha\sigma}(y)$, α being the row index, since $\boldsymbol{A}^{-1} = (a^{\alpha\beta}(y))$.

Hence the relations in (b) simply express that the sum and the product of the eigenvalues of the matrix $\boldsymbol{B}\boldsymbol{A}^{-1}$ are respectively equal to its trace and to its determinant, which may be also written as $\frac{\det(b_{\alpha\beta}(y))}{\det(a_{\alpha\beta}(y))}$ since $\boldsymbol{B}\boldsymbol{A}^{-1} = (b_\alpha^\beta(y))$. This proves (b).

(iii) Let $\boldsymbol{\eta}_1 = \begin{pmatrix}\eta_1^1\\ \eta_1^2\end{pmatrix} = \boldsymbol{C}\boldsymbol{\xi}_1$ and $\boldsymbol{\eta}_2 = \begin{pmatrix}\eta_2^1\\ \eta_2^2\end{pmatrix} = \boldsymbol{C}\boldsymbol{\xi}_2$, with $\boldsymbol{\xi}_1 = \begin{pmatrix}\xi_1^1\\ \xi_1^2\end{pmatrix}$ and $\boldsymbol{\xi}_2 = \begin{pmatrix}\xi_2^1\\ \xi_2^2\end{pmatrix}$, be two orthogonal ($\boldsymbol{\eta}_1^T\boldsymbol{\eta}_2 = 0$) eigenvectors of the symmetric matrix $\boldsymbol{C}^{-1}\boldsymbol{B}\boldsymbol{C}^{-1}$, corresponding to the eigenvalues $\frac{1}{R_1(y)}$ and $\frac{1}{R_2(y)}$, respectively. Hence

$$0 = \boldsymbol{\eta}_1^T\boldsymbol{\eta}_2 = \boldsymbol{\xi}_1^T\boldsymbol{C}^T\boldsymbol{C}\boldsymbol{\xi}_2 = \boldsymbol{\xi}_1^T\boldsymbol{A}\boldsymbol{\xi}_2 = 0,$$

since $\boldsymbol{C}^T = \boldsymbol{C}$. By (i), the corresponding lines $\Delta(P_1)$ and $\Delta(P_2)$ of the tangent plane are parallel to the vectors $\xi_1^\alpha\boldsymbol{a}_\alpha(y)$ and $\xi_2^\beta\boldsymbol{a}_\beta(y)$, which are orthogonal since

$$\{\xi_1^\alpha\boldsymbol{a}_\alpha(y)\}\cdot\{\xi_2^\beta\boldsymbol{a}_\beta(y)\} = a_{\alpha\beta}(y)\xi_1^\alpha\xi_2^\beta = \boldsymbol{\xi}_1^T\boldsymbol{A}\boldsymbol{\xi}_2.$$

If $\frac{1}{R_1(y)} \neq \frac{1}{R_2(y)}$, the directions of the vectors $\boldsymbol{\eta}_1$ and $\boldsymbol{\eta}_2$ are uniquely determined and the lines $\Delta(P_1)$ and $\Delta(P_2)$ are likewise uniquely determined. This proves (c). □

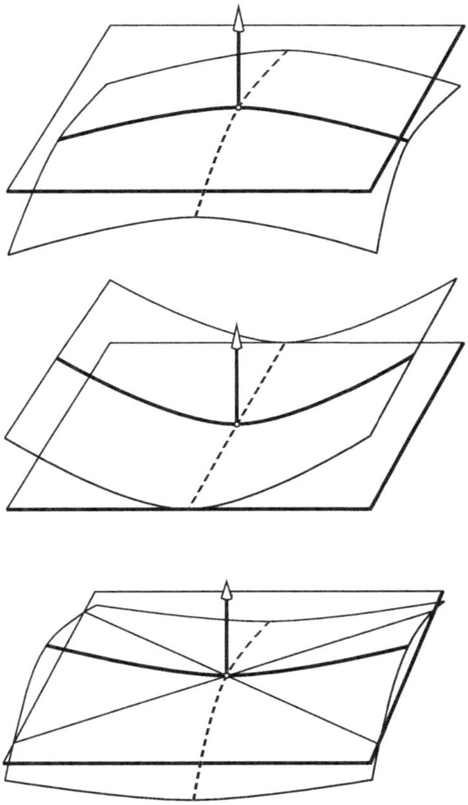

Figure 1.5.1. *Different kinds of points on a surface.* A point is elliptic if the Gaussian curvature is > 0 or equivalently, if the two principal radii of curvature are of the same sign; the surface is then locally on one side of its tangent plane. A point is parabolic if exactly one of the two principal radii of curvature is infinite; the surface is again locally on one side of its tangent plane. A point is hyperbolic if the Gaussian curvature is < 0 or equivalently, if the two principal radii of curvature are of different signs; the surface then intersects its tangent plane along two curves.

We are now in a position to state several *fundamental definitions*:

The elements $b_\alpha^\beta(y)$ of the (in general non-symmetric) matrix $(b_\alpha^\beta(y))$ defined in Theorem 1.5.1 are called the **mixed components of the second fundamental form** of the surface $\widehat{\omega} = \boldsymbol{\theta}(\omega)$ at $\widehat{y} = \boldsymbol{\theta}(y)$.

The real numbers $\dfrac{1}{R_1(y)}$ and $\dfrac{1}{R_2(y)}$ (one or both possibly equal to 0) found in Theorem 1.5.1 are called the **principal curvatures** of $\widehat{\omega}$ at $\widehat{y}$.

If $\dfrac{1}{R_1(y)} \neq 0$ *and* $\dfrac{1}{R_2(y)} \neq 0$, the real numbers $R_1(y)$ and $R_2(y)$ are called the **principal radii of curvature** of $\widehat{\omega}$ at $\widehat{y}$. If, e.g., $\dfrac{1}{R_1(y)} = 0$, the corresponding radius of curvature $R_1(y)$ is said to be *infinite*, according to the convention made in Section 1.4. While the principal radii of curvature may simultaneously change their signs in another system of curvilinear coordinates, the associated *centers of curvature* are intrinsically defined.

The numbers $\dfrac{1}{2}\left(\dfrac{1}{R_1(y)} + \dfrac{1}{R_2(y)}\right)$ and $\dfrac{1}{R_1(y)R_2(y)}$, which are the principal invariants of the matrix $(b_\alpha^\beta(y))$ (Theorem 1.5.1), are respectively called the **mean curvature** and the **Gaussian**, or **total, curvature** of the surface $\widehat{\omega}$ at $\widehat{y}$.

A point on a surface is an **elliptic, parabolic**, or **hyperbolic, point** according as its *Gaussian curvature* is $> 0, = 0$ but it is not a planar point, or < 0; see Figure 1.5.1.

1.6 Covariant Derivatives of a Vector Field Defined on a Surface; the Gauß and Weingarten Formulas

Consider a surface $\widehat{\omega} = \boldsymbol{\theta}(\omega)$ in $\boldsymbol{E}^3$, where $\boldsymbol{\theta} : \omega \subset \mathbb{R}^2 \to \boldsymbol{E}^3$ is a smooth enough injective immersion, let $\boldsymbol{a}_\alpha = \partial_\alpha \boldsymbol{\theta}$ as before, and let

$$\boldsymbol{a}_3(y) = \boldsymbol{a}^3(y) := \frac{\boldsymbol{a}_1(y) \wedge \boldsymbol{a}_2(y)}{|\boldsymbol{a}_1(y) \wedge \boldsymbol{a}_2(y)|}, \quad y \in \omega.$$

Then the vectors $\boldsymbol{a}_\alpha(y)$ (which form the covariant basis of the tangent plane to $\widehat{\omega}$ at $\widehat{y} = \boldsymbol{\theta}(y)$; see Figure 1.1.1) together with the vector $\boldsymbol{a}_3(y)$ (which is normal to $\widehat{\omega}$ and has Euclidean norm one) form the **covariant basis** *at* $\widehat{y}$.

Let the vectors $\boldsymbol{a}^\alpha(y)$ of the tangent plane to $\widehat{\omega}$ at $\widehat{y}$ be defined by the relations $\boldsymbol{a}^\alpha(y) \cdot \boldsymbol{a}_\beta(y) = \delta_\beta^\alpha$. Then the vectors $\boldsymbol{a}^\alpha(y)$ (which form the contravariant basis of the tangent plane at $\widehat{y}$; see again Figure 1.1.1) together with the vector $\boldsymbol{a}^3(y)$ form the **contravariant basis** *at* $\widehat{y}$; see Figure 1.6.1. Note that the vectors of the covariant and contravariant bases at $\widehat{y}$ satisfy

$$\boldsymbol{a}^i(y) \cdot \boldsymbol{a}_j(y) = \delta_j^i.$$

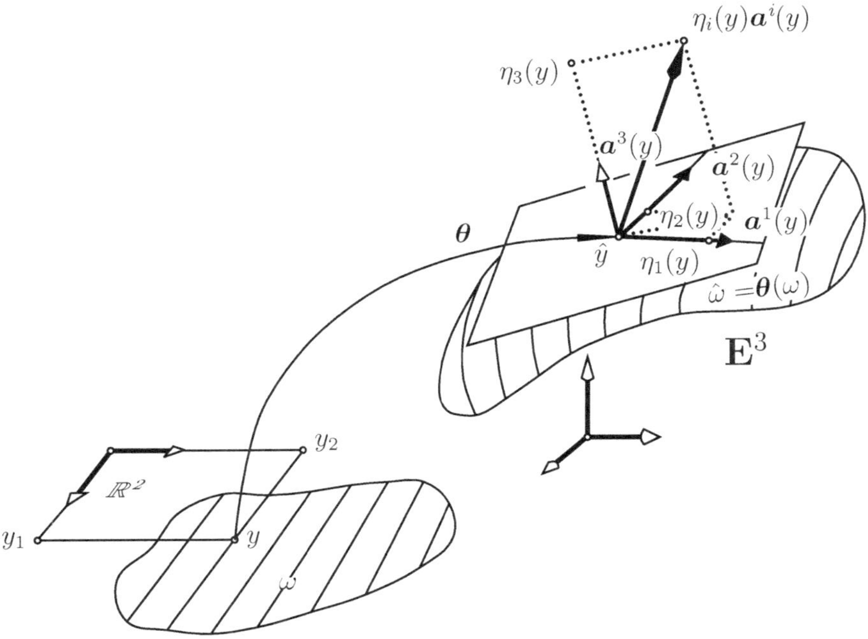

Figure 1.6.1. *Contravariant bases and vector fields along a surface.* At each point $\widehat{y} = \boldsymbol{\theta}(y) \in \widehat{\omega} = \boldsymbol{\theta}(\omega)$, the three vectors $\boldsymbol{a}^i(y)$, where $\boldsymbol{a}^\alpha(y)$ form the contravariant basis of the tangent plane to $\widehat{\omega}$ at $\widehat{y}$ (Figure 1.1.1) and $\boldsymbol{a}^3(y) = \dfrac{\boldsymbol{a}_1(y) \wedge \boldsymbol{a}_2(y)}{|\boldsymbol{a}_1(y) \wedge \boldsymbol{a}_2(y)|}$, form the contravariant basis at $\widehat{y}$. An arbitrary vector field defined on $\widehat{\omega}$ may then be defined by its covariant components $\eta_i : \omega \to \mathbb{R}$. This means that $\eta_i(y)\boldsymbol{a}^i(y)$ is the vector at the point $\widehat{y}$.

Suppose that a *vector field is defined on the surface* $\widehat{\omega}$. One way to define such a field in terms of the *curvilinear coordinates* used for defining the surface $\widehat{\omega}$ consists in writing it as $\eta_i \boldsymbol{a}^i : \omega \to \mathbb{R}^3$, i.e., in specifying its **covariant components** $\eta_i : \omega \to \mathbb{R}$ over the vector fields $\boldsymbol{a}^i$ formed by the *contravariant bases.* This means that this vector field is given by $\eta_i(y)\boldsymbol{a}^i(y)$ at each point $\widehat{y} = \boldsymbol{\theta}(y) \in \widehat{\omega}$ (Figure 1.6.1).

Our objective in this section is to compute the *partial derivatives* $\partial_\alpha(\eta_i \boldsymbol{a}^i)$ of such a vector field. These partial derivatives are found in the next theorem, as immediate consequences of two basic formulas, those of *Gauß* and *Weingarten.* The *Christoffel symbols "on a surface"* and the *covariant derivatives of a vector field defined on a surface* are also naturally introduced in this process.

Theorem 1.6.1. *Let ω be an open subset of $\mathbb{R}^2$ and let $\boldsymbol{\theta} \in \mathcal{C}^2(\omega; \boldsymbol{E}^3)$ be an immersion.*

(a) *The derivatives of the vectors of the covariant and contravariant bases are given by*

$$\partial_\alpha \boldsymbol{a}_\beta = \Gamma^\sigma_{\alpha\beta} \boldsymbol{a}_\sigma + b_{\alpha\beta} \boldsymbol{a}_3 \text{ and } \partial_\alpha \boldsymbol{a}^\beta = -\Gamma^\beta_{\alpha\sigma} \boldsymbol{a}^\sigma + b^\beta_\alpha \boldsymbol{a}^3,$$
$$\partial_\alpha \boldsymbol{a}_3 = \partial_\alpha \boldsymbol{a}^3 = -b_{\alpha\beta} \boldsymbol{a}^\beta = -b^\sigma_\alpha \boldsymbol{a}_\sigma,$$

where the covariant and mixed components $b_{\alpha\beta}$ and b^β_α of the second fundamental form of $\widehat{\omega}$ are defined in Theorems 1.4.3 *and* 1.5.1 *and*

$$\Gamma^\sigma_{\alpha\beta} := \boldsymbol{a}^\sigma \cdot \partial_\alpha \boldsymbol{a}_\beta.$$

(b) *Let there be given a vector field $\eta_i \boldsymbol{a}^i : \omega \to \mathbb{R}^3$ with covariant components $\eta_i \in \mathcal{C}^1(\omega)$. Then $\eta_i \boldsymbol{a}^i \in \mathcal{C}^1(\omega; \mathbb{R}^3)$ and the partial derivatives $\partial_\alpha(\eta_i \boldsymbol{a}^i) \in \mathcal{C}^0(\omega; \mathbb{R}^3)$ are given by*

$$\begin{aligned}\partial_\alpha(\eta_i \boldsymbol{a}^i) &= (\partial_\alpha \eta_\beta - \Gamma^\sigma_{\alpha\beta}\eta_\sigma - b_{\alpha\beta}\eta_3)\boldsymbol{a}^\beta + (\partial_\alpha \eta_3 + b^\beta_\alpha \eta_\beta)\boldsymbol{a}^3 \\ &= (\eta_{\beta|\alpha} - b_{\alpha\beta}\eta_3)\boldsymbol{a}^\beta + (\eta_{3|\alpha} + b^\beta_\alpha \eta_\beta)\boldsymbol{a}^3,\end{aligned}$$

where

$$\eta_{\beta|\alpha} := \partial_\alpha \eta_\beta - \Gamma^\sigma_{\alpha\beta}\eta_\sigma \text{ and } \eta_{3|\alpha} := \partial_\alpha \eta_3.$$

Proof. Since any vector $\boldsymbol{c}$ in the tangent plane can be expanded as $\boldsymbol{c} = (\boldsymbol{c} \cdot \boldsymbol{a}_\beta)\boldsymbol{a}^\beta = (\boldsymbol{c} \cdot \boldsymbol{a}^\sigma)\boldsymbol{a}_\sigma$, since $\partial_\alpha \boldsymbol{a}^3$ is in the tangent plane ($\partial_\alpha \boldsymbol{a}^3 \cdot \boldsymbol{a}^3 = \frac{1}{2}\partial_\alpha(\boldsymbol{a}^3 \cdot \boldsymbol{a}^3) = 0$), and since $\partial_\alpha \boldsymbol{a}^3 \cdot \boldsymbol{a}_\beta = -b_{\alpha\beta}$ (Theorem 1.4.3), it follows that

$$\partial_\alpha \boldsymbol{a}^3 = (\partial_\alpha \boldsymbol{a}^3 \cdot \boldsymbol{a}_\beta)\boldsymbol{a}^\beta = -b_{\alpha\beta}\boldsymbol{a}^\beta.$$

This formula, together with the definition of the functions b^β_α (Theorem 1.5.1), implies in turn that

$$\partial_\alpha \boldsymbol{a}_3 = (\partial_\alpha \boldsymbol{a}_3 \cdot \boldsymbol{a}^\sigma)\boldsymbol{a}_\sigma = -b_{\alpha\beta}(\boldsymbol{a}^\beta \cdot \boldsymbol{a}^\sigma)\boldsymbol{a}_\sigma = -b_{\alpha\beta}a^{\beta\sigma}\boldsymbol{a}_\sigma = -b^\sigma_\alpha \boldsymbol{a}_\sigma.$$

Any vector $\boldsymbol{c}$ can be expanded as $\boldsymbol{c} = (\boldsymbol{c} \cdot \boldsymbol{a}^i)\boldsymbol{a}_i = (\boldsymbol{c} \cdot \boldsymbol{a}_j)\boldsymbol{a}^j$. In particular,

$$\partial_\alpha \boldsymbol{a}_\beta = (\partial_\alpha \boldsymbol{a}_\beta \cdot \boldsymbol{a}^\sigma)\boldsymbol{a}_\sigma + (\partial_\alpha \boldsymbol{a}_\beta \cdot \boldsymbol{a}^3)\boldsymbol{a}_3 = \Gamma^\sigma_{\alpha\beta}\boldsymbol{a}_\sigma + b_{\alpha\beta}\boldsymbol{a}_3,$$

by definition of $\Gamma^\sigma_{\alpha\beta}$ and $b_{\alpha\beta}$. Finally,

$$\partial_\alpha \boldsymbol{a}^\beta = (\partial_\alpha \boldsymbol{a}^\beta \cdot \boldsymbol{a}_\sigma)\boldsymbol{a}^\sigma + (\partial_\alpha \boldsymbol{a}^\beta \cdot \boldsymbol{a}_3)\boldsymbol{a}^3 = -\Gamma^\beta_{\alpha\sigma}\boldsymbol{a}^\sigma + b^\beta_\alpha \boldsymbol{a}^3,$$

since

$$\partial_\alpha \boldsymbol{a}^\beta \cdot \boldsymbol{a}_3 = -\boldsymbol{a}^\beta \cdot \partial_\alpha \boldsymbol{a}_3 = b_\alpha^\sigma \boldsymbol{a}_\sigma \cdot \boldsymbol{a}^\beta = b_\alpha^\beta.$$

That $\eta_i \boldsymbol{a}^i \in \mathcal{C}^1(\omega; \mathbb{R}^3)$ if $\eta_i \in \mathcal{C}^1(\omega)$ is clear since $\boldsymbol{a}^i \in \mathcal{C}^1(\omega; \mathbb{R}^3)$ if $\boldsymbol{\theta} \in \mathcal{C}^2(\omega; \boldsymbol{E}^3)$. The formulas established *supra* immediately lead to the announced expression of $\partial_\alpha(\eta_i \boldsymbol{a}^i)$. □

The relations (found in Theorem 1.6.1)

$$\partial_\alpha \boldsymbol{a}_\beta = \Gamma_{\alpha\beta}^\sigma \boldsymbol{a}_\sigma + b_{\alpha\beta} \boldsymbol{a}_3 \text{ and } \partial_\alpha \boldsymbol{a}^\beta = -\Gamma_{\alpha\sigma}^\beta \boldsymbol{a}^\sigma + b_\alpha^\beta \boldsymbol{a}^3$$

and

$$\partial_\alpha \boldsymbol{a}_3 = \partial_\alpha \boldsymbol{a}^3 = -b_{\alpha\beta} \boldsymbol{a}^\beta = -b_\alpha^\sigma \boldsymbol{a}_\sigma,$$

respectively constitute the **formulas of Gauß** and **Weingarten**. The functions (also found in Theorem 1.6.1)

$$\eta_{\beta|\alpha} = \partial_\alpha \eta_\beta - \Gamma_{\alpha\beta}^\sigma \eta_\sigma \text{ and } \eta_{3|\alpha} = \partial_\alpha \eta_3$$

are the **first-order covariant derivatives of the vector field** $\eta_i \boldsymbol{a}^i : \omega \to \mathbb{R}^3$, and the functions

$$\Gamma_{\alpha\beta}^\sigma := \boldsymbol{a}^\sigma \cdot \partial_\alpha \boldsymbol{a}_\beta = -\partial_\alpha \boldsymbol{a}^\sigma \cdot \boldsymbol{a}_\beta$$

are the **Christoffel symbols of the second kind** (the Christoffel symbols of the first kind are introduced in the next section).

Remark 1.6.2. The Christoffel symbols $\Gamma_{\alpha\beta}^\sigma$ can be also defined solely in terms of the covariant components of the first fundamental form; see the proof of Theorem 1.7.1. □

1.7 The Gauß and Codazzi-Mainardi Equations

It is remarkable that *the components* $a_{\alpha\beta} = a_{\beta\alpha} : \omega \to \mathbb{R}$ *and* $b_{\alpha\beta} = b_{\beta\alpha} : \omega \to \mathbb{R}$ *of the first and second fundamental forms of a surface* $\boldsymbol{\theta}(\omega)$ *cannot be arbitrary functions.*

As shown in the next theorem, they must satisfy relations that take the form:

$$\begin{aligned}
\partial_\beta \Gamma_{\alpha\sigma\tau} - \partial_\sigma \Gamma_{\alpha\beta\tau} + \Gamma_{\alpha\beta}^\mu \Gamma_{\sigma\tau\mu} - \Gamma_{\alpha\sigma}^\mu \Gamma_{\beta\tau\mu} &= b_{\alpha\sigma} b_{\beta\tau} - b_{\alpha\beta} b_{\sigma\tau} \text{ in } \omega, \\
\partial_\beta b_{\alpha\sigma} - \partial_\sigma b_{\alpha\beta} + \Gamma_{\alpha\sigma}^\mu b_{\beta\mu} - \Gamma_{\alpha\beta}^\mu b_{\sigma\mu} &= 0 \text{ in } \omega,
\end{aligned}$$

where the functions $\Gamma_{\alpha\beta\tau}$ and $\Gamma_{\alpha\beta}^\sigma$ have simple expressions in terms of the functions $a_{\alpha\beta}$ and of some of their partial derivatives (as shown in the next proof, it so happens that the functions $\Gamma^\sigma_{\alpha\beta}$ as defined in Theorem 1.7.1 coincide with the Christoffel symbols introduced in the previous section; this explains why they are denoted by the same symbol).

These relations, which are meant to hold for all $\alpha, \beta, \sigma, \tau \in \{1, 2\}$, respectively constitute the **Gauß**, and **Codazzi-Mainardi, equations**.

Theorem 1.7.1. *Let ω be an open subset of $\mathbb{R}^2$, let $\boldsymbol{\theta} \in \mathcal{C}^3(\omega; \boldsymbol{E}^3)$ be an immersion, and let*

$$a_{\alpha\beta} := \partial_\alpha \boldsymbol{\theta} \cdot \partial_\beta \boldsymbol{\theta} \text{ and } b_{\alpha\beta} := \partial_{\alpha\beta} \boldsymbol{\theta} \cdot \left\{ \frac{\partial_1 \boldsymbol{\theta} \wedge \partial_2 \boldsymbol{\theta}}{|\partial_1 \boldsymbol{\theta} \wedge \partial_2 \boldsymbol{\theta}|} \right\}$$

denote the covariant components of the first and second fundamental forms of the surface $\boldsymbol{\theta}(\omega)$. Let the functions $\Gamma_{\alpha\beta\tau} \in \mathcal{C}^1(\omega)$ and $\Gamma^\sigma_{\alpha\beta} \in \mathcal{C}^1(\omega)$ be defined by

$$\Gamma_{\alpha\beta\tau} := \frac{1}{2}(\partial_\beta a_{\alpha\tau} + \partial_\alpha a_{\beta\tau} - \partial_\tau a_{\alpha\beta}),$$
$$\Gamma^\sigma_{\alpha\beta} := a^{\sigma\tau} \Gamma_{\alpha\beta\tau} \text{ where } (a^{\sigma\tau}) := (a_{\alpha\beta})^{-1}.$$

Then, necessarily,

$$\partial_\beta \Gamma_{\alpha\sigma\tau} - \partial_\sigma \Gamma_{\alpha\beta\tau} + \Gamma^\mu_{\alpha\beta} \Gamma_{\sigma\tau\mu} - \Gamma^\mu_{\alpha\sigma} \Gamma_{\beta\tau\mu} = b_{\alpha\sigma} b_{\beta\tau} - b_{\alpha\beta} b_{\sigma\tau} \text{ in } \omega,$$
$$\partial_\beta b_{\alpha\sigma} - \partial_\sigma b_{\alpha\beta} + \Gamma^\mu_{\alpha\sigma} b_{\beta\mu} - \Gamma^\mu_{\alpha\beta} b_{\sigma\mu} = 0 \text{ in } \omega.$$

Proof. Let $\boldsymbol{a}_\alpha = \partial_\alpha \boldsymbol{\theta}$. It is then immediately verified that the functions $\Gamma_{\alpha\beta\tau}$ are also given by

$$\Gamma_{\alpha\beta\tau} = \partial_\alpha \boldsymbol{a}_\beta \cdot \boldsymbol{a}_\tau.$$

Let $\boldsymbol{a}_3 = \dfrac{\boldsymbol{a}_1 \wedge \boldsymbol{a}_2}{|\boldsymbol{a}_1 \wedge \boldsymbol{a}_2|}$ and, for each $y \in \omega$, let the three vectors $\boldsymbol{a}^j(y)$ be defined by the relations $\boldsymbol{a}^j(y) \cdot \boldsymbol{a}_i(y) = \delta^j_i$. Since we also have $\boldsymbol{a}^\beta = a^{\alpha\beta} \boldsymbol{a}_\alpha$ and $\boldsymbol{a}^3 = \boldsymbol{a}_3$, the last relations imply that the functions $\Gamma^\sigma_{\alpha\beta}$ are also given by

$$\Gamma^\sigma_{\alpha\beta} = \partial_\alpha \boldsymbol{a}_\beta \cdot \boldsymbol{a}^\sigma.$$

Consequently,

$$\partial_\alpha \boldsymbol{a}_\beta = \Gamma^\sigma_{\alpha\beta} \boldsymbol{a}_\sigma + b_{\alpha\beta} \boldsymbol{a}_3,$$

since $\partial_\alpha \boldsymbol{a}_\beta = (\partial_\alpha \boldsymbol{a}_\beta \cdot \boldsymbol{a}^\sigma) \boldsymbol{a}_\sigma + (\partial_\alpha \boldsymbol{a}_\beta \cdot \boldsymbol{a}^3) \boldsymbol{a}_3$. Differentiating the same relations yields

$$\partial_\sigma \Gamma_{\alpha\beta\tau} = \partial_{\alpha\sigma} \boldsymbol{a}_\beta \cdot \boldsymbol{a}_\tau + \partial_\alpha \boldsymbol{a}_\beta \cdot \partial_\sigma \boldsymbol{a}_\tau,$$

so that the above relations together give

$$\partial_\alpha \boldsymbol{a}_\beta \cdot \partial_\sigma \boldsymbol{a}_\tau = \Gamma^\mu_{\alpha\beta} \boldsymbol{a}_\mu \cdot \partial_\sigma \boldsymbol{a}_\tau + b_{\alpha\beta} \boldsymbol{a}_3 \cdot \partial_\sigma \boldsymbol{a}_\tau = \Gamma^\mu_{\alpha\beta} \Gamma_{\sigma\tau\mu} + b_{\alpha\beta} b_{\sigma\tau}.$$

Consequently,

$$\partial_{\alpha\sigma} \boldsymbol{a}_\beta \cdot \boldsymbol{a}_\tau = \partial_\sigma \Gamma_{\alpha\beta\tau} - \Gamma^\mu_{\alpha\beta} \Gamma_{\sigma\tau\mu} - b_{\alpha\beta} b_{\sigma\tau}.$$

Since $\partial_{\alpha\sigma} \boldsymbol{a}_\beta = \partial_{\alpha\beta} \boldsymbol{a}_\sigma$, we also have

$$\partial_{\alpha\sigma} \boldsymbol{a}_\beta \cdot \boldsymbol{a}_\tau = \partial_\beta \Gamma_{\alpha\sigma\tau} - \Gamma^\mu_{\alpha\sigma} \Gamma_{\beta\tau\mu} - b_{\alpha\sigma} b_{\beta\tau}.$$

Hence the Gauß equations immediately follow.

Since $\partial_\alpha \boldsymbol{a}_3 = (\partial_\alpha \boldsymbol{a}_3 \cdot \boldsymbol{a}_\sigma) \boldsymbol{a}^\sigma + (\partial_\alpha \boldsymbol{a}_3 \cdot \boldsymbol{a}_3) \boldsymbol{a}^3$ and $\partial_\alpha \boldsymbol{a}_3 \cdot \boldsymbol{a}_\sigma = -b_{\alpha\sigma} = -\partial_\alpha \boldsymbol{a}_\sigma \cdot \boldsymbol{a}_3$, we have

$$\partial_\alpha \boldsymbol{a}_3 = -b_{\alpha\sigma} \boldsymbol{a}^\sigma.$$

Differentiating the relations $b_{\alpha\beta} = \partial_\alpha \boldsymbol{a}_\beta \cdot \boldsymbol{a}_3$, we obtain

$$\partial_\sigma b_{\alpha\beta} = \partial_{\alpha\sigma} \boldsymbol{a}_\beta \cdot \boldsymbol{a}_3 + \partial_\alpha \boldsymbol{a}_\beta \cdot \partial_\sigma \boldsymbol{a}_3.$$

This relation and the relations $\partial_\alpha \boldsymbol{a}_\beta = \Gamma^\sigma_{\alpha\beta} \boldsymbol{a}_\sigma + b_{\alpha\beta} \boldsymbol{a}_3$ and $\partial_\alpha \boldsymbol{a}_3 = -b_{\alpha\sigma} \boldsymbol{a}^\sigma$ together imply that

$$\partial_\alpha \boldsymbol{a}_\beta \cdot \partial_\sigma \boldsymbol{a}_3 = -\Gamma^\mu_{\alpha\beta} b_{\sigma\mu}.$$

Consequently,

$$\partial_{\alpha\sigma} \boldsymbol{a}_\beta \cdot \boldsymbol{a}_3 = \partial_\sigma b_{\alpha\beta} + \Gamma^\mu_{\alpha\beta} b_{\sigma\mu}.$$

Since $\partial_{\alpha\sigma} \boldsymbol{a}_\beta = \partial_{\alpha\beta} \boldsymbol{a}_\sigma$, we also have

$$\partial_{\alpha\sigma} \boldsymbol{a}_\beta \cdot \boldsymbol{a}_3 = \partial_\beta b_{\alpha\sigma} + \Gamma^\mu_{\alpha\sigma} b_{\beta\mu}.$$

Hence the Codazzi-Mainardi equations immediately follow. □

Remark 1.7.2. The vectors $\boldsymbol{a}_\alpha$ and $\boldsymbol{a}^\beta$ introduced above respectively form the covariant and contravariant bases of the tangent plane to the surface $\boldsymbol{\theta}(\omega)$, the unit vector $\boldsymbol{a}_3 = \boldsymbol{a}^3$ is normal to the surface, and the functions $a^{\alpha\beta}$ are the contravariant components of the first fundamental form (Sections 1.2 and 1.3). □

As shown in the above proof, the Gauß and Codazzi-Mainardi equations thus simply constitute a re-writing of the relations $\partial_{\alpha\sigma} \boldsymbol{a}_\beta = \partial_{\alpha\beta} \boldsymbol{a}_\sigma$ in the form of the equivalent relations $\partial_{\alpha\sigma} \boldsymbol{a}_\beta \cdot \boldsymbol{a}_\tau = \partial_{\alpha\beta} \boldsymbol{a}_\sigma \cdot \boldsymbol{a}_\tau$ and $\partial_{\alpha\sigma} \boldsymbol{a}_\beta \cdot \boldsymbol{a}_3 = \partial_{\alpha\beta} \boldsymbol{a}_\sigma \cdot \boldsymbol{a}_3$.

The functions

$$\Gamma_{\alpha\beta\tau} = \frac{1}{2}(\partial_\beta a_{\alpha\tau} + \partial_\alpha a_{\beta\tau} - \partial_\tau a_{\alpha\beta}) = \partial_\alpha \boldsymbol{a}_\beta \cdot \boldsymbol{a}_\tau = \Gamma_{\beta\alpha\tau}$$

and

$$\Gamma^\sigma_{\alpha\beta} = a^{\sigma\tau}\Gamma_{\alpha\beta\tau} = \partial_\alpha \boldsymbol{a}_\beta \cdot \boldsymbol{a}^\sigma = \Gamma^\sigma_{\beta\alpha}$$

are the **Christoffel symbols of the first**, and **second, kind**. We recall that the Christoffel symbols of the second kind also naturally appeared in a different context (that of covariant differentiation; cf. Section 1.6).

Finally, the functions

$$S_{\tau\alpha\beta\sigma} := \partial_\beta\Gamma_{\alpha\sigma\tau} - \partial_\sigma\Gamma_{\alpha\beta\tau} + \Gamma^\mu_{\alpha\beta}\Gamma_{\sigma\tau\mu} - \Gamma^\mu_{\alpha\sigma}\Gamma_{\beta\tau\mu}$$

are the **covariant components of the Riemann curvature tensor of the surface $\boldsymbol{\theta}(\omega)$**.

The definitions of the functions $\Gamma^\sigma_{\alpha\beta}$ and $\Gamma_{\alpha\beta\tau}$ imply that *the sixteen Gauß equations are satisfied if and only if they are satisfied for* $\alpha = 1$, $\beta = 2$, $\sigma = 1$, $\tau = 2$ and that *the Codazzi-Mainardi equations are satisfied if and only if they are satisfied* for $\alpha = 1$, $\beta = 2$, $\sigma = 1$ and $\alpha = 1$, $\beta = 2$, $\sigma = 2$ (other choices of indices with the same properties are clearly possible).

In other words, the Gauß equations and the Codazzi-Mainardi equations in fact respectively reduce to *one* and *two* equations.

Letting $\alpha = 2$, $\beta = 1$, $\sigma = 2$, $\tau = 1$ in the Gauß equations gives in particular

$$S_{1212} = \det(b_{\alpha\beta}).$$

Consequently, the *Gaussian curvature* at each point $\boldsymbol{\Theta}(y)$ of the surface $\boldsymbol{\theta}(\omega)$ can be written as

$$\frac{1}{R_1(y)R_2(y)} = \frac{S_{1212}(y)}{\det(a_{\alpha\beta}(y))}, \; y \in \omega,$$

since $\dfrac{1}{R_1(y)R_2(y)} = \dfrac{\det(b_{\alpha\beta}(y)}{\det(a_{\alpha\beta}(y))}$ (Theorem 1.5.1). By inspection of the function S_{1212}, we thus reach the astonishing conclusion that, *at each point of the surface, a notion involving the "curvature" of the surface, viz., the Gaussian curvature, is entirely determined by the knowledge of the "metric" of the surface at the same point*, viz., the components of the first fundamental forms and their partial derivatives of order ≤ 2 at the same point! This startling conclusion naturally deserves a theorem:

Theorem 1.7.3. *Let ω be an open subset of $\mathbb{R}^2$, let $\boldsymbol{\theta} \in \mathcal{C}^3(\omega; \boldsymbol{E}^3)$ be an immersion, let $a_{\alpha\beta} = \partial_\alpha\boldsymbol{\theta} \cdot \partial_\beta\boldsymbol{\theta}$ denote the covariant components of the first* *fundamental form of the surface* $\boldsymbol{\theta}(\omega)$, *and let the functions* $\Gamma_{\alpha\beta\tau}$ *and* S_{1212} *be defined by*

$$\Gamma_{\alpha\beta\tau} := \frac{1}{2}(\partial_\beta a_{\alpha\tau} + \partial_\alpha a_{\beta\tau} - \partial_\tau a_{\alpha\beta}),$$
$$S_{1212} := \frac{1}{2}(2\partial_{12}a_{12} - \partial_{11}a_{22} - \partial_{22}a_{11}) + a^{\alpha\beta}(\Gamma_{12\alpha}\Gamma_{12\beta} - \Gamma_{11\alpha}\Gamma_{22\beta}).$$

Then, at each point $\boldsymbol{\theta}(y)$ *of the surface* $\boldsymbol{\theta}(\omega)$, *the principal curvatures* $\frac{1}{R_1(y)}$ *and* $\frac{1}{R_2(y)}$ *satisfy*

$$\frac{1}{R_1(y)R_2(y)} = \frac{S_{1212}(y)}{\det(a_{\alpha\beta}(y))},\ y \in \omega.$$

Theorem 1.7.3 constitutes the famed **Theorema Egregium** of Gauß [1828], so named by Gauß who had been himself astounded by his discovery.

1.8 The Fundamental Theorem of Surface Theory

Let $\mathbb{M}^2, \mathbb{S}^2$, and $\mathbb{S}^2_>$ denote the sets of all square matrices of order two, of all symmetric matrices of order two, and of all symmetric, positive definite matrices of order two.

So far, we have considered that we are given an open set $\omega \subset \mathbb{R}^2$ and a smooth enough immersion $\boldsymbol{\theta} : \omega \to \boldsymbol{E}^3$, thus allowing us to define the fields $(a_{\alpha\beta}) : \omega \to \mathbb{S}^2_>$ and $(b_{\alpha\beta}) : \omega \to \mathbb{S}^2$, where $a_{\alpha\beta} : \omega \to \mathbb{R}$ and $b_{\alpha\beta} : \omega \to \mathbb{R}$ are the covariant components of the *first* and *second fundamental forms* of the surface $\boldsymbol{\theta}(\omega) \subset \boldsymbol{E}^3$.

Remark 1.8.1. *The immersion* $\boldsymbol{\theta}$ *need not be injective* in order that these matrix fields be well defined. □

We now turn to the *reciprocal questions*:

Given an open subset ω of $\mathbb{R}^2$ and two smooth enough matrix fields $(a_{\alpha\beta}) : \omega \to \mathbb{S}^2_>$ and $(b_{\alpha\beta}) : \omega \to \mathbb{S}^2$, when are they the first and second fundamental forms of a surface $\boldsymbol{\theta}(\omega) \subset \boldsymbol{E}^3$, i.e., when does there exist an immersion $\boldsymbol{\theta} : \omega \to \boldsymbol{E}^3$ such that

$$a_{\alpha\beta} = \partial_\alpha\boldsymbol{\theta} \cdot \partial_\beta\boldsymbol{\theta} \text{ and } b_{\alpha\beta} = \partial_{\alpha\beta}\boldsymbol{\theta} \cdot \left\{\frac{\partial_1\boldsymbol{\theta} \wedge \partial_2\boldsymbol{\theta}}{|\partial_1\boldsymbol{\theta} \wedge \partial_2\boldsymbol{\theta}|}\right\} \text{ in } \omega?$$

If such an immersion exists, to what extent is it unique?

The answers to these questions turn out to be remarkably simple: *If* ω *is simply-connected, the necessary conditions of* Theorem 1.7.1, i.e., *the Gauß and Codazzi-Mainardi equations, are also sufficient for the existence of such an immersion. If* ω *is connected, this immersion is unique up to isometries in* $\boldsymbol{E}^3$.

Remark 1.8.2. Whether an immersion found in this fashion is *injective* is a different issue, which accordingly should be resolved by different means. □

These existence and uniqueness results together constitute the **fundamental theorem of surface theory**. This theorem comprises two essentially distinct parts, a *global existence result* (Theorem 1.8.3) and a *uniqueness result* (Theorem 1.8.4), the latter being also called *rigidity theorem*. Note that these two results are established under *different assumptions* on the set ω and on the smoothness of the fields $(a_{\alpha\beta})$ and $(b_{\alpha\beta})$. We begin with the issue of *existence*.

A direct proof of the fundamental theorem of surface theory is given in Klingenberg [1973, Theorem 3.8.8]. See also Ciarlet & Larsonneur [2001] or Ciarlet [2005] for a self-contained, and essentially elementary, proof. A proof of the "local" version of this theorem, which constitutes *Bonnet's theorem*, is found in, e.g., do Carmo [1976] or Kühnel [2002].

Theorem 1.8.3. *Let ω be a connected and simply-connected open subset of $\mathbb{R}^2$ and let $(a_{\alpha\beta}) \in \mathcal{C}^2(\omega; \mathbb{S}^2_>)$ and $(b_{\alpha\beta}) \in \mathcal{C}^1(\omega; \mathbb{S}^2)$ be two matrix fields that satisfy the Gauß and Codazzi-Mainardi equations, viz.,*

$$\partial_\beta \Gamma_{\alpha\sigma\tau} - \partial_\sigma \Gamma_{\alpha\beta\tau} + \Gamma^\mu_{\alpha\beta}\Gamma_{\sigma\tau\mu} - \Gamma^\mu_{\alpha\sigma}\Gamma_{\beta\tau\mu} = b_{\alpha\sigma}b_{\beta\tau} - b_{\alpha\beta}b_{\sigma\tau} \text{ in } \omega,$$
$$\partial_\beta b_{\alpha\sigma} - \partial_\sigma b_{\alpha\beta} + \Gamma^\mu_{\alpha\sigma} b_{\beta\mu} - \Gamma^\mu_{\alpha\beta} b_{\sigma\mu} = 0 \text{ in } \omega,$$

where

$$\Gamma_{\alpha\beta\tau} := \frac{1}{2}(\partial_\beta a_{\alpha\tau} + \partial_\alpha a_{\beta\tau} - \partial_\tau a_{\alpha\beta}),$$
$$\Gamma^\sigma_{\alpha\beta} := a^{\sigma\tau}\Gamma_{\alpha\beta\tau} \text{ where } (a^{\sigma\tau}) := (a_{\alpha\beta})^{-1}.$$

Then there exists an immersion $\boldsymbol{\theta} \in \mathcal{C}^3(\omega; \boldsymbol{E}^3)$ such that

$$a_{\alpha\beta} = \partial_\alpha \boldsymbol{\theta} \cdot \partial_\beta \boldsymbol{\theta} \text{ and } b_{\alpha\beta} = \partial_{\alpha\beta}\boldsymbol{\theta} \cdot \left\{ \frac{\partial_1 \boldsymbol{\theta} \wedge \partial_2 \boldsymbol{\theta}}{|\partial_1 \boldsymbol{\theta} \wedge \partial_2 \boldsymbol{\theta}|} \right\} \text{ in } \omega.$$

□

The *regularity assumptions* made in Theorem 1.8.3 on the matrix fields $(a_{\alpha\beta})$ and $(b_{\alpha\beta})$ can be significantly relaxed in several ways.

In this direction, Hartman & Wintner [1950] first showed that the existence theorem still holds if $(a_{\alpha\beta}) \in \mathcal{C}^1(\omega; \mathbb{S}^2_>)$ and $(b_{\alpha\beta}) \in \mathcal{C}^0(\omega; \mathbb{S}^2)$, with a resulting mapping $\boldsymbol{\theta}$ in the space $\mathcal{C}^2(\omega; \boldsymbol{E}^3)$. Then S. Mardare [2003b] established that if $(a_{\alpha\beta}) \in W^{1,\infty}_{\text{loc}}(\omega; \mathbb{S}^2_>)$ and $(b_{\alpha\beta}) \in L^\infty_{\text{loc}}(\omega; \mathbb{S}^2)$ are two matrix fields that satisfy the Gauß and Codazzi-Mainardi equations in the sense of distributions, there exists a mapping $\boldsymbol{\theta} \in W^{2,\infty}_{\mathrm{loc}}(\omega; \boldsymbol{E}^3)$ such that $(a_{\alpha\beta})$ and $(b_{\alpha\beta})$ are the fundamental forms of the surface $\boldsymbol{\theta}(\omega)$. The last word seems to belong to S. Mardare [2005], who was able to further reduce these regularities, to those of the spaces $W^{1,p}_{\mathrm{loc}}(\omega; \mathbb{S}^2_>)$ for the field $(a_{\alpha\beta})$ and $L^p_{\mathrm{loc}}(\omega; \mathbb{S}^2)$ for the field $(b_{\alpha\beta})$ for any $p > 2$, with a resulting mapping $\boldsymbol{\theta}$ in the space $W^{2,p}_{\mathrm{loc}}(\omega; \boldsymbol{E}^3)$.

Theorem 1.8.3 establishes the *existence* of an immersion $\boldsymbol{\theta} : \omega \subset \mathbb{R}^2 \to \boldsymbol{E}^3$ giving rise to a surface $\boldsymbol{\theta}(\omega)$ with prescribed first and second fundamental forms, provided these forms satisfy *ad hoc* sufficient conditions. We now turn to the question of *uniqueness* of such immersions.

This is the object of the next theorem, which is called the **rigidity theorem for surfaces**. This result asserts that, if two immersions $\widetilde{\boldsymbol{\theta}} \in \mathcal{C}^2(\omega; \boldsymbol{E}^3)$ and $\boldsymbol{\theta} \in \mathcal{C}^2(\omega; \boldsymbol{E}^3)$ share the same fundamental forms, then the surface $\boldsymbol{\theta}(\omega)$ is obtained by subjecting the surface $\widetilde{\boldsymbol{\theta}}(\omega)$ to a *rotation* (represented by an orthogonal matrix $\boldsymbol{Q}$ with $\det \boldsymbol{Q} = 1$), then by subjecting the rotated surface to a *translation* (represented by a vector $\mathbf{c}$). Note in passing that the converse clearly holds.

Such a geometric transformation of the surface $\widetilde{\boldsymbol{\theta}}(\omega)$ is sometimes called a "*rigid transformation*", to remind that it corresponds to the idea of a "rigid" one in $\boldsymbol{E}^3$. This observation motivates the terminology "rigidity theorem".

Theorem 1.8.4. *Let ω be a connected open subset of $\mathbb{R}^2$ and let $\boldsymbol{\theta} \in \mathcal{C}^2(\omega; \boldsymbol{E}^3)$ and $\widetilde{\boldsymbol{\theta}} \in \mathcal{C}^2(\omega; \boldsymbol{E}^3)$ be two immersions such that their associated first and second fundamental forms satisfy* (with self-explanatory notations)

$$a_{\alpha\beta} = \widetilde{a}_{\alpha\beta} \text{ and } b_{\alpha\beta} = \widetilde{b}_{\alpha\beta} \text{ in } \omega.$$

Then there exist a vector $\mathbf{c} \in \boldsymbol{E}^3$ and a matrix $\boldsymbol{Q} \in \mathbb{O}^3_+$ such that

$$\boldsymbol{\theta}(y) = \mathbf{c} + \boldsymbol{Q}\widetilde{\boldsymbol{\theta}}(y) \text{ for all } y \in \omega.$$

□

More details about the various notions of classical Differential Geometry considered in Chapter 1 are found in classic texts such as Stoker [1969], Klingenberg [1973], do Carmo [1976], Berger & Gostiaux [1987], Spivak [1999], or Kühnel [2002].

2 An Introduction to Shell Theory

In this chapter, which heavily relies on Chapter 1, we begin by a detailed description of the nonlinear and linear equations proposed by W.T. Koiter for modeling thin elastic shells. These equations are "two-dimensional", in the sense that they are expressed in terms of two curvilinear coordinates used for defining the middle surface of the shell. The existence, uniqueness, and regularity of solutions to the linear Koiter equations are then established, thanks to a fundamental "Korn inequality on a surface" and to an "infinitesimal rigid displacement lemma on a surface".

2.1 What is a Two-Dimensional Shell Problem?

A *domain* ω in $\mathbb{R}^2$ is an open, bounded, connected subset of $\mathbb{R}^2$, whose boundary is Lipschitz-continuous, the set ω being locally on one side of its boundary. To begin with, we briefly recapitulate some important notions introduced and studied in Chapter 1. Note in this respect that we shall extend *without further notice* all the definitions given, or properties studied, on arbitrary *open subsets of* $\mathbb{R}^2$ in Chapter 1 to their analogs on *domains in* $\mathbb{R}^2$.

Greek indices and exponents (except ν in the notation ∂_ν) range in the set $\{1, 2\}$, Latin indices and exponents range in the set $\{1, 2, 3\}$ (save when they are used for indexing sequences), and the summation convention with respect to repeated indices and exponents is systematically used. The notation $\boldsymbol{E}^3$ denotes a three-dimensional Euclidean space, the Euclidean scalar product and the exterior product of $\boldsymbol{a}, \boldsymbol{b} \in \boldsymbol{E}^3$ are noted $\boldsymbol{a} \cdot \boldsymbol{b}$ and $\boldsymbol{a} \wedge \boldsymbol{b}$, and the Euclidean norm of $\boldsymbol{a} \in \boldsymbol{E}^3$ is noted $|\boldsymbol{a}|$.

Let ω be a domain in $\mathbb{R}^2$. Let $y = (y_\alpha)$ denote a generic point in the set $\overline{\omega}$, and let $\partial_\alpha := \partial/\partial y_\alpha$. Let there be given an *immersion* $\boldsymbol{\theta} \in \mathcal{C}^3(\overline{\omega}; \boldsymbol{E}^3)$, i.e., a mapping such that the two vectors

$$\boldsymbol{a}_\alpha(y) := \partial_\alpha \boldsymbol{\theta}(y)$$

are linearly independent at all points $y \in \overline{\omega}$. These two vectors thus span the tangent plane to the *surface*

$$S := \boldsymbol{\theta}(\overline{\omega})$$

at the point $\boldsymbol{\theta}(y)$, and the unit vector

$$\boldsymbol{a}_3(y) := \frac{\boldsymbol{a}_1(y) \wedge \boldsymbol{a}_2(y)}{|\boldsymbol{a}_1(y) \wedge \boldsymbol{a}_2(y)|}$$

is normal to S at the point $\boldsymbol{\theta}(y)$. The three vectors $\boldsymbol{a}_i(y)$ constitute the *covariant basis* at the point $\boldsymbol{\theta}(y)$, while the three vectors $\boldsymbol{a}^i(y)$ defined by the relations

$$\boldsymbol{a}^i(y) \cdot \boldsymbol{a}_j(y) = \delta^i_j,$$

where δ^i_j is the Kronecker symbol, constitute the *contravariant basis* at the point $\boldsymbol{\theta}(y) \in S$ (recall that $\boldsymbol{a}^3(y) = \boldsymbol{a}_3(y)$ and that the vectors $\boldsymbol{a}^\alpha(y)$ are also in the tangent plane to S at $\boldsymbol{\theta}(y)$).

The covariant and contravariant components $a_{\alpha\beta}$ and $a^{\alpha\beta}$ of the *first fundamental form* of S, the *Christoffel symbols* $\Gamma^\sigma_{\alpha\beta}$, and the covariant and mixed components $b_{\alpha\beta}$ and b^β_α of the *second fundamental form* of S are then defined by letting:

$$a_{\alpha\beta} := \boldsymbol{a}_\alpha \cdot \boldsymbol{a}_\beta, \quad a^{\alpha\beta} := \boldsymbol{a}^\alpha \cdot \boldsymbol{a}^\beta, \quad \Gamma^\sigma_{\alpha\beta} := \boldsymbol{a}^\sigma \cdot \partial_\beta \boldsymbol{a}_\alpha,$$
$$b_{\alpha\beta} := \boldsymbol{a}^3 \cdot \partial_\beta \boldsymbol{a}_\alpha, \quad b^\beta_\alpha := a^{\beta\sigma} b_{\sigma\alpha}.$$

The *area* element along S is $\sqrt{a}\,\mathrm{d}y$, where

$$a := \det(a_{\alpha\beta}).$$

Note that one also has $\sqrt{a} = |\boldsymbol{a}_1 \wedge \boldsymbol{a}_2|$.

Let $\Omega := \omega \times \,]-\varepsilon, \varepsilon[$, let $x = (x_i)$ denote a generic point in the set $\overline{\Omega}$ (hence $x_\alpha = y_\alpha$), and let $\partial_i := \partial/\partial x_i$. Consider an *elastic shell* with *middle surface* $S = \boldsymbol{\theta}(\overline{\omega})$ and *thickness* $2\varepsilon > 0$, i.e., an elastic body whose *reference configuration* is the set $\boldsymbol{\Theta}(\overline{\omega} \times [-\varepsilon, \varepsilon])$, where (cf. Figure 2.1.1)

$$\boldsymbol{\Theta}(y, x_3) := \boldsymbol{\theta}(y) + x_3 \boldsymbol{a}_3(y) \text{ for all } (y, x_3) \in \overline{\omega} \times [-\varepsilon, \varepsilon].$$

Naturally, this definition makes sense physically only if the mapping $\boldsymbol{\Theta}$ is globally injective on the set $\overline{\Omega}$. Fortunately, this is indeed the case if the immersion $\boldsymbol{\theta}$ is itself globally injective on the set $\overline{\omega}$ and ε is small enough, according to the following result (due to Ciarlet [2000a, Theorem 3.1-1]).

Theorem 2.1.1. *Let ω be a domain in $\mathbb{R}^2$ and let $\boldsymbol{\theta} \in \mathcal{C}^3(\overline{\omega}; \boldsymbol{E}^3)$ be an injective immersion. Then there exists $\varepsilon > 0$ such that the mapping $\boldsymbol{\Theta} : \overline{\Omega} \to \boldsymbol{E}^3$ defined by*

$$\boldsymbol{\Theta}(y, x_3) := \boldsymbol{\theta}(y) + x_3 \boldsymbol{a}_3(y) \text{ for all } (y, x_3) \in \overline{\Omega},$$

where $\Omega := \omega \times \,]-\varepsilon, \varepsilon[$, is a $\mathcal{C}^2$-diffeomorphism from $\overline{\Omega}$ onto $\boldsymbol{\Theta}(\overline{\Omega})$ and $\det(\boldsymbol{g}_1, \boldsymbol{g}_2, \boldsymbol{g}_3) > 0$ in $\overline{\Omega}$, where $\boldsymbol{g}_i := \partial_i \boldsymbol{\Theta}$. □

Proof. The assumed regularity on $\boldsymbol{\theta}$ implies that $\boldsymbol{\Theta} \in \mathcal{C}^2(\overline{\omega} \times [-\varepsilon, \varepsilon]; \boldsymbol{E}^3)$ for any $\varepsilon > 0$. The relations

$$\boldsymbol{g}_\alpha = \partial_\alpha \boldsymbol{\Theta} = \boldsymbol{a}_\alpha + x_3 \partial_\alpha \boldsymbol{a}_3 \text{ and } \boldsymbol{g}_3 = \partial_3 \boldsymbol{\Theta} = \boldsymbol{a}_3$$

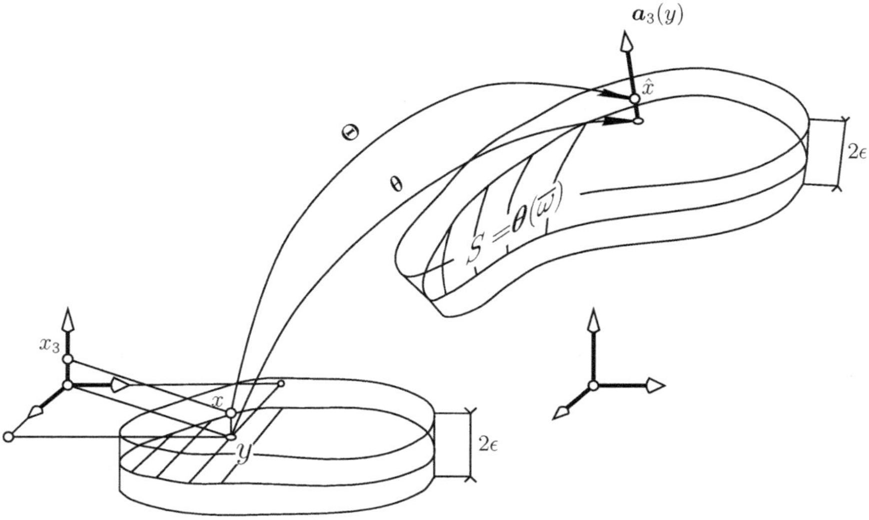

Figure 2.1.1. *The reference configuration of an elastic shell.* Let ω be a domain in $\mathbb{R}^2$, let $\Omega = \omega \times]-\varepsilon, \varepsilon[> 0$, let $\boldsymbol{\theta} \in \mathcal{C}^3(\overline{\omega}; \boldsymbol{E}^3)$ be an immersion, and let the mapping $\boldsymbol{\Theta} : \overline{\Omega} \to \boldsymbol{E}^3$ be defined by $\boldsymbol{\Theta}(y, x_3) = \boldsymbol{\theta}(y) + x_3 \boldsymbol{a}_3(y)$ for all $(y, x_3) \in \overline{\Omega}$. Then the mapping $\boldsymbol{\Theta}$ is globally injective on $\overline{\Omega}$ if the immersion $\boldsymbol{\theta}$ is globally injective on $\overline{\omega}$ and $\varepsilon > 0$ is small enough (Theorem 2.1.1). In this case, the set $\boldsymbol{\Theta}(\overline{\Omega})$ may be viewed as the reference configuration of an elastic shell with thickness 2ε and middle surface $S = \theta(\overline{\omega})$. The coordinates (y_1, y_2, x_3) of an arbitrary point $x \in \overline{\Omega}$ are then viewed as curvilinear coordinates of the point $\widehat{x} = \boldsymbol{\Theta}(x)$ of the reference configuration of the shell.

imply that

$$\det(\boldsymbol{g}_1, \boldsymbol{g}_2, \boldsymbol{g}_3)|_{x_3=0} = \det(\boldsymbol{a}_1, \boldsymbol{a}_2, \boldsymbol{a}_3) > 0 \text{ in } \overline{\omega}.$$

Hence $\det(\boldsymbol{g}_1, \boldsymbol{g}_2, \boldsymbol{g}_3) > 0$ on $\overline{\omega} \times [-\varepsilon, \varepsilon]$ if $\varepsilon > 0$ is small enough.

Therefore, the *implicit function theorem* can be applied if ε is small enough: It shows that, *locally*, the mapping $\boldsymbol{\Theta}$ is a $\mathcal{C}^2$-diffeomorphism: Given any $y \in \overline{\omega}$, there exist a neighborhood $U(y)$ of y in $\overline{\omega}$ and $\varepsilon(y) > 0$ such that $\boldsymbol{\Theta}$ is a $\mathcal{C}^2$-diffeomorphism from the set $U(y) \times [-\varepsilon(y), \varepsilon(y)]$ onto $\boldsymbol{\Theta}(U(y) \times [-\varepsilon(y), \varepsilon(y)])$. See, e.g., Schwartz [1992, Chapter 3] (the proof of the implicit function theorem, which is almost invariably given for functions defined over open sets, can be easily extended to functions defined over *closures of domains*, such as the sets $\overline{\omega} \times [-\varepsilon, \varepsilon]$; see, e.g., Stein [1970]).

To establish that the mapping $\boldsymbol{\Theta} : \overline{\omega} \times [-\varepsilon, \varepsilon] \to \boldsymbol{E}^3$ is injective provided $\varepsilon > 0$ is small enough, we proceed by contradiction: If this property is false, there exist $\varepsilon_n > 0$, (y^n, x_3^n), and $(\widetilde{y}^n, \widetilde{x}_3^n)$, $n \geq 0$, such that

$$\varepsilon_n \to 0 \text{ as } n \to \infty, \quad y^n \in \overline{\omega}, \quad \widetilde{y}^n \in \overline{\omega}, \quad |x_3^n| \leq \varepsilon_n, \quad |\widetilde{x}_3^n| \leq \varepsilon_n,$$
$$(y^n, x_3^n) \neq (\widetilde{y}^n, \widetilde{x}_3^n) \text{ and } \boldsymbol{\Theta}(y^n, x_3^n) = \boldsymbol{\Theta}(\widetilde{y}^n, \widetilde{x}_3^n).$$

Since the set $\overline{\omega}$ is compact, there exist $y \in \overline{\omega}$ and $\widetilde{y} \in \overline{\omega}$, and there exists a subsequence, still indexed by n for convenience, such that

$$y^n \to y, \quad \widetilde{y}^n \to \widetilde{y}, \quad x_3^n \to 0, \quad \widetilde{x}_3^n \to 0 \text{ as } n \to \infty.$$

Hence

$$\boldsymbol{\theta}(y) = \lim_{n\to\infty} \boldsymbol{\Theta}(y^n, x_3^n) = \lim_{n\to\infty} \boldsymbol{\Theta}(\widetilde{y}^n, \widetilde{x}_3^n) = \boldsymbol{\theta}(\widetilde{y}),$$

by the continuity of the mapping $\boldsymbol{\Theta}$ and thus $y = \widetilde{y}$ since the mapping $\boldsymbol{\theta}$ is injective by assumption. But these properties contradict the local injectivity (noted above) of the mapping $\boldsymbol{\Theta}$. Hence there exists $\varepsilon > 0$ such that $\boldsymbol{\Theta}$ is injective on the set $\overline{\Omega} = \overline{\omega} \times [-\varepsilon, \varepsilon]$. □

In what follows, we assume that $\varepsilon > 0$ is small enough so that the conclusions of Theorem 2.1.1 hold. In particular then, the coordinates y_1, y_2, x_3 of the points $x \in \overline{\Omega}$ constitute a system of "three-dimensional" *curvilinear coordinates* for describing the reference configuration $\boldsymbol{\Theta}(\overline{\Omega})$ of the shell.

Then for each $x \in \overline{\Omega}$, the three linearly independent vectors $\boldsymbol{g}_i(x) := \partial_i \boldsymbol{\Theta}(x)$ constitute the *covariant basis* at the point $\boldsymbol{\Theta}(x)$, and the three (likewise linearly independent) vectors $\boldsymbol{g}^j(x)$ defined by the relations $\boldsymbol{g}^j(x) \cdot \boldsymbol{g}_i(x) = \delta_i^j$ constitute the *contravariant basis* at the same point.

Let γ_0 be a measurable subset of the boundary $\gamma := \partial\omega$ that satisfies *length* $\gamma_0 > 0$. It is assumed that the shell is subjected to a *homogeneous boundary condition of place* along the portion $\boldsymbol{\Theta}(\gamma_0 \times [-\varepsilon, \varepsilon])$ of its lateral face $\boldsymbol{\Theta}(\gamma \times [-\varepsilon, \varepsilon])$. This means that the displacement field of the shell vanishes on the set $\boldsymbol{\Theta}(\gamma_0 \times [-\varepsilon, \varepsilon])$.

The shell is subjected to *applied body forces* in its interior $\boldsymbol{\Theta}(\Omega)$ and to *applied surface forces* on its "upper" and "lower" faces $\boldsymbol{\Theta}(\Gamma_+)$ and $\boldsymbol{\Theta}(\Gamma_-)$, where $\Gamma_\pm := \omega \times \{\pm\varepsilon\}$. Such applied forces are given by the "contravariant" components $f^i \in L^2(\Omega)$ and $h^i \in L^2(\Gamma_+ \cup \Gamma_-)$ of their densities per unit volume and per unit area, respectively (this means that $f^i(x)\boldsymbol{g}_i(x)\sqrt{g(x)}\,\mathrm{d}x$ is the body force applied to the volume $\sqrt{g(x)}\,\mathrm{d}x$ at each $x \in \overline{\Omega}$, where $g(x) := \det(\boldsymbol{g}_i(x) \cdot \boldsymbol{g}_j(x))$; a similar interpretation holds for the surface forces).

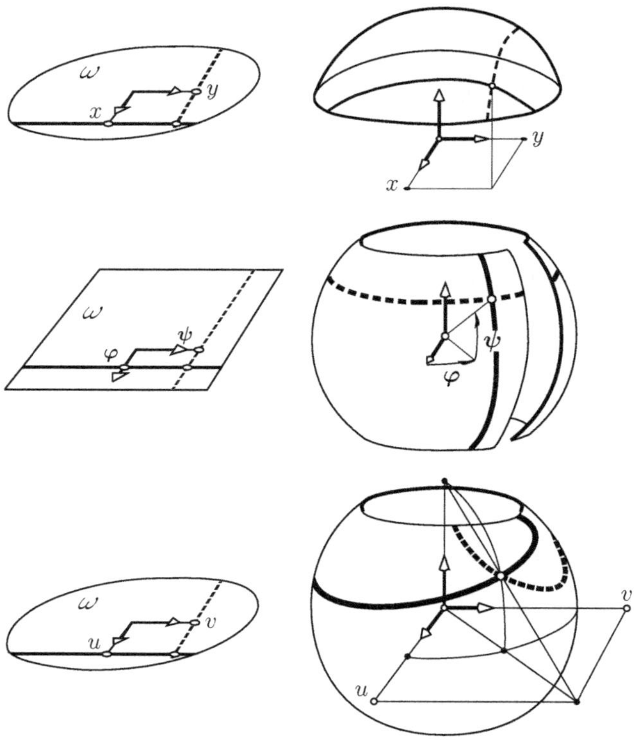

Figure 2.1.2. *An elastic shell modeled as a three-dimensional problem.* Let $\Omega = \omega \times \,]-\varepsilon, \varepsilon[$. The set $\boldsymbol{\Theta}(\overline{\Omega})$, where $\boldsymbol{\Theta}(y, x_3) = \boldsymbol{\theta}(y) + x_3 \boldsymbol{a}_3(y)$ for all $x = (y, x_3) \in \overline{\Omega}$, is the reference configuration of a shell, with thickness 2ε and middle surface $S = \boldsymbol{\theta}(\overline{\omega})$ (Figure 2.1.1), which is subjected to a boundary condition of place along the portion $\boldsymbol{\Theta}(\Gamma_0)$ of its lateral face (i.e., the displacement vanishes on $\boldsymbol{\Theta}(\Gamma_0)$), where $\Gamma_0 = \gamma_0 \times [-\varepsilon, \varepsilon]$ and $\gamma_0 \subset \gamma = \partial\omega$. The shell is subjected to applied body forces in its interior $\boldsymbol{\Theta}(\Omega)$ and to applied surface forces on its upper and lower faces $\boldsymbol{\Theta}(\Gamma_+)$ and $\boldsymbol{\Theta}(\Gamma_-)$ where $\Gamma_\pm = \overline{\omega} \times \{\pm\varepsilon\}$. Under the influence of these forces, a point $\boldsymbol{\Theta}(x)$ undergoes a displacement $u_i(x)\boldsymbol{g}^i(x)$, where the three vectors $\boldsymbol{g}^i(x)$ form the contravariant basis at the point $\boldsymbol{\Theta}(x)$. The unknowns of the problem are the three covariant components $u_i : \overline{\Omega} \to \mathbb{R}$ of the displacement field $u_i \boldsymbol{g}^i : \overline{\Omega} \to \mathbb{R}^3$ of the points of $\boldsymbol{\Theta}(\overline{\Omega})$, which thus satisfy the boundary conditions $u_i = 0$ on Γ_0. The objective consists in finding *ad hoc* conditions affording the "replacement" of this three-dimensional problem by a "two-dimensional problem posed over the middle surface S" if ε is "small enough"; see Figure 2.1.3.

Note that, for the sake of visual clarity, the thickness is overly exaggerated.

Finally, the elastic material constituting the shell is assumed to be *homogeneous* and *isotropic* and the reference configuration $\boldsymbol{\Theta}(\overline{\Omega})$ of the shell is assumed to be a *natural state*. Hence the material is characterized by two *Lamé constants* λ and μ satisfying $3\lambda + 2\mu > 0$ and $\mu > 0$ (for a detailed description of elastic materials, see, e.g., Ciarlet [1988, Chapter 3]).

Such a shell, endowed with its curvilinear coordinates, $(y_1, y_2, x_3) \in \overline{\Omega}$, can thus be modeled by means of the *equations of three-dimensional, linear or nonlinear, elasticity in curvilinear coordinates* (for a detailed description of these equations, see Ciarlet [2005]). In such equations, the "natural" *unknowns* are the *three "covariant" components* $u_i : \overline{\Omega} \to \mathbb{R}$ *of the displacement field* $u_i \boldsymbol{g}^i : \overline{\Omega} \to \mathbb{R}$ *of the points of the reference configuration* $\boldsymbol{\Theta}(\overline{\Omega})$, where the vector fields $\boldsymbol{g}^i$ denote the contravariant bases; cf. Figure 2.1.2.

In a "*two-dimensional approach*", the above three-dimensional problem is "replaced" by a, presumably much simpler, *"two-dimensional" problem*, this time "*posed over the middle surface* S *of the shell*". This means that the new unknowns should be now the *three covariant components* $\zeta_i : \overline{\omega} \to \mathbb{R}$ *of the displacement field* $\zeta_i \boldsymbol{a}^i : \overline{\omega} \to \boldsymbol{E}^3$ *of the points of the middle surface* $S = \boldsymbol{\theta}(\overline{\omega})$; cf. Figure 2.1.3.

During the past decades, considerable progress has been made towards a rigorous justification of such a "replacement". The central idea is that of *asymptotic analysis:* It consists in showing that, if the data are of *ad hoc* orders of magnitude, *the three-dimensional displacement vector field* (once properly "scaled") *converges in an appropriate function space as* $\varepsilon \to 0$ *to a "limit" vector field that can be entirely computed by solving a two-dimensional problem.*

In this direction, see Ciarlet [2000a, Part A] for a thorough overview in the *linear* case and the key contributions of Le Dret & Raoult [1996] and Friesecke, James, Mora & Müller [2003] in the *nonlinear* case.

2.2 The Nonlinear Koiter Shell Equations

We now describe the **nonlinear Koiter shell equations**, so named after Koiter [1966], and since then a two-dimensional nonlinear model of choice in computational mechanics (its relation to an asymptotic analysis as $\varepsilon \to 0$ is briefly discussed at the end of this section).

To begin with, we need to define two important tensor fields. Given an arbitrary displacement field $\eta_i \boldsymbol{a}^i : \overline{\omega} \to \mathbb{R}^3$ of the surface S with smooth enough components $\eta_i : \overline{\omega} \to \mathbb{R}$, define the vector field $\boldsymbol{\eta} := (\eta_i) : \overline{\omega} \to \mathbb{R}^3$ and let

$$a_{\alpha\beta}(\boldsymbol{\eta}) := \boldsymbol{a}_\alpha(\boldsymbol{\eta}) \cdot \boldsymbol{a}_\beta(\boldsymbol{\eta}), \text{ where } \boldsymbol{a}_\alpha(\boldsymbol{\eta}) := \partial_\alpha(\boldsymbol{\theta} + \eta_i \boldsymbol{a}^i),$$

denote the covariant components of the first fundamental form of the *de-*

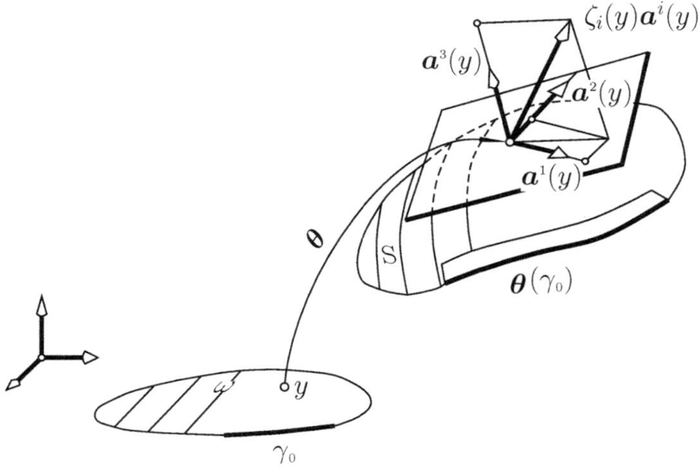

Figure 2.1.3. *An elastic shell modeled as a two-dimensional problem.* For $\varepsilon > 0$ "small enough" and data of *ad hoc* orders of magnitude, the three-dimensional shell problem (Figure 2.1.2) is "replaced" by a "two-dimensional shell problem". This means that the new unknowns are the three covariant components $\zeta_i : \overline{\omega} \to \mathbb{R}$ of the displacement field $\zeta_i \boldsymbol{a}^i : \overline{\omega} \to \mathbb{R}^3$ of the points of the middle surface $S = \boldsymbol{\theta}(\overline{\omega})$. In this process, the "three-dimensional" boundary conditions on Γ_0 need to be replaced by *ad hoc* "two-dimensional" boundary conditions on γ_0. For instance, the "boundary conditions of clamping" $\zeta_i = \partial_\gamma \zeta_3 = 0$ on γ_0 (used in Koiter's linear equations; cf. Section 2.3) mean that the points of, and the tangent spaces to, the deformed and undeformed middle surfaces coincide along the set $\boldsymbol{\theta}(\gamma_0)$.

formed surface $(\boldsymbol{\theta} + \eta_i \boldsymbol{a}^i)(\overline{\omega})$. Then the functions

$$G_{\alpha\beta}(\boldsymbol{\eta}) := \frac{1}{2}(a_{\alpha\beta}(\boldsymbol{\eta}) - a_{\alpha\beta})$$

denote the **covariant components of the change of metric tensor** associated with the displacement field $\eta_i \boldsymbol{a}^i$ of S.

If the two vectors $a_\alpha(\boldsymbol{\eta})$ are linearly independent at all points of ω, let

$$b_{\alpha\beta}(\boldsymbol{\eta}) := \frac{1}{\sqrt{a(\boldsymbol{\eta})}} \partial_{\alpha\beta}(\boldsymbol{\theta} + \eta_i \boldsymbol{a}^i) \cdot \{\boldsymbol{a}_1(\boldsymbol{\eta}) \wedge \boldsymbol{a}_2(\boldsymbol{\eta})\},$$

where

$$a(\boldsymbol{\eta}) := \det(a_{\alpha\beta}(\boldsymbol{\eta})),$$

denote the covariant components of the second fundamental form of the *deformed surface* $(\boldsymbol{\theta} + \eta_i \boldsymbol{a}^i)(\overline{\omega})$. Then the functions

$$R_{\alpha\beta}(\boldsymbol{\eta}) := b_{\alpha\beta}(\boldsymbol{\eta}) - b_{\alpha\beta}$$

denote the **covariant components of the change of curvature tensor field** associated with the displacement field $\eta_i \boldsymbol{a}^i$ of S. Note that $\sqrt{a(\boldsymbol{\eta})} = |\boldsymbol{a}_1(\boldsymbol{\eta}) \wedge \boldsymbol{a}_2(\boldsymbol{\eta})|$.

Note that *both* surfaces $\boldsymbol{\theta}(\overline{\omega})$ and $(\boldsymbol{\theta} + \eta_i \boldsymbol{a}^i)(\overline{\omega})$ are equipped with the *same* curvilinear coordinates y_1, y_2.

As a point of departure, consider an elastic shell modeled as a three-dimensional problem (Section 2.1). The nonlinear two-dimensional equations proposed by Koiter [1966] for modeling an elastic shell are then derived from those of nonlinear three-dimensional elasticity on the basis of two *a priori* assumptions: One assumption, of a *geometrical* nature, is the *Kirchhoff-Love assumption.* It asserts that any point situated on a normal to the middle surface remains on the normal to the deformed middle surface after the deformation has taken place and that, in addition, the distance between such a point and the middle surface remains constant. The other assumption, of a *mechanical* nature, asserts that the state of stress inside the shell is planar and parallel to the middle surface (this second assumption is itself based on delicate *a priori* estimates due to John [1965, 1971]).

Taking these *a priori* assumptions into account, W.T. Koiter then reached the conclusion that the unknown vector field $\boldsymbol{\zeta} = (\zeta_i)$ should be a *stationary point*, in particular a *minimizer*, over a set of smooth enough vector fields $\boldsymbol{\eta} = (\eta_i) : \overline{\omega} \to \mathbb{R}^3$ satisfying *ad hoc* boundary conditions on γ_0, of the functional j defined by (cf. Koiter [1966, eqs. (4.2), (8.1), and (8.3)]):

$$\begin{aligned} j(\boldsymbol{\eta}) = \frac{1}{2} \int_\omega \Big\{ \varepsilon a^{\alpha\beta\sigma\tau} G_{\sigma\tau}(\boldsymbol{\eta}) G_{\alpha\beta}(\boldsymbol{\eta}) + \frac{\varepsilon^3}{3} a^{\alpha\beta\sigma\tau} R_{\sigma\tau}(\boldsymbol{\eta}) R_{\alpha\beta}(\boldsymbol{\eta}) \Big\} \sqrt{a}\, \mathrm{d}y \\ - \int_\omega p^i \eta_i \sqrt{a}\, \mathrm{d}y, \end{aligned}$$

where the functions

$$a^{\alpha\beta\sigma\tau} := \frac{4\lambda\mu}{\lambda + 2\mu} a^{\alpha\beta} a^{\sigma\tau} + 2\mu (a^{\alpha\sigma} a^{\beta\tau} + a^{\alpha\tau} a^{\beta\sigma})$$

denote the **contravariant components of the shell elasticity tensor** and the functions $p^i \in L^2(\omega)$ are defined by

$$p^i := \int_{-\varepsilon}^{\varepsilon} f^i \, \mathrm{d}x_3 + h^i(\cdot, +\varepsilon) + h^i(\cdot, -\varepsilon).$$

The above functional j is called **Koiter's energy for a nonlinear elastic shell.**

The *stored energy function* w_K found in Koiter's energy j is thus defined by

$$w_K(\boldsymbol{\eta}) = \frac{\varepsilon}{2} a^{\alpha\beta\sigma\tau} G_{\sigma\tau}(\boldsymbol{\eta}) G_{\alpha\beta}(\boldsymbol{\eta}) + \frac{\varepsilon^3}{6} a^{\alpha\beta\sigma\tau} R_{\sigma\tau}(\boldsymbol{\eta}) R_{\alpha\beta}(\boldsymbol{\eta})$$

for *ad hoc* vector fields $\boldsymbol{\eta}$. This expression is the sum of the *"membrane" part*

$$w_M(\boldsymbol{\eta}) = \frac{\varepsilon}{2} a^{\alpha\beta\sigma\tau} G_{\sigma\tau}(\boldsymbol{\eta}) G_{\alpha\beta}(\boldsymbol{\eta})$$

and of the *"flexural" part*

$$w_F(\boldsymbol{\eta}) = \frac{\varepsilon^3}{6} a^{\alpha\beta\sigma\tau} R_{\sigma\tau}(\boldsymbol{\eta}) R_{\alpha\beta}(\boldsymbol{\eta}).$$

The long-standing question of how to rigorously identify and justify the nonlinear two-dimensional equations of elastic shells from three-dimensional elasticity was finally settled in two key contributions, one by Le Dret & Raoult [1996] and one by Friesecke, James, Mora & Müller [2003], who respectively justified the equations of a *nonlinearly elastic membrane shell* and those of a *nonlinearly elastic flexural shell* by means of Γ-*convergence theory* (a nonlinearly elastic shell is a membrane shell if there are no nonzero admissible displacements of its middle surface S that preserve the metric of S; it is a flexural shell otherwise).

The stored energy function $w_M^\sharp$ of a nonlinearly elastic membrane shell is an *ad hoc* quasiconvex envelope, which turns out to be only a function of the covariant components $a_{\alpha\beta}(\boldsymbol{\eta})$ of the first fundamental form of the unknown deformed middle surface (the notion of quasiconvexity, which plays a central role in the calculus of variations, is due to Morrey [1952]; an excellent introduction to this notion is provided in Dacorogna [1989, Chapter 5]). The function $w_M^\sharp$ reduces to the above "membrane" part w_M in Koiter's stored energy function w_K only for a restricted class of displacement fields $\eta_i \boldsymbol{a}^i$ of the middle surface. By contrast, the stored energy function of a nonlinearly elastic flexural shell is always equal to the above "flexural" part w_F in Koiter's stored energy function w_K.

Remark 2.2.1. Interestingly, a *formal* asymptotic analysis of the three-dimensional equations is only capable of delivering the above "restricted" expression $w_M(\boldsymbol{\eta})$, but otherwise fails to provide the general expression, i.e., valid for all types of displacements, found by Le Dret & Raoult [1996]. By contrast, the same formal approach yields the correct expression $w_F(\boldsymbol{\eta})$. For details, see Miara [1998], Lods & Miara [1998], and Ciarlet [2000a, Part B]. □

Another closely related set of nonlinear shell equations "of Koiter's type" has been proposed by Ciarlet [2000b]. In these equations, the denominator $\sqrt{a(\boldsymbol{\eta})}$ that appears in the functions $R_{\alpha\beta}(\boldsymbol{\eta}) = b_{\alpha\beta}(\boldsymbol{\eta}) - b_{\alpha\beta}$ is simply replaced by $\sqrt{a}$, thereby avoiding the possibility of a vanishing denominator in the expression $w_K(\boldsymbol{\eta})$. Then Ciarlet & Roquefort [2001] have shown that the leading term of a formal asymptotic expansion of a solution to this two-dimensional model, with the thickness 2ε as the "small" parameter, coincides with that found by a formal asymptotic analysis of the three-dimensional equations. This result thus raises hopes that a rigorous justification, again by means of Γ-convergence theory, of either types of nonlinear Koiter's models might be possible.

2.3 The Linear Koiter Shell Equations

Consider the Koiter energy j for a nonlinearly elastic shell, defined by (cf. Section 2.1)

$$j(\boldsymbol{\eta}) = \frac{1}{2}\int_\omega \Big\{\varepsilon a^{\alpha\beta\sigma\tau} G_{\sigma\tau}(\boldsymbol{\eta}) G_{\alpha\beta}(\boldsymbol{\eta}) + \frac{\varepsilon^3}{3} a^{\alpha\beta\sigma\tau} R_{\sigma\tau}(\boldsymbol{\eta}) R_{\alpha\beta}(\boldsymbol{\eta})\Big\}\sqrt{a}\,\mathrm{d}y \\ - \int_\omega p^i \eta_i \sqrt{a}\,\mathrm{d}y,$$

for smooth enough vector fields $\boldsymbol{\eta} = (\eta_i) : \overline{\omega} \to \mathbb{R}^3$. One of its virtues is that the integrands of the first two integrals are *quadratic* expressions in terms of the covariant components $G_{\alpha\beta}(\boldsymbol{\eta})$ and $R_{\alpha\beta}(\boldsymbol{\eta})$ of the change of metric, and change of curvature, tensors associated with a displacement field $\eta_i \boldsymbol{a}^i$ of the middle surface $S = \boldsymbol{\theta}(\overline{\omega})$ of the shell. In order to obtain the energy corresponding to the *linear* equations of Koiter [1970], which we are about to describe, it suffices, "*by definition*", to replace the covariant components

$$G_{\alpha\beta}(\boldsymbol{\eta}) = \frac{1}{2}(a_{\alpha\beta}(\boldsymbol{\eta}) - a_{\alpha\beta}) \text{ and } R_{\alpha\beta}(\boldsymbol{\eta}) = b_{\alpha\beta}(\boldsymbol{\eta}) - b_{\alpha\beta},$$

of these tensors by their *linear parts with respect to* $\boldsymbol{\eta} = (\eta_i)$, respectively denoted $\gamma_{\alpha\beta}(\boldsymbol{\eta})$ and $\rho_{\alpha\beta}(\boldsymbol{\eta})$ below. Accordingly, our first task consists in finding explicit expressions of such linearized tensors. To begin with, we compute the components $\gamma_{\alpha\beta}(\boldsymbol{\eta})$.

A word of caution. *The vector fields*

$$\boldsymbol{\eta} = (\eta_i) \text{ and } \widetilde{\boldsymbol{\eta}} := \eta_i \boldsymbol{a}^i,$$

which are both defined on $\overline{\omega}$, *must be carefully distinguished!* While the latter has an *intrinsic character*, the former has not; it only provides a means of recovering the field $\widetilde{\boldsymbol{\eta}}$ *via* its covariant components η_i. □

Theorem 2.3.1. *Let ω be a domain in $\mathbb{R}^2$ and let $\boldsymbol{\theta} \in \mathcal{C}^2(\overline{\omega}; \boldsymbol{E}^3)$ be an immersion. Given a displacement field $\widetilde{\boldsymbol{\eta}} := \eta_i \boldsymbol{a}^i$ of the surface $S = \boldsymbol{\theta}(\overline{\omega})$ with smooth enough covariant components $\eta_i : \overline{\omega} \to \mathbb{R}$, let the function $\gamma_{\alpha\beta}(\boldsymbol{\eta}) : \overline{\omega} \to \mathbb{R}$ be defined by*

$$\gamma_{\alpha\beta}(\boldsymbol{\eta}) := \frac{1}{2}\left[a_{\alpha\beta}(\boldsymbol{\eta}) - a_{\alpha\beta}\right]^{\text{lin}},$$

where $a_{\alpha\beta}$ and $a_{\alpha\beta}(\boldsymbol{\eta})$ are the covariant components of the first fundamental form of the surfaces $\boldsymbol{\theta}(\overline{\omega})$ and $(\boldsymbol{\theta} + \eta_i \boldsymbol{a}^i)(\overline{\omega})$, and $[\cdots]^{\text{lin}}$ denotes the linear part with respect to $\boldsymbol{\eta} = (\eta_i)$ in the expression $[\cdots]$. Then

$$\begin{aligned}\gamma_{\alpha\beta}(\boldsymbol{\eta}) &= \frac{1}{2}\left(\partial_\beta \widetilde{\boldsymbol{\eta}} \cdot \boldsymbol{a}_\alpha + \partial_\alpha \widetilde{\boldsymbol{\eta}} \cdot \boldsymbol{a}_\beta\right) = \gamma_{\beta\alpha}(\boldsymbol{\eta}) \\ &= \frac{1}{2}(\eta_{\alpha|\beta} + \eta_{\beta|\alpha}) - b_{\alpha\beta}\eta_3 \\ &= \frac{1}{2}(\partial_\beta \eta_\alpha + \partial_\alpha \eta_\beta) - \Gamma^\sigma_{\alpha\beta}\eta_\sigma - b_{\alpha\beta}\eta_3,\end{aligned}$$

where the covariant derivatives $\eta_{\alpha|\beta}$ are defined by $\eta_{\alpha|\beta} = \partial_\beta \eta_\alpha - \Gamma^\sigma_{\alpha\beta}\eta_\sigma$ (Theorem 1.6.1). *In particular then,*

$$\eta_\alpha \in H^1(\omega) \text{ and } \eta_3 \in L^2(\omega) \Rightarrow \gamma_{\alpha\beta}(\boldsymbol{\eta}) \in L^2(\omega).$$

Proof. The covariant components $a_{\alpha\beta}(\boldsymbol{\eta})$ of the metric tensor of the surface $(\boldsymbol{\theta} + \eta_i \boldsymbol{a}^i)(\overline{\omega})$ are by definition given by

$$a_{\alpha\beta}(\boldsymbol{\eta}) = \partial_\alpha(\boldsymbol{\theta} + \widetilde{\boldsymbol{\eta}}) \cdot \partial_\beta(\boldsymbol{\theta} + \widetilde{\boldsymbol{\eta}}).$$

The relations

$$\partial_\alpha(\boldsymbol{\theta} + \widetilde{\boldsymbol{\eta}}) = \boldsymbol{a}_\alpha + \partial_\alpha \widetilde{\boldsymbol{\eta}}$$

then show that

$$\begin{aligned}a_{\alpha\beta}(\boldsymbol{\eta}) &= (\boldsymbol{a}_\alpha + \partial_\alpha \widetilde{\boldsymbol{\eta}}) \cdot (\boldsymbol{a}_\beta + \partial_\beta \widetilde{\boldsymbol{\eta}}) \\ &= a_{\alpha\beta} + \partial_\beta \widetilde{\boldsymbol{\eta}} \cdot \boldsymbol{a}_\alpha + \partial_\alpha \widetilde{\boldsymbol{\eta}} \cdot \boldsymbol{a}_\beta + \partial_\alpha \widetilde{\boldsymbol{\eta}} \cdot \partial_\beta \widetilde{\boldsymbol{\eta}},\end{aligned}$$

hence that

$$\gamma_{\alpha\beta}(\boldsymbol{\eta}) = \frac{1}{2}[a_{\alpha\beta}(\boldsymbol{\eta}) - a_{\alpha\beta}]^{\text{lin}} = \frac{1}{2}(\partial_\beta \widetilde{\boldsymbol{\eta}} \cdot \boldsymbol{a}_\alpha + \partial_\alpha \widetilde{\boldsymbol{\eta}} \cdot \boldsymbol{a}_\beta).$$

The other expressions of $\gamma_{\alpha\beta}(\boldsymbol{\eta})$ immediately follow from the expression of $\partial_\alpha \widetilde{\boldsymbol{\eta}} = \partial_\alpha(\eta_i \boldsymbol{a}^i)$ given in Theorem 1.6.1. □

The functions $\gamma_{\alpha\beta}(\boldsymbol{\eta})$ are called the **covariant components of the linearized change of metric tensor** associated with a displacement $\eta_i \boldsymbol{a}^i$ of the surface S.

We next compute the components $\rho_{\alpha\beta}(\boldsymbol{\eta})$.

Theorem 2.3.2. *Let ω be a domain in $\mathbb{R}^2$ and let $\boldsymbol{\theta} \in \mathcal{C}^3(\overline{\omega}; \boldsymbol{E}^3)$ be an immersion. Given a displacement field $\widetilde{\boldsymbol{\eta}} := \eta_i \boldsymbol{a}^i$ of the surface $S = \boldsymbol{\theta}(\overline{\omega})$ with smooth enough and "small enough" covariant components $\eta_i : \overline{\omega} \to \mathbb{R}$, let the functions $\rho_{\alpha\beta}(\boldsymbol{\eta}) : \overline{\omega} \to \mathbb{R}$ be defined by*

$$\rho_{\alpha\beta}(\boldsymbol{\eta}) := [b_{\alpha\beta}(\boldsymbol{\eta}) - b_{\alpha\beta}]^{\text{lin}},$$

where $b_{\alpha\beta}$ and $b_{\alpha\beta}(\boldsymbol{\eta})$ are the covariant components of the second fundamental form of the surfaces $\boldsymbol{\theta}(\overline{\omega})$ and $(\boldsymbol{\theta} + \eta_i \boldsymbol{a}^i)(\overline{\omega})$, and $[\cdots]^{\text{lin}}$ denotes the linear part with respect to $\boldsymbol{\eta} = (\eta_i)$ in the expression $[\cdots]$. Then

$$\begin{aligned}
\rho_{\alpha\beta}(\boldsymbol{\eta}) &= (\partial_{\alpha\beta}\widetilde{\boldsymbol{\eta}} - \Gamma^{\sigma}_{\alpha\beta}\partial_\sigma \widetilde{\boldsymbol{\eta}}) \cdot \boldsymbol{a}_3 = \rho_{\beta\alpha}(\boldsymbol{\eta}) \\
&= \eta_{3|\alpha\beta} - b^{\sigma}_{\alpha} b_{\sigma\beta}\eta_3 + b^{\sigma}_{\alpha}\eta_{\sigma|\beta} + b^{\tau}_{\beta}\eta_{\tau|\alpha} + b^{\tau}_{\beta}|_{\alpha}\eta_\tau \\
&= \partial_{\alpha\beta}\eta_3 - \Gamma^{\sigma}_{\alpha\beta}\partial_\sigma\eta_3 - b^{\sigma}_{\alpha} b_{\sigma\beta}\eta_3 \\
&\quad + b^{\sigma}_{\alpha}(\partial_\beta\eta_\sigma - \Gamma^{\tau}_{\beta\sigma}\eta_\tau) + b^{\tau}_{\beta}(\partial_\alpha\eta_\tau - \Gamma^{\sigma}_{\alpha\tau}\eta_\sigma) \\
&\quad + (\partial_\alpha b^{\tau}_{\beta} + \Gamma^{\tau}_{\alpha\sigma} b^{\sigma}_{\beta} - \Gamma^{\sigma}_{\alpha\beta} b^{\tau}_{\sigma})\eta_\tau,
\end{aligned}$$

where the covariant derivatives $\eta_{\alpha|\beta}$ are defined by $\eta_{\alpha|\beta} = \partial_\beta\eta_\alpha - \Gamma^{\sigma}_{\alpha\beta}\eta_\sigma$ (Theorem 1.6.1) *and*

$$\eta_{3|\alpha\beta} := \partial_{\alpha\beta}\eta_3 - \Gamma^{\sigma}_{\alpha\beta}\partial_\sigma\eta_3 \text{ and } b^{\tau}_{\beta}|_{\alpha} := \partial_\alpha b^{\tau}_{\beta} + \Gamma^{\tau}_{\alpha\sigma} b^{\sigma}_{\beta} - \Gamma^{\sigma}_{\alpha\beta} b^{\tau}_{\sigma}.$$

In particular then,

$$\eta_\alpha \in H^1(\omega) \text{ and } \eta_3 \in H^2(\omega) \Rightarrow \rho_{\alpha\beta}(\boldsymbol{\eta}) \in L^2(\omega).$$

The functions $b^{\tau}_{\beta}|_{\alpha}$ satisfy the symmetry relations

$$b^{\tau}_{\beta}|_{\alpha} = b^{\tau}_{\alpha}|_{\beta}.$$

Proof. For convenience, the proof is divided into five parts. In parts (i) and (ii), we establish elementary relations satisfied by the vectors $\boldsymbol{a}_i$ and $\boldsymbol{a}^i$ of the covariant and contravariant bases along S.

(i) *The two vectors $\boldsymbol{a}_\alpha = \partial_\alpha\boldsymbol{\theta}$ satisfy $|\boldsymbol{a}_1 \wedge \boldsymbol{a}_2| = \sqrt{a}$, where $a = \det(a_{\alpha\beta})$.*

Let $\boldsymbol{A}$ denote the matrix of order three with $\boldsymbol{a}_1, \boldsymbol{a}_2, \boldsymbol{a}_3$ as its column vectors. Consequently,

$$\det \boldsymbol{A} = (\boldsymbol{a}_1 \wedge \boldsymbol{a}_2) \cdot \boldsymbol{a}_3 = (\boldsymbol{a}_1 \wedge \boldsymbol{a}_2) \cdot \frac{\boldsymbol{a}_1 \wedge \boldsymbol{a}_2}{|\boldsymbol{a}_1 \wedge \boldsymbol{a}_2|} = |\boldsymbol{a}_1 \wedge \boldsymbol{a}_2|.$$

Besides,

$$(\det \boldsymbol{A})^2 = \det(\boldsymbol{A}^T \boldsymbol{A}) = \det(a_{\alpha\beta}) = a,$$

since $\boldsymbol{a}_\alpha \cdot \boldsymbol{a}_\beta = a_{\alpha\beta}$ and $\boldsymbol{a}_\alpha \cdot \boldsymbol{a}_3 = \delta_{\alpha 3}$. Hence $|\boldsymbol{a}_1 \wedge \boldsymbol{a}_2| = \sqrt{a}$.

(ii) *The vectors* $\boldsymbol{a}_i$ *and* $\boldsymbol{a}^\alpha$ *are related by* $\boldsymbol{a}_1 \wedge \boldsymbol{a}_3 = -\sqrt{a}\boldsymbol{a}^2$ *and* $\boldsymbol{a}_3 \wedge \boldsymbol{a}_2 = -\sqrt{a}\boldsymbol{a}^1$.

To prove that two vectors $\boldsymbol{c}$ and $\boldsymbol{d}$ coincide, it suffices to prove that $\boldsymbol{c} \cdot \boldsymbol{a}_i = \boldsymbol{d} \cdot \boldsymbol{a}_i$ for $i \in \{1, 2, 3\}$. In the present case,

$$(\boldsymbol{a}_1 \wedge \boldsymbol{a}_3) \cdot \boldsymbol{a}_1 = 0 \text{ and } (\boldsymbol{a}_1 \wedge \boldsymbol{a}_3) \cdot \boldsymbol{a}_3 = 0,$$
$$(\boldsymbol{a}_1 \wedge \boldsymbol{a}_3) \cdot \boldsymbol{a}_2 = -(\boldsymbol{a}_1 \wedge \boldsymbol{a}_2) \cdot \boldsymbol{a}_3 = -\sqrt{a},$$

since $\sqrt{a}\boldsymbol{a}_3 = \boldsymbol{a}_1 \wedge \boldsymbol{a}_2$ by (i), on the one hand; on the other hand,

$$-\sqrt{a}\boldsymbol{a}^2 \cdot \boldsymbol{a}_1 = -\sqrt{a}\boldsymbol{a}^2 \cdot \boldsymbol{a}_3 = 0 \text{ and } -\sqrt{a}\boldsymbol{a}^2 \cdot \boldsymbol{a}_2 = -\sqrt{a},$$

since $\boldsymbol{a}^i \cdot \boldsymbol{a}_j = \delta^i_j$. Hence $\boldsymbol{a}_1 \wedge \boldsymbol{a}_3 = -\sqrt{a}\boldsymbol{a}^2$. The other relation is similarly established.

(iii) *The covariant components* $b_{\alpha\beta}(\boldsymbol{\eta})$ *satisfy*

$$b_{\alpha\beta}(\boldsymbol{\eta}) = b_{\alpha\beta} + (\partial_{\alpha\beta}\tilde{\boldsymbol{\eta}} - \Gamma^\sigma_{\alpha\beta}\partial_\sigma\tilde{\boldsymbol{\eta}}) \cdot \boldsymbol{a}_3 + \text{ h.o.t.},$$

where "h.o.t." *stands for "higher-order terms"*, i.e., terms of order higher than linear with respect to $\boldsymbol{\eta} = (\eta_i)$. Consequently,

$$\rho_{\alpha\beta}(\boldsymbol{\eta}) := [b_{\alpha\beta}(\boldsymbol{\eta}) - b_{\alpha\beta}]^{\text{lin}} = \left(\partial_{\alpha\beta}\tilde{\boldsymbol{\eta}} - \Gamma^\sigma_{\alpha\beta}\partial_\sigma\tilde{\boldsymbol{\eta}}\right) \cdot \boldsymbol{a}_3 = \rho_{\beta\alpha}(\boldsymbol{\eta}).$$

Since the vectors $\boldsymbol{a}_\alpha = \partial_\alpha\boldsymbol{\theta}$ are linearly independent in $\overline{\omega}$ and the fields $\boldsymbol{\eta} = (\eta_i)$ are smooth enough by assumption, *the vectors* $\partial_\alpha(\boldsymbol{\theta} + \eta_i\boldsymbol{a}^i)$ *are also linearly independent in* $\overline{\omega}$ *provided the fields* $\boldsymbol{\eta}$ *are "small enough"*, e.g., with respect to the norm of the space $\mathcal{C}^1(\overline{\omega}; \mathbb{R}^3)$. The following computations are therefore licit as they apply to a linearization around $\boldsymbol{\eta} = \boldsymbol{0}$.

Let

$$\boldsymbol{a}_\alpha(\boldsymbol{\eta}) := \partial_\alpha(\boldsymbol{\theta} + \tilde{\boldsymbol{\eta}}) = \boldsymbol{a}_\alpha + \partial_\alpha\tilde{\boldsymbol{\eta}} \text{ and } \boldsymbol{a}_3(\boldsymbol{\eta}) := \frac{\boldsymbol{a}_1(\boldsymbol{\eta}) \wedge \boldsymbol{a}_2(\boldsymbol{\eta})}{\sqrt{a(\boldsymbol{\eta})}},$$

where

$$a(\boldsymbol{\eta}) := \det(a_{\alpha\beta}(\boldsymbol{\eta})) \text{ and } a_{\alpha\beta}(\boldsymbol{\eta}) := \boldsymbol{a}_\alpha(\boldsymbol{\eta}) \cdot \boldsymbol{a}_\beta(\eta).$$

Then

$$\begin{aligned}
b_{\alpha\beta}(\boldsymbol{\eta}) &= \partial_\alpha \boldsymbol{a}_\beta(\boldsymbol{\eta}) \cdot \boldsymbol{a}_3(\boldsymbol{\eta}) \\
&= \frac{1}{\sqrt{a(\boldsymbol{\eta})}}(\partial_\alpha \boldsymbol{a}_\beta + \partial_{\alpha\beta}\widetilde{\boldsymbol{\eta}}) \cdot (\boldsymbol{a}_1 \wedge \boldsymbol{a}_2 + \boldsymbol{a}_1 \wedge \partial_2 \widetilde{\boldsymbol{\eta}} + \partial_1 \widetilde{\boldsymbol{\eta}} \wedge \boldsymbol{a}_2 + \text{ h.o.t.}) \\
&= \frac{1}{\sqrt{a(\boldsymbol{\eta})}} \left\{ \sqrt{a}(b_{\alpha\beta} + \partial_{\alpha\beta}\widetilde{\boldsymbol{\eta}} \cdot \boldsymbol{a}_3) \right\} \\
&\quad + \frac{1}{\sqrt{a(\boldsymbol{\eta})}} \left\{ (\Gamma^\sigma_{\alpha\beta} \boldsymbol{a}_\sigma + b_{\alpha\beta}\boldsymbol{a}_3) \cdot (\boldsymbol{a}_1 \wedge \partial_2 \widetilde{\boldsymbol{\eta}} + \partial_1 \widetilde{\boldsymbol{\eta}} \wedge \boldsymbol{a}_2) + \text{ h.o.t. } \right\},
\end{aligned}$$

since $b_{\alpha\beta} = \partial_\alpha \boldsymbol{a}_\beta \cdot \boldsymbol{a}_3$ and $\partial_\alpha \boldsymbol{a}_\beta = \Gamma^\sigma_{\alpha\beta}\boldsymbol{a}_\sigma + b_{\alpha\beta}\boldsymbol{a}_3$ by the formula of Gauß (Theorem 1.6.1 (a)). Next,

$$\begin{aligned}
&(\Gamma^\sigma_{\alpha\beta}\boldsymbol{a}_\sigma + b_{\alpha\beta}\boldsymbol{a}_3) \cdot (\boldsymbol{a}_1 \wedge \partial_2 \widetilde{\boldsymbol{\eta}}) \\
&\quad = \Gamma^2_{\alpha\beta}\boldsymbol{a}_2 \cdot (\boldsymbol{a}_1 \wedge \partial_2 \widetilde{\boldsymbol{\eta}}) - b_{\alpha\beta}\partial_2 \widetilde{\boldsymbol{\eta}} \cdot (\boldsymbol{a}_1 \wedge \boldsymbol{a}_3) \\
&\quad = \sqrt{a}\left(-\Gamma^2_{\alpha\beta}\partial_2 \widetilde{\boldsymbol{\eta}} \cdot \boldsymbol{a}_3 + b_{\alpha\beta}\partial_2 \widetilde{\boldsymbol{\eta}} \cdot \boldsymbol{a}^2\right),
\end{aligned}$$

since, by (ii), $\boldsymbol{a}_2 \cdot (\boldsymbol{a}_1 \wedge \partial_2 \widetilde{\boldsymbol{\eta}}) = -\partial_2 \widetilde{\boldsymbol{\eta}} \cdot (\boldsymbol{a}_1 \wedge \boldsymbol{a}_2) = -\sqrt{a}\partial_2 \widetilde{\boldsymbol{\eta}} \cdot \boldsymbol{a}_3$ and $\boldsymbol{a}_1 \wedge \boldsymbol{a}_3 = -\sqrt{a}\boldsymbol{a}^2$; likewise,

$$(\Gamma^\sigma_{\alpha\beta}\boldsymbol{a}_\sigma + b_{\alpha\beta}\boldsymbol{a}_3) \cdot (\partial_1 \widetilde{\boldsymbol{\eta}} \wedge \boldsymbol{a}_2) = \sqrt{a}(-\Gamma^1_{\alpha\beta}\partial_1 \widetilde{\boldsymbol{\eta}} \cdot \boldsymbol{a}_3 + b_{\alpha\beta}\partial_1 \boldsymbol{\eta} \cdot \boldsymbol{a}^1).$$

Consequently,

$$b_{\alpha\beta}(\boldsymbol{\eta}) = \sqrt{\frac{a}{a(\boldsymbol{\eta})}} \left\{ b_{\alpha\beta}(1 + \partial_\sigma \tilde{\boldsymbol{\eta}} \cdot \boldsymbol{a}^\sigma) + (\partial_{\alpha\beta}\widetilde{\boldsymbol{\eta}} - \Gamma^\sigma_{\alpha\beta}\partial_\sigma \widetilde{\boldsymbol{\eta}}) \cdot \boldsymbol{a}_3 + \text{ h.o.t.} \right\}.$$

There remains to find the linear term with respect to $\boldsymbol{\eta} = (\eta_i)$ in the expansion $\dfrac{1}{\sqrt{a(\boldsymbol{\eta})}} = \dfrac{1}{\sqrt{a}}(1 + \cdots)$. To this end, we note that

$$\det(\boldsymbol{A} + \boldsymbol{H}) = (\det \boldsymbol{A})(1 + \operatorname{tr} \boldsymbol{A}^{-1}\boldsymbol{H} + o(\boldsymbol{H})),$$

with $\boldsymbol{A} := (a_{\alpha\beta})$ and $\boldsymbol{A} + \boldsymbol{H} := (a_{\alpha\beta}(\boldsymbol{\eta}))$. Hence

$$\boldsymbol{H} = (\partial_\beta \widetilde{\boldsymbol{\eta}} \cdot \boldsymbol{a}_\alpha + \partial_\alpha \widetilde{\boldsymbol{\eta}} \cdot \boldsymbol{a}_\beta + \text{ h.o.t.}),$$

since $[a_{\alpha\beta}(\boldsymbol{\eta}) - a_{\alpha\beta}]^{\text{lin}} = \partial_\beta \widetilde{\boldsymbol{\eta}} \cdot \boldsymbol{a}_\alpha + \partial_\alpha \widetilde{\boldsymbol{\eta}} \cdot \boldsymbol{a}_\beta$ (Theorem 2.3.1). Therefore,

$$a(\boldsymbol{\eta}) = \det(a_{\alpha\beta}(\boldsymbol{\eta})) = \det(a_{\alpha\beta})(1 + 2\partial_\alpha \widetilde{\boldsymbol{\eta}} \cdot \boldsymbol{a}^\alpha + \text{ h.o.t.}),$$

since $\boldsymbol{A}^{-1} = (a^{\alpha\beta})$; consequently,

$$\frac{1}{\sqrt{a(\boldsymbol{\eta})}} = \frac{1}{\sqrt{a}}(1 - \partial_\alpha \widetilde{\boldsymbol{\eta}} \cdot \boldsymbol{a}^\alpha + \text{ h.o.t.}).$$

Noting that there are no linear terms with respect to $\boldsymbol{\eta} = (\eta_i)$ in the product $(1 - \partial_\alpha \widetilde{\boldsymbol{\eta}} \cdot \boldsymbol{a}^\alpha)(1 + \partial_\sigma \widetilde{\boldsymbol{\eta}} \cdot \boldsymbol{a}^\sigma)$, we find the announced expansion, viz.,

$$b_{\alpha\beta}(\boldsymbol{\eta}) = b_{\alpha\beta} + (\partial_{\alpha\beta}\widetilde{\boldsymbol{\eta}} - \Gamma^\sigma_{\alpha\beta}\partial_\sigma \widetilde{\boldsymbol{\eta}}) \cdot \boldsymbol{a}_3 + \text{ h.o.t.}$$

(iv) *The components* $\rho_{\alpha\beta}(\boldsymbol{\eta})$ *can be also written as*

$$\rho_{\alpha\beta}(\boldsymbol{\eta}) = \eta_{3|\alpha\beta} - b^\sigma_\alpha b_{\sigma\beta}\eta_3 + b^\sigma_\alpha \eta_{\sigma|\beta} + b^\tau_\beta \eta_{\tau|\alpha} + b^\tau_\beta|_\alpha \eta_\tau,$$

where the functions $\eta_{3|\alpha\beta}$ *and* $b^\tau_\beta|_\alpha$ *are defined as in the statement of the theorem.*

By Theorem 1.6.1 (b),

$$\partial_\sigma \widetilde{\boldsymbol{\eta}} = (\partial_\sigma \eta_\beta - \Gamma^\tau_{\sigma\beta}\eta_\tau - b_{\sigma\beta}\eta_3)\boldsymbol{a}^\beta + (\partial_\sigma \eta_3 + b^\tau_\sigma \eta_\tau)\boldsymbol{a}^3.$$

Hence

$$-\Gamma^\sigma_{\alpha\beta}\partial_\sigma \widetilde{\boldsymbol{\eta}} \cdot \boldsymbol{a}_3 = -\Gamma^\sigma_{\alpha\beta}(\partial_\sigma \eta_3 + b^\tau_\sigma \eta_\tau),$$

since $\boldsymbol{a}^i \cdot \boldsymbol{a}_3 = \delta^i_3$. Again by Theorem 1.6.1 (b),

$$\begin{aligned}
\partial_{\alpha\beta}\widetilde{\boldsymbol{\eta}} \cdot \boldsymbol{a}_3 &= \partial_\alpha \left\{(\partial_\beta \eta_\sigma - \Gamma^\tau_{\beta\sigma}\eta_\tau - b_{\beta\sigma}\eta_3)\boldsymbol{a}^\sigma + (\partial_\beta \eta_3 + b^\tau_\beta \eta_\tau)\boldsymbol{a}^3\right\} \cdot \boldsymbol{a}_3 \\
&= (\partial_\beta \eta_\sigma - \Gamma^\tau_{\beta\sigma}\eta_\tau - b_{\beta\sigma}\eta_3)\partial_\alpha \boldsymbol{a}^\sigma \cdot \boldsymbol{a}_3 \\
&\quad + (\partial_{\alpha\beta}\eta_3 + (\partial_\alpha b^\tau_\beta)\eta_\tau + b^\tau_\beta \partial_\alpha \eta_\tau)\boldsymbol{a}^3 \cdot \boldsymbol{a}_3 + (\partial_\beta \eta_3 + b^\tau_\beta \eta_\tau)\partial_\alpha \boldsymbol{a}^3 \cdot \boldsymbol{a}_3 \\
&= b^\sigma_\alpha(\partial_\beta \eta_\sigma - \Gamma^\tau_{\beta\sigma}\eta_\tau) - b^\sigma_\alpha b_{\sigma\beta}\eta_3 + \partial_{\alpha\beta}\eta_3 + (\partial_\alpha b^\tau_\beta)\eta_\tau + b^\tau_\beta \partial_\alpha \eta_\tau,
\end{aligned}$$

since

$$\begin{aligned}
\partial_\alpha \boldsymbol{a}^\sigma \cdot \boldsymbol{a}_3 &= (-\Gamma^\sigma_{\alpha\tau}\boldsymbol{a}^\tau + b^\sigma_\alpha \boldsymbol{a}^3) \cdot \boldsymbol{a}_3 = b^\sigma_\alpha, \\
\partial_\alpha \boldsymbol{a}^3 \cdot \boldsymbol{a}_3 &= -b_{\alpha\sigma}\boldsymbol{a}^\sigma \cdot \boldsymbol{a}_3 = 0,
\end{aligned}$$

by the formulas of Gauß and Weingarten (Theorem 1.6.1 (a)). We thus obtain

$$\begin{aligned}
\rho_{\alpha\beta}(\boldsymbol{\eta}) &= (\partial_{\alpha\beta}\widetilde{\boldsymbol{\eta}} - \Gamma^\sigma_{\alpha\beta}\partial_\sigma \widetilde{\boldsymbol{\eta}}) \cdot \boldsymbol{a}_3 \\
&= b^\sigma_\alpha(\partial_\beta \eta_\sigma - \Gamma^\tau_{\beta\sigma}\eta_\tau) - b^\sigma_\alpha b_{\sigma\beta}\eta_3 + \partial_{\alpha\beta}\eta_3 + (\partial_\alpha b^\tau_\beta)\eta_\tau + b^\tau_\beta \partial_\alpha \eta_\tau \\
&\quad - \Gamma^\sigma_{\alpha\beta}(\partial_\sigma \eta_3 + b^\tau_\sigma \eta_\tau).
\end{aligned}$$

While this relation seemingly involves only the covariant derivatives $\eta_{3|\alpha\beta}$ and $\eta_{\sigma|\beta}$, it may be easily rewritten so as to involve in addition the functions $\eta_{\tau|\alpha}$ and $b^\tau_{\beta|\alpha}$. The stratagem simply consists in using the relation $\Gamma^\tau_{\alpha\sigma} b^\sigma_\beta \eta_\tau - \Gamma^\sigma_{\alpha\tau} b^\tau_\beta \eta_\sigma = 0$! This gives

$$\begin{aligned}
\rho_{\alpha\beta}(\boldsymbol{\eta}) &= (\partial_{\alpha\beta}\eta_3 - \Gamma^\sigma_{\alpha\beta}\partial_\tau\eta_3) - b^\sigma_\alpha b_{\sigma\beta}\eta_3 \\
&\quad + b^\sigma_\alpha(\partial_\beta\eta_\sigma - \Gamma^\tau_{\beta\sigma}\eta_\tau) + b^\tau_\beta(\partial_\alpha\eta_\tau - \Gamma^\sigma_{\alpha\tau}\eta_\tau) \\
&\quad + (\partial_\alpha b^\tau_\beta + \Gamma^\tau_{\alpha\sigma} b^\sigma_\beta b^\sigma_\beta - \Gamma^\sigma_{\alpha\beta} b^\tau_\sigma)\eta_\tau.
\end{aligned}$$

(v) *The functions $b^\tau_\beta|_\alpha$ are symmetric with respect to the indices α and β.*

Again, because of the formulas of Gauß and Weingarten, we can write

$$\begin{aligned}
\mathbf{0} &= \partial_{\alpha\beta}\boldsymbol{a}^\tau - \partial_{\beta\alpha}\boldsymbol{a}^\tau = \partial_\alpha\left(-\Gamma^\tau_{\beta\sigma}\boldsymbol{a}^\sigma + b^\tau_\beta\boldsymbol{a}^3\right) - \partial_\beta\left(-\Gamma^\tau_{\alpha\sigma}\boldsymbol{a}^\sigma + b^\tau_\alpha\boldsymbol{a}^3\right) \\
&= -(\partial_\alpha\Gamma^\tau_{\beta\sigma})\boldsymbol{a}^\sigma + \Gamma^\tau_{\beta\sigma}\Gamma^\sigma_{\alpha\nu}\boldsymbol{a}^\nu - \Gamma^\tau_{\beta\sigma}b^\sigma_\alpha\boldsymbol{a}^3 + (\partial_\alpha b^\tau_\beta)\boldsymbol{a}^3 - b^\tau_\beta b_{\alpha\sigma}\boldsymbol{a}^\sigma \\
&\quad + (\partial_\beta\Gamma^\tau_{\alpha\sigma})\boldsymbol{a}^\sigma - \Gamma^\tau_{\alpha\sigma}\Gamma^\sigma_{\beta\mu}\boldsymbol{a}^\mu + \Gamma^\tau_{\alpha\sigma}b^\sigma_\beta\boldsymbol{a}^3 - (\partial_\beta b^\tau_\alpha)\boldsymbol{a}^3 + b^\tau_\alpha b_{\beta\sigma}\boldsymbol{a}^\sigma.
\end{aligned}$$

Consequently,

$$0 = (\partial_{\alpha\beta}\boldsymbol{a}^\tau - \partial_{\beta\alpha}\boldsymbol{a}^\tau)\cdot\boldsymbol{a}^3 = \partial_\alpha b^\tau_\beta - \partial_\beta b^\tau_\alpha + \Gamma^\tau_{\alpha\sigma}b^\sigma_\beta - \Gamma^\tau_{\beta\sigma}b^\sigma_\alpha,$$

on the one hand. On the other hand, we immediately infer from the definition of the functions $b^\tau_\beta|_\alpha$ that we also have

$$b^\tau_\beta|_\alpha - b^\tau_\alpha|_\beta = \partial_\alpha b^\tau_\beta - \partial_\beta b^\tau_\alpha + \Gamma^\tau_{\alpha\sigma}b^\sigma_\beta - \Gamma^\tau_{\beta\sigma}b^\sigma_\alpha,$$

and thus the proof is complete. □

The functions $\rho_{\alpha\beta}(\boldsymbol{\eta})$ are called the **covariant components of the linearized change of curvature tensor** associated with a displacement $\eta_i\boldsymbol{a}^i$ of the surface S. The functions

$$\eta_{3|\alpha\beta} = \partial_{\alpha\beta}\eta_3 - \Gamma^\sigma_{\alpha\beta}\partial_\sigma\eta_3 \text{ and } b^\tau_\beta|_\alpha = \partial_\alpha b^\tau_\beta + \Gamma^\tau_{\alpha\sigma}b^\sigma_\beta - \Gamma^\sigma_{\alpha\beta}b^\tau_\sigma$$

respectively represent a **second-order covariant derivative of the vector field $\eta_i\boldsymbol{a}^i$** and a **first-order covariant derivative of the second fundamental form** of S, defined here by means of its *mixed* components b^τ_β.

Remarks 2.3.3. (1) Covariant derivatives $b_{\alpha\beta|\sigma}$ can be likewise defined. More specifically, each function

$$b_{\alpha\beta|\sigma} := \partial_\sigma b_{\alpha\beta} - \Gamma^\tau_{\alpha\sigma}b_{\tau\beta} - \Gamma^\tau_{\beta\sigma}b_{\alpha\tau}$$

represents a *first-order covariant derivatives of the second fundamental form*, defined here by means of its *covariant* components $b_{\alpha\beta}$. By a proof analogous to that given in Theorem 2.3.2 for establishing the symmetry relations $b^\tau_\beta|_\alpha = b^\tau_\alpha|_\beta$, one can then show that these covariant derivatives likewise satisfy the symmetry relations

$$b_{\alpha\beta|\sigma} = b_{\alpha\sigma|\beta},$$

which are themselves equivalent to the relations

$$\partial_\sigma b_{\alpha\beta} - \partial_\beta b_{\alpha\sigma} + \Gamma^\tau_{\alpha\beta} b_{\tau\sigma} - \Gamma^\tau_{\alpha\sigma} b_{\tau\beta} = 0,$$

i.e., the familiar Codazzi-Mainardi equations (Theorem 1.7.1)!

(2) The functions $c_{\alpha\beta} := b^\sigma_\alpha b_{\sigma\beta} = c_{\beta\alpha}$ appearing in the expression of $\rho_{\alpha\beta}(\boldsymbol{\eta})$ are the covariant components of the *third fundamental form* of S. For details, see, e.g., Stoker [1969, p. 98] or Klingenberg [1973, p. 48].

(3) The functions $b_{\alpha\beta}(\boldsymbol{\eta})$ are not always well defined (in order that they be, the vectors $a_\alpha(\boldsymbol{\eta})$ must be linearly independent in $\overline{\omega}$), but the functions $\rho_{\alpha\beta}(\boldsymbol{\eta})$ are *always* well defined.

(4) The symmetry $\rho_{\alpha\beta}(\boldsymbol{\eta}) = \rho_{\beta\alpha}(\boldsymbol{\eta})$ follows immediately by inspection of the expression $\rho_{\alpha\beta}(\boldsymbol{\eta}) = (\partial_{\alpha\beta}\widetilde{\boldsymbol{\eta}} - \Gamma^\sigma_{\alpha\beta}\partial_\sigma\widetilde{\boldsymbol{\eta}}) \cdot \boldsymbol{a}_3$ found there. By contrast, deriving the same symmetry from the other expression of $\rho_{\alpha\beta}(\boldsymbol{\eta})$ requires proving first that the covariant derivatives $b^\sigma_\beta|_\alpha$ are themselves symmetric with respect to the indices α and β (cf. part (v) of the proof of Theorem 2.3.1). □

While the expression of the components $\rho_{\alpha\beta}(\boldsymbol{\eta})$ in terms of the covariant components η_i of the displacement field is fairly complicated but well known (see, e.g., Koiter [1970]), that in terms of $\widetilde{\boldsymbol{\eta}} = \eta_i \boldsymbol{a}^i$ is *remarkably simple* but seems to have been mostly ignored, although it already appeared in Bamberger [1981]. Together with the expression of the components $\gamma_{\alpha\beta}(\boldsymbol{\eta})$ in terms of $\widetilde{\boldsymbol{\eta}}$ (Theorem 2.3.1), this simpler expression was efficiently put to use by Blouza & Le Dret [1999], who showed that their principal merit is to afford the definition of the components $\gamma_{\alpha\beta}(\boldsymbol{\eta})$ and $\rho_{\alpha\beta}(\boldsymbol{\eta})$ under substantially weaker regularity assumptions on the mapping $\boldsymbol{\theta}$.

More specifically, we were led to assume that $\boldsymbol{\theta} \in \mathcal{C}^3(\overline{\omega}; \boldsymbol{E}^3)$ in Theorem 2.3.1 in order to insure that $\rho_{\alpha\beta}(\boldsymbol{\eta}) \in L^2(\omega)$ if $\boldsymbol{\eta} \in H^1(\omega) \times H^1(\omega) \times H^2(\omega)$. The culprits responsible for this regularity are the functions $b^\tau_\beta|_\alpha$ appearing in the functions $\rho_{\alpha\beta}(\boldsymbol{\eta})$. Otherwise Blouza & Le Dret [1999] have shown how this regularity assumption on $\boldsymbol{\theta}$ can be weakened if only the expressions of $\gamma_{\alpha\beta}(\boldsymbol{\eta})$ and $\rho_{\alpha\beta}(\boldsymbol{\eta})$ in terms of the field $\widetilde{\boldsymbol{\eta}}$ are considered.

We are now in a position to describe the **linear Koiter shell equations**. Let γ_0 be a measurable subset of $\gamma = \partial\omega$ that satisfies *length* $\gamma_0 > 0$, let ∂_ν

denote the outer normal derivative operator along $\partial\omega$ (since ω is a domain, a unit outer normal vector (ν_α) exists dγ-almost everywhere along γ, and thus $\partial_\nu = \nu_\alpha\partial_\alpha$), and let the space $\boldsymbol{V}(\omega)$ be defined by

$$\boldsymbol{V}(\omega) := \{\boldsymbol{\eta} = (\eta_i) \in H^1(\omega) \times H^1(\omega) \times H^2(\omega); \eta_i = \partial_\nu\eta_3 = 0 \text{ on } \gamma_0\}.$$

Then the unknown vector field $\boldsymbol{\zeta} = (\zeta_i) : \overline{\omega} \to \mathbb{R}^3$, where the functions ζ_i are the covariant components of the displacement field $\zeta_i\boldsymbol{a}^i$ of the middle surface $S = \boldsymbol{\theta}(\overline{\omega})$ of the shell, should be a *stationary point over the space* $\boldsymbol{V}(\omega)$ of the **functional** j defined by

$$j(\boldsymbol{\eta}) = \frac{1}{2}\int_\omega \Big\{\varepsilon a^{\alpha\beta\sigma\tau}\gamma_{\sigma\tau}(\boldsymbol{\eta})\gamma_{\alpha\beta}(\boldsymbol{\eta}) + \frac{\varepsilon^3}{3}a^{\alpha\beta\sigma\tau}\rho_{\sigma\tau}(\boldsymbol{\eta})\rho_{\alpha\beta}(\boldsymbol{\eta})\Big\}\sqrt{a}\,\mathrm{d}y$$
$$- \int_\omega p^i\eta_i\sqrt{a}\,\mathrm{d}y$$

for all $\boldsymbol{\eta} \in \boldsymbol{V}(\omega)$. This functional j is called **Koiter's energy for a linearly elastic shell.**

Equivalently, the vector field $\boldsymbol{\zeta} \in \boldsymbol{V}(\omega)$ should satisfy the *variational equations*

$$\int_\omega \Big\{\varepsilon a^{\alpha\beta\sigma\tau}\gamma_{\sigma\tau}(\boldsymbol{\zeta})\gamma_{\alpha\beta}(\boldsymbol{\eta}) + \frac{\varepsilon^3}{3}a^{\alpha\beta\sigma\tau}\rho_{\sigma\tau}(\boldsymbol{\zeta})\rho_{\alpha\beta}(\boldsymbol{\eta})\Big\}\sqrt{a}\,\mathrm{d}y$$
$$= \int_\omega p^i\eta_i\sqrt{a}\,\mathrm{d}y \text{ for all } \boldsymbol{\eta} = (\eta_i) \in \boldsymbol{V}(\omega).$$

We recall that the functions

$$a^{\alpha\beta\sigma\tau} := \frac{4\lambda\mu}{\lambda + 2\mu}a^{\alpha\beta}a^{\sigma\tau} + 2\mu(a^{\alpha\sigma}a^{\beta\tau} + a^{\alpha\tau}a^{\beta\sigma})$$

denote the *contravariant components of the shell elasticity tensor* (λ and μ are the Lamé constants of the elastic material constituting the shell), $\gamma_{\alpha\beta}(\boldsymbol{\eta})$ and $\rho_{\alpha\beta}(\boldsymbol{\eta})$ denote the covariant components of the *linearized change of metric*, and *change of curvature, tensors* associated with a displacement field $\eta_i\boldsymbol{a}^i$ of S, and the given functions $p^i \in L^2(\omega)$ account for the applied forces. Finally, the boundary conditions $\eta_i = \partial_\nu\eta_3 = 0$ on γ_0 express that the shell is *clamped* along the portion $\boldsymbol{\theta}(\gamma_0)$ of its middle surface (see Figure 2.1.3).

The choice of the function spaces $H^1(\omega)$ and $H^2(\omega)$ for the tangential components η_α and normal components η_3 of the displacement fields $\eta_i\boldsymbol{a}^i$ is guided by the natural requirement that the functions $\gamma_{\alpha\beta}(\boldsymbol{\eta})$ and $\rho_{\alpha\beta}(\boldsymbol{\eta})$ be both in $L^2(\omega)$, so that the energy is in turn well defined for $\boldsymbol{\eta} = (\eta_i) \in \boldsymbol{V}(\omega)$. Otherwise these choices can be weakened to accommodate shells whose middle surfaces have little regularity (see Blouza & Le Dret [1999]).

Remark 2.3.4. Koiter's linear equations can be fully justified by means of an asymptotic analysis of the "three-dimensional" equations of linearized elasticity as $\varepsilon \to 0$; see Ciarlet [2000a, Chapter 7] and the references therein. For a brief introduction to other linear shell equations, see Ciarlet [2005, Section 4.5]; for an in-depth study and an extensive list of references, see Ciarlet [2000a, Part A].

2.4 A Fundamental Lemma of J.L. Lions

We first review some essential definitions and notations, together with a fundamental *lemma of J.L. Lions* (Theorem 2.4.1). This lemma will play a key rôle in the proofs of the Korn inequalities on a surface (Section 2.5).

A *domain* Ω in $\mathbb{R}^d$ is an *open, bounded, connected* subset of $\mathbb{R}^d$ with a *Lipschitz-continuous boundary* Γ, the set Ω being *locally on one side of* Γ. As Γ is Lipschitz-continuous, a *measure* $\mathrm{d}\Gamma$ can be defined along Γ and a *unit outer normal vector* $\boldsymbol{\nu} = (\nu_i)_{i=1}^d$ ("unit" means that its Euclidean norm is one) exists $\mathrm{d}\Gamma$-almost everywhere along Γ.

Let Ω be a domain in $\mathbb{R}^d$. For each integer $m \geq 1, H^m(\Omega)$ and $H_0^m(\Omega)$ denote the usual *Sobolev spaces*. In particular,

$$H^1(\Omega) := \{v \in L^2(\Omega);\ \partial_i v \in L^2(\Omega),\ 1 \leq i \leq d\},$$
$$H^2(\Omega) := \{v \in H^1(\Omega);\ \partial_{ij} v \in L^2(\Omega),\ 1 \leq i, j \leq d\},$$

where $\partial_i v$ and $\partial_{ij} v$ denote partial derivatives in the sense of distributions, and

$$H_0^1(\Omega) := \{v \in H^1(\Omega);\ v = 0 \text{ on } \Gamma\},$$

where the relation $v = 0$ on Γ is to be understood in the sense of trace. The norm in $L^2(\Omega)$ is noted $\|\cdot\|_{0,\Omega}$ and the norm in $H^m(\Omega)$, $m \geq 1$, is noted $\|\cdot\|_{m,\Omega}$. In particular then,

$$\|v\|_{0,\Omega} := \left\{ \int_\Omega |v|^2 \,\mathrm{d}x \right\}^{1/2} \text{ if } v \in L^2(\Omega),$$

$$\|v\|_{1,\Omega} := \left\{ \|v\|_{0,\Omega}^2 + \sum_{i=1}^d \|\partial_i v\|_{0,\Omega}^2 \right\}^{1/2} \text{ if } v \in H^1(\Omega).$$

We also consider the Sobolev space

$$H^{-1}(\Omega) := \text{dual space of } H_0^1(\Omega).$$

Another possible definition of the space $H_0^1(\Omega)$ being

$$H_0^1(\Omega) = \text{closure of } \mathcal{D}(\Omega) \text{ with respect to } \|\cdot\|_{1,\Omega},$$

where $\mathcal{D}(\Omega)$ denotes the space of infinitely differentiable real-valued functions defined over Ω whose support is a compact subset of Ω, it is clear that

$$v \in L^2(\Omega) \implies v \in H^{-1}(\Omega) \text{ and } \partial_i v \in H^{-1}(\Omega),\ 1 \le i \le n,$$

since (the duality between the spaces $\mathcal{D}(\Omega)$ and $\mathcal{D}'(\Omega)$ is denoted by $\langle \cdot, \cdot \rangle$):

$$|\langle v, \varphi \rangle| = \left| \int_\Omega v\varphi \, \mathrm{d}x \right| \le \|v\|_{0,\Omega} \|\varphi\|_{1,\Omega},$$
$$|\langle \partial_i v, \varphi \rangle| = |-\langle v, \partial_i \varphi \rangle| = \left| -\int_\Omega v\partial_i \varphi \, \mathrm{d}x \right| \le \|v\|_{0,\Omega} \|\varphi\|_{1,\Omega}$$

for all $\varphi \in \mathcal{D}(\Omega)$. It is remarkable (but also remarkably difficult to prove!) that the converse implication holds:

Theorem 2.4.1. *Let Ω be a domain in $\mathbb{R}^d$ and let v be a distribution on Ω. Then*

$$\{v \in H^{-1}(\Omega) \text{ and } \partial_i v \in H^{-1}(\Omega),\ 1 \le i \le d\} \implies v \in L^2(\Omega).$$

□

This implication was first proved by J.L. Lions, as stated in Magenes & Stampacchia [1958, p. 320, Note $(^{27})$]; for this reason, it will be henceforth referred to as the **lemma of J.L. Lions**. Its first published proof for domains with smooth boundaries appeared in Duvaut & Lions [1972, p. 111]; another proof was also given by Tartar [1978]. Various extensions to "genuine" domains, i.e., with Lipschitz-continuous boundaries, are given in Bolley & Camus [1976], Geymonat & Suquet [1986], and Borchers & Sohr [1990]; Amrouche & Girault [1994, Proposition 2.10] even proved that the more general implication

$$\{v \in \mathcal{D}'(\Omega) \text{ and } \partial_i v \in H^m(\Omega),\ 1 \le i \le n\} \implies v \in H^{m+1}(\Omega)$$

holds for *arbitrary* integers $m \in \mathbb{Z}$.

Our objective in the next sections is to study the existence and uniqueness of the solution to the above variational equations. To this end, we shall see (Theorem 2.6.1) that, under the assumptions $3\lambda + 2\mu > 0$ and $\mu > 0$, there exists a constant $c_e > 0$ such that

$$\sum_{\alpha,\beta} |t_{\alpha\beta}|^2 \le c_e a^{\alpha\beta\sigma\tau}(y) t_{\sigma\tau} t_{\alpha\beta}$$

for all $y \in \overline{\omega}$ and all symmetric matrices $(t_{\alpha\beta})$. When *length* $\gamma_0 > 0$, the existence and uniqueness of a solution to this variational problem by means of the *Lax-Milgram lemma* will then be a consequence of the existence of a constant c such that

$$\Big\{\sum_\alpha \|\eta_\alpha\|^2_{1,\omega} + \|\eta_3\|^2_{2,\omega}\Big\}^{1/2}$$
$$\leq c\Big\{\sum_{\alpha,\beta} \|\gamma_{\alpha\beta}(\boldsymbol{\eta})\|^2_{0,\omega} + \sum_{\alpha,\beta} \|\rho_{\alpha\beta}(\boldsymbol{\eta})\|^2_{0,\omega}\Big\}^{1/2} \text{ for all } \boldsymbol{\eta} \in \boldsymbol{V}(\omega).$$

The objective of the next section precisely consists in showing that such a fundamental *"Korn's inequality on a surface" indeed holds* (Theorem 2.5.3).

2.5 Korn's Inequalities on a Surface

To begin with, we establish a **Korn's inequality on a surface, "without boundary conditions"**, as a consequence of the *lemma of J.L. Lions* (cf. Theorem 2.4.1). We follow here Ciarlet & Miara [1992] (see also Bernadou, Ciarlet & Miara [1994]).

Theorem 2.5.1. *Let ω be a domain in $\mathbb{R}^2$ and let $\boldsymbol{\theta} \in \mathcal{C}^3(\overline{\omega}; \boldsymbol{E}^3)$ be an injective immersion. Given $\boldsymbol{\eta} = (\eta_i) \in H^1(\omega) \times H^1(\omega) \times H^2(\omega)$, let*

$$\gamma_{\alpha\beta}(\boldsymbol{\eta}) := \Big\{\frac{1}{2}(\partial_\beta \widetilde{\boldsymbol{\eta}} \cdot \boldsymbol{a}_\alpha + \partial_\alpha \widetilde{\boldsymbol{\eta}} \cdot \boldsymbol{a}_\beta)\Big\} \in L^2(\omega),$$
$$\rho_{\alpha\beta}(\boldsymbol{\eta}) := \Big\{(\partial_{\alpha\beta} \widetilde{\boldsymbol{\eta}} - \Gamma^\sigma_{\alpha\beta} \partial_\sigma \widetilde{\boldsymbol{\eta}}) \cdot \boldsymbol{a}_3\Big\} \in L^2(\omega)$$

denote the covariant components of the linearized change of metric, and linearized change of curvature, tensors associated with the displacement field $\widetilde{\boldsymbol{\eta}} := \eta_i \boldsymbol{a}^i$ of the surface $S = \boldsymbol{\theta}(\overline{\omega})$. Then there exists a constant $c_0 = c_0(\omega, \boldsymbol{\theta})$ such that

$$\Big\{\sum_\alpha \|\eta_\alpha\|^2_{1,\omega} + \|\eta_3\|^2_{2,\omega}\Big\}^{1/2}$$
$$\leq c_0\Big\{\sum_\alpha \|\eta_\alpha\|^2_{0,\omega} + \|\eta_3\|^2_{1,\omega} + \sum_{\alpha,\beta} \|\gamma_{\alpha\beta}(\boldsymbol{\eta})\|^2_{0,\omega} + \sum_{\alpha,\beta} \|\rho_{\alpha\beta}(\boldsymbol{\eta})\|^2_{0,\omega}\Big\}^{1/2}$$

for all $\boldsymbol{\eta} = (\eta_i) \in H^1(\omega) \times H^1(\omega) \times H^2(\omega)$.

Proof. The "fully explicit" expressions of the functions $\gamma_{\alpha\beta}(\boldsymbol{\eta})$ and $\rho_{\alpha\beta}(\boldsymbol{\eta})$, as found in Theorems 2.3.1 and 2.3.2, are used in this proof, simply because they are more convenient for its purposes.

(i) *Define the space*

$$\boldsymbol{W}(\omega) := \{\boldsymbol{\eta} = (\eta_i) \in L^2(\omega) \times L^2(\omega) \times H^1(\omega);\\ \gamma_{\alpha\beta}(\boldsymbol{\eta}) \in L^2(\omega),\ \rho_{\alpha\beta}(\boldsymbol{\eta}) \in L^2(\omega)\}.$$

Then, equipped with the norm $\|\cdot\|_{\boldsymbol{W}(\omega)}$ *defined by*

$$\|\boldsymbol{\eta}\|_{\boldsymbol{W}(\omega)} := \Big\{\sum_\alpha \|\eta_\alpha\|^2_{0,\omega} + \|\eta_3\|^2_{1,\omega} + \sum_{\alpha,\beta} \|\gamma_{\alpha\beta}(\boldsymbol{\eta})\|^2_{0,\omega} + \sum_{\alpha,\beta} \|\rho_{\alpha\beta}(\boldsymbol{\eta})\|^2_{0,\omega}\Big\}^{1/2},$$

the space $\boldsymbol{W}(\omega)$ *is a Hilbert space.*

The relations "$\gamma_{\alpha\beta}(\boldsymbol{\eta}) \in L^2(\omega)$" and "$\rho_{\alpha\beta}(\boldsymbol{\eta}) \in L^2(\omega)$" appearing in the definition of the space $\boldsymbol{W}(\omega)$ are to be understood in the sense of distributions. They mean that a vector field $\boldsymbol{\eta} \in L^2(\omega) \times L^2(\omega) \times H^1(\omega)$ belongs to $\boldsymbol{W}(\omega)$ if there exist functions in $L^2(\omega)$, *denoted* $\gamma_{\alpha\beta}(\boldsymbol{\eta})$ and $\rho_{\alpha\beta}(\boldsymbol{\eta})$, such that for all $\varphi \in \mathcal{D}(\omega)$,

$$\begin{aligned}
\int_\omega \gamma_{\alpha\beta}(\boldsymbol{\eta})\varphi \,\mathrm{d}y &= -\int_\omega \Big\{\frac{1}{2}(\eta_\beta\partial_\alpha\varphi + \eta_\alpha\partial_\beta\varphi) + \Gamma^\sigma_{\alpha\beta}\eta_\sigma\varphi + b_{\alpha\beta}\eta_3\varphi\Big\}\,\mathrm{d}y,\\
\int_\omega \rho_{\alpha\beta}(\boldsymbol{\eta})\varphi \,\mathrm{d}y &= -\int_\omega \{\partial_\alpha\eta_3\partial_\beta\varphi + \Gamma^\sigma_{\alpha\beta}\partial_\sigma\eta_3\varphi + b^\sigma_\alpha b_{\sigma\beta}\eta_3\varphi\\
&\qquad + \eta_\sigma\partial_\beta(b^\sigma_\alpha\varphi) + b^\sigma_\alpha\Gamma^\tau_{\beta\sigma}\eta_\tau\varphi\\
&\qquad + \eta_\tau\partial_\alpha(b^\tau_\beta\varphi) + b^\tau_\beta\Gamma^\sigma_{\alpha\tau}\eta_\sigma\varphi\\
&\qquad - \big(\partial_\alpha b^\tau_\beta + \Gamma^\tau_{\alpha\sigma}b^\sigma_\beta - \Gamma^\sigma_{\alpha\beta}b^\tau_\sigma\big)\eta_\tau\varphi\}\,\mathrm{d}y.
\end{aligned}$$

Let there be given a Cauchy sequence $(\boldsymbol{\eta}^k)_{k=1}^\infty$ with elements $\boldsymbol{\eta}^k = (\eta_i^k) \in \boldsymbol{W}(\omega)$. The definition of the norm $\|\cdot\|_{\boldsymbol{W}(\omega)}$ shows that there exist $\eta_\alpha \in L^2(\omega)$, $\eta_3 \in H^1(\omega)$, $\gamma_{\alpha\beta} \in L^2(\omega)$, and $\rho_{\alpha\beta} \in L^2(\omega)$ such that

$$\begin{gathered}
\eta^k_\alpha \to \eta_\alpha \text{ in } L^2(\omega), \quad \eta^k_3 \to \eta_3 \text{ in } H^1(\omega),\\
\gamma_{\alpha\beta}(\boldsymbol{\eta}^k) \to \gamma_{\alpha\beta} \text{ in } L^2(\omega), \quad \rho_{\alpha\beta}(\boldsymbol{\eta}^k) \to \rho_{\alpha\beta} \text{ in } L^2(\omega)
\end{gathered}$$

as $k \to \infty$. Given a function $\varphi \in \mathcal{D}(\omega)$, letting $k \to \infty$ in the relations $\int_\omega \gamma_{\alpha\beta}(\boldsymbol{\eta}^k)\varphi\,\mathrm{d}\omega = \dots$ and $\int_\omega \rho_{\alpha\beta}(\boldsymbol{\eta}^k)\varphi\,\mathrm{d}\omega = \dots$ then shows that $\gamma_{\alpha\beta} = \gamma_{\alpha\beta}(\boldsymbol{\eta})$ and $\rho_{\alpha\beta} = \rho_{\alpha\beta}(\boldsymbol{\eta})$.

(ii) *The spaces* $\boldsymbol{W}(\omega)$ *and* $H^1(\omega) \times H^1(\omega) \times H^2(\omega)$ *coincide.*

Clearly, $H^1(\omega) \times H^1(\omega) \times H^2(\omega) \subset \boldsymbol{W}(\omega)$. To prove the other inclusion, let $\boldsymbol{\eta} = (\eta_i) \in \boldsymbol{W}(\omega)$. The relations

$$s_{\alpha\beta}(\boldsymbol{\eta}) := \frac{1}{2}(\partial_\alpha\eta_\beta + \partial_\beta\eta_\alpha) = \gamma_{\alpha\beta}(\boldsymbol{\eta}) + \Gamma^\sigma_{\alpha\beta}\eta_\sigma + b_{\alpha\beta}\eta_3$$

then imply that $s_{\alpha\beta}(\boldsymbol{\eta}) \in L^2(\omega)$ since the functions $\Gamma^\sigma_{\alpha\beta}$ and $b_{\alpha\beta}$ are continuous on $\overline{\omega}$. Therefore,

$$\partial_\sigma \eta_\alpha \in H^{-1}(\omega),$$
$$\partial_\beta(\partial_\sigma \eta_\alpha) = \{\partial_\beta s_{\alpha\sigma}(\boldsymbol{\eta}) + \partial_\sigma s_{\alpha\beta}(\boldsymbol{\eta}) - \partial_\alpha s_{\beta\sigma}(\boldsymbol{\eta})\} \in H^{-1}(\omega),$$

since $\chi \in L^2(\omega)$ implies $\partial_\sigma \chi \in H^{-1}(\omega)$. Hence $\partial_\sigma \eta_\alpha \in L^2(\omega)$ by the *lemma of J.L. Lions* (Theorem 2.4.1) and thus $\eta_\alpha \in H^1(\omega)$.

The definition of the functions $\rho_{\alpha\beta}(\boldsymbol{\eta})$, the continuity over $\overline{\omega}$ of the functions $\Gamma^\sigma_{\alpha\beta}, b_{\sigma\beta}, b^\sigma_\alpha$, and $\partial_\alpha b^\tau_\beta$, and the relations $\rho_{\alpha\beta}(\boldsymbol{\eta}) \in L^2(\omega)$ then imply that $\partial_{\alpha\beta}\eta_3 \in L^2(\omega)$, hence that $\eta_3 \in H^2(\omega)$.

(iii) *Korn's inequality without boundary conditions.*

The identity mapping ι from the space $H^1(\omega) \times H^1(\omega) \times H^2(\omega)$ equipped with its product norm $\boldsymbol{\eta} = (\eta_i) \to \{\sum_\alpha \|\eta_\alpha\|^2_{1,\omega} + \|\eta_3\|^2_{2,\omega}\}^{1/2}$ into the space $\boldsymbol{W}(\omega)$ equipped with $\|\cdot\|_{\boldsymbol{W}(\omega)}$ is injective, continuous, and surjective by (ii). Since both spaces are complete (cf. (i)), the *closed graph theorem* then shows that the inverse mapping ι^{-1} is also continuous or equivalently, that the inequality of Korn's type without boundary conditions holds. □

In order to establish a Korn's inequality "with boundary conditions", we have to identify classes of boundary conditions to be imposed on the fields $\boldsymbol{\eta} = (\eta_i) \in H^1(\omega) \times H^1(\omega) \times H^2(\omega)$ in order that we can "get rid" of the norms $\|\eta_\alpha\|_{0,\omega}$ and $\|\eta_3\|_{1,\omega}$ in the right-hand side of the above inequality, i.e., situations where the *semi-norm*

$$\boldsymbol{\eta} = (\eta_i) \to \Big\{\sum_{\alpha,\beta} \|\gamma_{\alpha\beta}(\boldsymbol{\eta})\|^2_{0,\omega} + \sum_{\alpha,\beta} \|\rho_{\alpha\beta}(\boldsymbol{\eta})\|^2_{0,\omega}\Big\}^{1/2}$$

becomes a *norm*, which should be in addition *equivalent* to the product norm.

To this end, we need an *infinitesimal rigid displacement lemma "on a surface"* (the adjective "infinitesimal" reminds that only the *linearized* parts $\gamma_{\alpha\beta}(\boldsymbol{\eta})$ and $\rho_{\alpha\beta}(\boldsymbol{\eta})$ of the "full" change of metric and curvature tensors $\frac{1}{2}(a_{\alpha\beta}(\boldsymbol{\eta}) - a_{\alpha\beta})$ and $(b_{\alpha\beta}(\boldsymbol{\eta}) - b_{\alpha\beta})$ are required to vanish in ω), which is due to Bernadou & Ciarlet [1976, Theorems 5.1-1 and 5.2-1]; see also Bernadou, Ciarlet & Miara [1994, Lemmas 2.5 and 2.6] and Blouza & Le Dret [1999, Theorem 6].

Part (a) in the next theorem is an **infinitesimal rigid displacement lemma on a surface, "without boundary conditions"**, while part (b) is an **infinitesimal rigid displacement lemma on a surface, "with boundary conditions"**.

Any proof of this theorem is, at least to some extent, delicate. The one given here is not the shortest, but it is a natural one: It relies on the classical, and much easier to prove, "three-dimensional infinitesimal rigid displacement lemma in Cartesian coordinates" used in part (iii) of the next proof (a "direct" proof, such as the one originally found by Bernadou & Ciarlet [1976], is surprisingly "technical").

In the next proof, the functions $\widehat{e}_{ij}(\widehat{\boldsymbol{v}})$ and $e_{ij}(\boldsymbol{v})$ are the *Cartesian* and *covariant components* of the *"three-dimensional" linearized strain tensor* respectively associated with a displacement field $\widehat{v}_i\widehat{\boldsymbol{e}}^i : \{\widehat{\Omega}\}^- \to \mathbb{R}^3$ and a displacement field $v_i\boldsymbol{g}^i : \overline{\Omega} \to \mathbb{R}^3$; the functions Γ^p_{ij} are the *Christoffel symbols* (of the second kind) associated with the mapping $\boldsymbol{\Theta} : \overline{\Omega} \to \{\widehat{\Omega}\}^-$; finally, the functions $v_{i\|j}$ are the *covariant derivatives* of the vector field $v_i\boldsymbol{g}^i : \overline{\Omega} \to \mathbb{R}^3$. □

Theorem 2.5.2. *Let there be given a domain ω in $\mathbb{R}^2$ and an injective immersion $\boldsymbol{\theta} \in \mathcal{C}^3(\overline{\omega}; \boldsymbol{E}^3)$.*

(a) *Let $\boldsymbol{\eta} = (\eta_i) \in H^1(\omega) \times H^1(\omega) \times H^2(\omega)$ be such that*

$$\gamma_{\alpha\beta}(\boldsymbol{\eta}) = \rho_{\alpha\beta}(\boldsymbol{\eta}) = 0 \text{ in } \omega.$$

Then there exist two vectors $\boldsymbol{a}, \boldsymbol{b} \in \mathbb{R}^3$ such that

$$\eta_i(y)\boldsymbol{a}^i(y) = \boldsymbol{a} + \boldsymbol{b} \wedge \boldsymbol{\theta}(y) \text{ for all } y \in \overline{\omega}.$$

(b) *Let γ_0 be a $d\gamma$-measurable subset of $\gamma = \partial\omega$ that satisfies* length $\gamma_0 >$ 0 *and let a vector field $\boldsymbol{\eta} = (\eta_i) \in H^1(\omega) \times H^1(\omega) \times H^2(\omega)$ be such that*

$$\gamma_{\alpha\beta}(\boldsymbol{\eta}) = \rho_{\alpha\beta}(\boldsymbol{\eta}) = 0 \text{ in } \omega \text{ and } \eta_i = \partial_\nu\eta_3 = 0 \text{ on } \gamma_0.$$

Then $\boldsymbol{\eta} = \boldsymbol{0}$ in ω.

Proof. The proof is divided into five parts, numbered (i) to (v).

(i) *In parts* (i) *to* (iii), *Ω denotes a domain in $\mathbb{R}^3$ and $\boldsymbol{\Theta} : \overline{\Omega} \to \boldsymbol{E}^3$ denotes a $\mathcal{C}^2$-diffeomorphism from $\overline{\Omega}$ onto its image $\boldsymbol{\Theta}(\overline{\Omega})$.* Consequently, the three vectors $\boldsymbol{g}_i(x) := \partial_i\boldsymbol{\Theta}(x)$, where $\partial_i = \partial/\partial x_i$, are linearly independent at all points $x = (x_i) \in \overline{\Omega}$ and the three vectors $\boldsymbol{g}^i(x)$ defined by $\boldsymbol{g}^i(x)\boldsymbol{g}_j(x) = \delta^i_j$ are likewise linearly independent at all points $x \in \overline{\Omega}$. Note also that $\boldsymbol{g}_i \in \mathcal{C}^1(\overline{\Omega}; \mathbb{R}^3)$ and $\boldsymbol{g}^i \in \mathcal{C}^1(\overline{\Omega}; \mathbb{R}^3)$.

Let $\widehat{\boldsymbol{e}}_i$ denote the basis of $\mathbb{E}^3$, let $\widehat{x}_i$ denote the Cartesian coordinates of a point $\widehat{x} \in \mathbb{E}^3$, let $\widehat{\partial}_i := \partial/\partial\widehat{x}_i$, and let $\widehat{\Omega} := \boldsymbol{\Theta}(\Omega)$. *With any vector field*

$v_i \boldsymbol{g}^i : \overline{\Omega} \to \mathbb{R}^3$ with $\boldsymbol{v} := (v_i) \in H^1(\Omega; \mathbb{R}^3)$, we then associate a vector field $\widehat{v}_i \widehat{\boldsymbol{e}}^i : \{\widehat{\Omega}\}^- \to \mathbb{R}^3$ by letting

$$\widehat{v}_i(\widehat{x})\widehat{\boldsymbol{e}}^i = v_i(x)\boldsymbol{g}^i(x) \text{ for all } \widehat{x} = \boldsymbol{\Theta}(x),\ x \in \overline{\Omega}.$$

It is then clear that $\widehat{\boldsymbol{v}} := (\widehat{v}_i) \in H^1(\widehat{\Omega}; \mathbb{R}^3)$.

We now show that, *for all $x \in \Omega$,*

$$\widehat{\partial}_j \widehat{v}_i(\widehat{x}) = \big(v_{k\|\ell}[\boldsymbol{g}^k]_i[\boldsymbol{g}^\ell]_j\big)(x),\ \widehat{x} = \boldsymbol{\Theta}(x),$$

where

$$v_{i\|j} := \partial_j v_i - \Gamma^p_{ij} v_p \text{ and } \Gamma^p_{ij} := \boldsymbol{g}^p \cdot \partial_i \boldsymbol{g}_j,$$

and

$$[\boldsymbol{g}^k(x)]_i := \boldsymbol{g}^k(x) \cdot \widehat{\boldsymbol{e}}_i$$

denotes the i-th component of $\boldsymbol{g}^k(x)$ over the basis $\{\widehat{\boldsymbol{e}}_1, \widehat{\boldsymbol{e}}_2, \widehat{\boldsymbol{e}}_3\}$.

In what follows, the simultaneous appearance of $\widehat{x}$ and x in an equality means that they are related by $\widehat{x} = \boldsymbol{\Theta}(x)$ and that the equality in question holds for all $x \in \Omega$.

Let $\boldsymbol{\Theta}(x) = \Theta^k(x)\widehat{\boldsymbol{e}}_k$ and $\widehat{\boldsymbol{\Theta}}(\widehat{x}) = \widehat{\Theta}^i(\widehat{x})\boldsymbol{e}_i$, where $\widehat{\boldsymbol{\Theta}} : \widehat{\Omega} \to \boldsymbol{E}^3$ denotes the inverse mapping of $\boldsymbol{\Theta} : \Omega \to \boldsymbol{E}^3$. Since $\widehat{\boldsymbol{\Theta}}(\boldsymbol{\Theta}(x)) = x$ for all $x \in \Omega$, the chain rule shows that the matrices $\boldsymbol{\nabla\Theta}(x) := (\partial_j \Theta^k(x))$ (the row index is k) and $\widehat{\boldsymbol{\nabla}}\widehat{\boldsymbol{\Theta}}(\widehat{x}) := (\widehat{\partial}_k \widehat{\Theta}^i(\widehat{x}))$ (the row index is i) satisfy

$$\widehat{\boldsymbol{\nabla}}\widehat{\boldsymbol{\Theta}}(\widehat{x})\boldsymbol{\nabla\Theta}(x) = \boldsymbol{I},$$

or equivalently,

$$\widehat{\partial}_k \widehat{\Theta}^i(\widehat{x})\partial_j \Theta^k(x) = \begin{pmatrix} \widehat{\partial}_1 \widehat{\Theta}^i(\widehat{x}) & \widehat{\partial}_2 \widehat{\Theta}^i(\widehat{x}) & \widehat{\partial}_3 \widehat{\Theta}^i(\widehat{x}) \end{pmatrix} \begin{pmatrix} \partial_j \Theta^1(x) \\ \partial_j \Theta^2(x) \\ \partial_j \Theta^3(x) \end{pmatrix} = \delta^i_j.$$

The components of the above column vector being precisely those of the vector $\boldsymbol{g}_j(x)$, the components of the above row vector must be those of the vector $\boldsymbol{g}^i(x)$ since $\boldsymbol{g}^i(x)$ is uniquely defined for each exponent i by the three relations $\boldsymbol{g}^i(x) \cdot \boldsymbol{g}_j(x) = \delta^i_j$, $j = 1, 2, 3$. Hence the k-th component of $\boldsymbol{g}^i(x)$ over the basis $\{\widehat{\boldsymbol{e}}_1, \widehat{\boldsymbol{e}}_2, \widehat{\boldsymbol{e}}_3\}$ can be also expressed in terms of the inverse mapping $\widehat{\boldsymbol{\Theta}}$, as:

$$[\boldsymbol{g}^i(x)]_k = \widehat{\partial}_k \widehat{\Theta}^i(\widehat{x}).$$

We next compute the derivatives $\partial_\ell \boldsymbol{g}^q(x)$ (it is easily seen that the fields $\boldsymbol{g}^q = g^{qr}\boldsymbol{g}_r$ are of class $\mathcal{C}^1$ on Ω since $\boldsymbol{\Theta}$ is assumed to be of class $\mathcal{C}^2$). These derivatives will be needed below for expressing the derivatives $\widehat{\partial}_j \widehat{u}_i(\widehat{x})$ as functions of x (recall that $\widehat{u}_i(\widehat{x}) = u_k(x)[\boldsymbol{g}^k(x)]_i$). Recalling that the vectors $\boldsymbol{g}^k(x)$ form a basis, we may write *a priori*

$$\partial_\ell \boldsymbol{g}^q(x) = -\Gamma^q_{\ell k}(x)\boldsymbol{g}^k(x),$$

thereby unambiguously defining functions $\Gamma^q_{\ell k} : \Omega \to \mathbb{R}$. To find their expressions in terms of the mappings $\boldsymbol{\Theta}$ and $\widehat{\boldsymbol{\Theta}}$, we observe that

$$\Gamma^q_{\ell k}(x) = \Gamma^q_{\ell m}(x)\delta^m_k = \Gamma^q_{\ell m}(x)\boldsymbol{g}^m(x) \cdot \boldsymbol{g}_k(x) = -\partial_\ell \boldsymbol{g}^q(x) \cdot \boldsymbol{g}_k(x).$$

Hence, noting that $\partial_\ell(\boldsymbol{g}^q(x) \cdot \boldsymbol{g}_k(x)) = 0$ and $[\boldsymbol{g}^q(x)]_p = \widehat{\partial}_p \widehat{\Theta}^q(\widehat{x})$, we obtain

$$\Gamma^q_{\ell k}(x) = \boldsymbol{g}^q(x) \cdot \partial_\ell \boldsymbol{g}_k(x) = \widehat{\partial}_p \widehat{\Theta}^q(\widehat{x})\partial_{\ell k}\Theta^p(x) = \Gamma^q_{k\ell}(x).$$

Since $\boldsymbol{\Theta} \in \mathcal{C}^2(\Omega; \boldsymbol{E}^3)$ and $\widehat{\boldsymbol{\Theta}} \in \mathcal{C}^1(\widehat{\Omega}; \mathbb{R}^3)$ by assumption, the last relations show that $\Gamma^q_{\ell k} \in \mathcal{C}^0(\Omega)$.

We are now in a position to compute the partial derivatives $\widehat{\partial}_j \widehat{v}_i(\widehat{x})$ as functions of x, by means of the relation $\widehat{v}_i(\widehat{x}) = v_k(x)[\boldsymbol{g}^k(x)]_i$. To this end, we first note that a differentiable function $w : \Omega \to \mathbb{R}$ satisfies

$$\widehat{\partial}_j w(\widehat{\boldsymbol{\Theta}}(\widehat{x})) = \partial_\ell w(x)\widehat{\partial}_j \widehat{\Theta}^\ell(\widehat{x}) = \partial_\ell w(x)[\boldsymbol{g}^\ell(x)]_j,$$

by the chain rule. In particular then,

$$\begin{aligned}
\widehat{\partial}_j \widehat{v}_i(\widehat{x}) &= \widehat{\partial}_j v_k(\widehat{\boldsymbol{\Theta}}(\widehat{x}))[\boldsymbol{g}^k(x)]_i + v_q(x)\widehat{\partial}_j [\boldsymbol{g}^q(\widehat{\boldsymbol{\Theta}}(\widehat{x}))]_i \\
&= \partial_\ell v_k(x)[\boldsymbol{g}^\ell(x)]_j [\boldsymbol{g}^k(x)]_i + v_q(x)\big(\partial_\ell [\boldsymbol{g}^q(x)]_i\big)[\boldsymbol{g}^\ell(x)]_j \\
&= \left(\partial_\ell v_k(x) - \Gamma^q_{\ell k}(x)v_q(x)\right)[\boldsymbol{g}^k(x)]_i [\boldsymbol{g}^\ell(x)]_j,
\end{aligned}$$

since $\partial_\ell \boldsymbol{g}^q(x) = -\Gamma^q_{\ell k}(x)\boldsymbol{g}^k(x)$. We have therefore shown that

$$\widehat{\partial}_j \widehat{v}_i(\widehat{x}) = v_{k\|\ell}(x)[\boldsymbol{g}^k(x)]_i [\boldsymbol{g}^\ell(x)]_j,$$

where

$$v_{k\|\ell}(x) := \partial_\ell v_k(x) - \Gamma^q_{\ell k}(x)v_q(x),$$

and $[\boldsymbol{g}^k(x)]_i$ and $\Gamma^q_{\ell k}(x)$ are defined as above.

(ii) With any vector field $v_i g^i : \overline{\Omega} \to \mathbb{R}^3$ with $\boldsymbol{v} := (v_i) \in H^1(\Omega; \mathbb{R}^3)$, we next associate the functions $e_{ij}(\boldsymbol{v}) \in L^2(\Omega)$ defined by

$$e_{ij}(\boldsymbol{v}) := \frac{1}{2}(v_{i\|j} + v_{j\|i}) = \frac{1}{2}(\partial_j v_i + \partial_i v_j) - \Gamma^p_{ij} v_p,$$

and, with any vector field $\widehat{v}_i\widehat{e}^i : \overline{\widehat{\Omega}} \to \mathbb{R}^3$ with $\widehat{\boldsymbol{v}} := (\widehat{v}_i) \in H^1(\overline{\widehat{\Omega}}; \mathbb{R}^3)$ we associate the functions $\widehat{e}_{ij}(\widehat{\boldsymbol{v}}) \in L^2(\widehat{\Omega})$ defined by

$$\widehat{e}_{ij}(\widehat{\boldsymbol{v}}) := \frac{1}{2}(\widehat{\partial}_j\widehat{v}_i + \widehat{\partial}_i\widehat{v}_j).$$

If the fields $\boldsymbol{v} \in H^1(\Omega; \mathbb{R}^3)$ and $\widehat{\boldsymbol{v}} \in H^1(\widehat{\Omega}; \mathbb{R}^3)$ are related as in (i), *it then immediately follows from* (i) *that*

$$\widehat{e}_{ij}(\widehat{\boldsymbol{v}})(\widehat{x}) = \left(e_{k\ell}(\boldsymbol{v})[\boldsymbol{g}^k]_i[\boldsymbol{g}^\ell]_j\right)(x) \text{ for all } \widehat{x} = \boldsymbol{\Theta}(x),\ x \in \Omega.$$

(iii) *Let a vector field $\boldsymbol{v} = (v_i) \in \boldsymbol{H}^1(\Omega)$ be such that*

$$e_{ij}(\boldsymbol{v}) = 0 \text{ in } \Omega.$$

Then there exist two vectors $\boldsymbol{a}, \boldsymbol{b} \in \mathbb{R}^3$ such that the associated vector field $v_i\boldsymbol{g}^i$ is of the form

$$v_i(x)\boldsymbol{g}^i(x) = \boldsymbol{a} + \boldsymbol{b} \wedge \boldsymbol{\Theta}(x) \text{ for all } x \in \overline{\Omega}.$$

Next, let Γ_0 be a $\mathrm{d}\Gamma$*-measurable subset of the boundary $\partial\Omega$ that satisfies* area $\Gamma_0 > 0$, *and let a vector field $\boldsymbol{v} = (v_i) \in \boldsymbol{H}^1(\Omega)$ be such that*

$$e_{ij}(\boldsymbol{v}) = 0 \text{ in } \Omega \textit{ and } \boldsymbol{v} = \boldsymbol{0} \text{ on } \Gamma_0.$$

Then $\boldsymbol{v} = \boldsymbol{0}$ in Ω.

It follows from part (ii) that

$$e_{ij}(\boldsymbol{v}) = 0 \text{ in } \Omega \text{ implies } \widehat{e}_{ij}(\widehat{\boldsymbol{v}}) = 0 \text{ in } \widehat{\Omega}.$$

Then the identity

$$\widehat{\partial}_j(\widehat{\partial}_k\widehat{v}_i) = \widehat{\partial}_j\widehat{e}_{ik}(\widehat{\boldsymbol{v}}) + \widehat{\partial}_k\widehat{e}_{ij}(\widehat{\boldsymbol{v}}) - \widehat{\partial}_i\widehat{e}_{jk}(\widehat{\boldsymbol{v}}) \text{ in } \mathcal{D}'(\widehat{\Omega})$$

further shows that

$$\widehat{e}_{ij}(\widehat{\boldsymbol{v}}) = 0 \text{ in } \widehat{\Omega} \text{ implies } \widehat{\partial}_j(\widehat{\partial}_k\widehat{v}_i) = 0 \text{ in } \mathcal{D}'(\widehat{\Omega}).$$

By a classical result from distribution theory (Schwartz [1966, p. 60]), each function $\widehat{v}_i$ is therefore a polynomial of degree ≤ 1 in the variables $\widehat{x}_j$, since the set $\widehat{\Omega}$ is connected. There thus exist constants a_i and b_{ij} such that

$$\widehat{v}_i(\widehat{x}) = a_i + b_{ij}\widehat{x}_j \text{ for all } \widehat{x} = (\widehat{x}_i) \in \widehat{\Omega}.$$

But $\widehat{e}_{ij}(\widehat{\boldsymbol{v}}) = 0$ also implies that $b_{ij} = -b_{ji}$. Hence there exist two vectors $\boldsymbol{a}, \boldsymbol{b} \in \mathbb{R}^3$ such that ($\boldsymbol{o}\widehat{\boldsymbol{x}}$ denotes the column vector with components x_i)

$$\widehat{\boldsymbol{v}}_i(\widehat{x})\widehat{\boldsymbol{e}}^i = \boldsymbol{a} + \boldsymbol{b} \wedge \boldsymbol{o}\widehat{\boldsymbol{x}} \text{ for all } \widehat{x} \in \widehat{\Omega},$$

or equivalently, such that

$$v_i(x)\boldsymbol{g}^i(x) = \boldsymbol{a} + \boldsymbol{b} \wedge \boldsymbol{\Theta}(x) \text{ for all } x \in \Omega.$$

Since the set where such a vector field $\widehat{v}_i\widehat{\boldsymbol{e}}^i$ vanishes is always of zero area unless $\boldsymbol{a} = \boldsymbol{b} = \boldsymbol{0}$ (as is easily proved; see, e.g., Ciarlet [1988, Theorem 6.3-4]), the assumption $\mathit{area}\,\Gamma_0 > 0$ implies that $\widehat{\boldsymbol{v}} = \boldsymbol{0}$.

(iv) *We now let*

$$\Omega := \omega \times \left]-\varepsilon, \varepsilon\right[,$$

and we let the mapping $\boldsymbol{\Theta} : \overline{\Omega} \to \boldsymbol{E}^3$ *be defined by*

$$\boldsymbol{\Theta}(y, x_3) := \boldsymbol{\theta}(y) + x_3\boldsymbol{a}_3(y) \text{ for all } x = (x_i) := (y, x_3) \in \overline{\Omega},$$

where $\varepsilon > 0$ has been chosen in such a way that the mapping $\boldsymbol{\Theta}$ is a $\mathcal{C}^2$-diffeomorphism from $\overline{\Omega}$ into its image $\boldsymbol{\Theta}(\overline{\Omega})$ (Theorem 2.1.1). *With any vector field* $\eta_i\boldsymbol{a}^i$ *with covariant components* η_α *in* $H^1(\omega)$ *and* η_3 *in* $H^2(\omega)$, *let there be associated the vector field* $v_i\boldsymbol{g}^i$ *defined on* $\overline{\Omega}$ *by*

$$v_i(y, x_3)\boldsymbol{g}^i(y, x_3) = \eta_i(y)\boldsymbol{a}^i(y) - x_3(\partial_\alpha\eta_3 + b_\alpha^\sigma\eta_\sigma)(y)\boldsymbol{a}^\alpha(y)$$

for all $(y, x_3) \in \overline{\Omega}$, *where the vectors* $\boldsymbol{g}^i$ *are defined by* $\boldsymbol{g}^i \cdot \boldsymbol{g}_j = \delta_j^i$.

Then the covariant components v_i *of the vector field* $v_i\boldsymbol{g}^i$ *are in* $H^1(\Omega)$ *and the corresponding functions* $e_{ij}(\boldsymbol{v}) \in L^2(\Omega)$ *defined as in part* (ii) *are given by*

$$\begin{aligned} e_{\alpha\beta}(\boldsymbol{v}) &= \gamma_{\alpha\beta}(\boldsymbol{\eta}) - x_3\rho_{\alpha\beta}(\boldsymbol{\eta}) \\ &\quad + \frac{x_3^2}{2}\{b_\alpha^\sigma\rho_{\beta\sigma}(\boldsymbol{\eta}) + b_\beta^\tau\rho_{\alpha\tau}(\boldsymbol{\eta}) - 2b_\alpha^\sigma b_\beta^\tau\gamma_{\sigma\tau}(\boldsymbol{\eta})\}, \\ e_{i3}(\boldsymbol{v}) &= 0. \end{aligned}$$

Note that, as in the above expressions of the functions $e_{\alpha\beta}(\boldsymbol{v})$, the dependence on x_3 is explicit, but the dependence with respect to $y \in \overline{\omega}$ is omitted, throughout the proof. The explicit expressions of the functions $\gamma_{\alpha\beta}(\boldsymbol{\eta})$ and $\rho_{\alpha\beta}(\boldsymbol{\eta})$ in terms of the functions η_i (Theorems 2.3.1 and 2.3.2) are used in this part of the proof.

To prove the above assertion, we proceed in two stages. First, *given functions* $\eta_\alpha, \mathcal{X}_\alpha \in H^1(\omega)$ *and* $\eta_3 \in H^2(\omega)$, *let the vector field* $v_i \boldsymbol{g}^i$ *be defined on* $\overline{\Omega}$ *by*

$$v_i \boldsymbol{g}^i = \eta_i \boldsymbol{a}^i + x_3 \mathcal{X}_\alpha \boldsymbol{a}^\alpha.$$

Then the functions v_i *are in* $H^1(\Omega)$. *Besides, the functions* $e_{i\|j}(\boldsymbol{v})$ *defined as in part* (ii) *are given by*

$$\begin{aligned}
e_{\alpha\beta}(\boldsymbol{v}) &= \Big\{\frac{1}{2}(\eta_{\alpha|\beta} + \eta_{\beta|\alpha}) - b_{\alpha\beta}\eta_3\Big\} \\
&\quad + \frac{x_3}{2}\{\mathcal{X}_{\alpha|\beta} + \mathcal{X}_{\beta|\alpha} - b_\alpha^\sigma(\eta_{\sigma|\beta} - b_{\beta\sigma}\eta_3) - b_\beta^\tau(\eta_{\tau|\alpha} - b_{\alpha\tau}\eta_3)\} \\
&\quad + \frac{x_3^2}{2}\{-b_\alpha^\sigma \mathcal{X}_{\sigma|\beta} - b_\beta^\tau \mathcal{X}_{\tau|\alpha}\}, \\
e_{\alpha 3}(\boldsymbol{v}) &= \frac{1}{2}(\mathcal{X}_\alpha + \partial_\alpha \eta_3 + b_\alpha^\sigma \eta_\sigma), \\
e_{33}(\boldsymbol{v}) &= 0,
\end{aligned}$$

where $\eta_{\alpha|\beta} = \partial_\beta \eta_\alpha - \Gamma_{\alpha\beta}^\sigma \eta_\sigma$ and $\mathcal{X}_{\alpha|\beta} = \partial_\beta \mathcal{X}_\alpha - \Gamma_{\alpha\beta}^\sigma \mathcal{X}_\sigma$ designate the covariant derivatives of the fields $\eta_i \boldsymbol{a}^i$ and $\mathcal{X}_i \boldsymbol{a}^i$ with $\mathcal{X}_3 = 0$ (Section 1.6).

To see this, we note that

$$\partial_\alpha \boldsymbol{a}_3 = -b_\alpha^\sigma \boldsymbol{a}_\sigma$$

by the formula of Weingarten (Theorem 1.6.1). Hence, the vectors of the covariant basis associated with the mapping $\boldsymbol{\Theta} = \boldsymbol{\theta} + x_3 \boldsymbol{a}_3$ are given by

$$\boldsymbol{g}_\alpha = \boldsymbol{a}_\alpha - x_3 b_\alpha^\sigma \boldsymbol{a}_\sigma \text{ and } \boldsymbol{g}_3 = \boldsymbol{a}_3.$$

The assumed regularities of the functions η_i and $\mathcal{X}_\alpha$ imply that

$$v_i = (v_j \boldsymbol{g}^j) \cdot \boldsymbol{g}_i = (\eta_j \boldsymbol{a}^j + x_3 \mathcal{X}_\alpha \boldsymbol{a}^\alpha) \cdot \boldsymbol{g}_i \in H^1(\Omega)$$

since $\boldsymbol{g}_i \in \boldsymbol{C}^1(\overline{\Omega})$. The announced expressions for the functions $e_{ij}(\boldsymbol{v})$ are then obtained by simple computations, based on the relations $v_{i\|j} = \{\partial_j(v_k \boldsymbol{g}^k)\} \cdot \boldsymbol{g}_i$ (part (i)) and $e_{ij}(\boldsymbol{v}) = \frac{1}{2}(v_{i\|j} + v_{j\|i})$.

Second, we show that, *when*

$$\mathcal{X}_\alpha = -(\partial_\alpha \eta_3 + b_\alpha^\sigma \eta_\sigma),$$

the functions $e_{ij}(\boldsymbol{v})$ *above take the expressions announced in the statement of part* (iv).

We first note that $\mathcal{X}_\alpha \in H^1(\omega)$ (since $b_\alpha^\sigma \in \mathcal{C}^1(\overline{\omega})$) and that $e_{\alpha 3}(\boldsymbol{v}) = 0$ when $\mathcal{X}_\alpha = -(\partial_\alpha \eta_3 + b_\alpha^\sigma \eta_\sigma)$. It thus remains to find the explicit forms of the functions $e_{\alpha\beta}(\boldsymbol{v})$ in this case. Replacing the functions $\mathcal{X}_\alpha$ by their expressions and using the symmetry relations $b^\sigma_\alpha|_\beta = b^\sigma_\beta|_\alpha$ (Theorem 2.3.2), we find that

$$\frac{1}{2}\{\mathcal{X}_{\alpha|\beta} + \mathcal{X}_{\beta|\alpha} - b^\sigma_\alpha(\eta_{\sigma|\beta} - b_{\beta\sigma}\eta_3) - b^\tau_\beta(\eta_{\tau|\alpha} - b_{\alpha\tau}\eta_3)\}$$
$$= -\eta_{3|\alpha\beta} - b^\sigma_\alpha\eta_{\sigma|\beta} - b^\tau_\beta\eta_{\tau|\alpha} - b^\tau_{\beta|\alpha}\eta_\tau + b^\sigma_\alpha b_{\sigma\beta}\eta_3,$$

i.e., the factor of x_3 in $e_{\alpha\beta}(\boldsymbol{v})$ is equal to $-\rho_{\alpha\beta}(\boldsymbol{\eta})$. Finally,

$$\begin{aligned} -b^\sigma_\alpha\mathcal{X}_{\sigma|\beta} - b^\tau_\beta\mathcal{X}_{\tau|\alpha} &= b^\sigma_\alpha\big(\eta_{3|\beta\sigma} + b^\tau_{\sigma|\beta}\eta_\tau + b^\tau_\sigma\eta_{\tau|\beta}\big) + b^\tau_\beta\big(\eta_{3|\alpha\tau} + b^\sigma_{\tau|\alpha}\eta_\sigma + b^\sigma_\tau\eta_{\sigma|\alpha}\big)\\ &= b^\sigma_\alpha\big(\rho_{\beta\sigma}(\boldsymbol{\eta}) - b^\tau_\beta\eta_{\tau|\sigma} + b^\tau_\beta b_{\tau\sigma}\eta_3\big) + b^\tau_\beta\big(\rho_{\alpha\tau}(\boldsymbol{\eta}) - b^\sigma_\alpha\eta_{\sigma|\tau} + b^\sigma_\alpha b_{\sigma\tau}\eta_3\big)\\ &= b^\sigma_\alpha\rho_{\beta\sigma}(\boldsymbol{\eta}) + b^\tau_\beta\rho_{\alpha\tau}(\boldsymbol{\eta}) - 2b^\sigma_\alpha b^\tau_\beta\gamma_{\sigma\tau}(\boldsymbol{\eta}),\end{aligned}$$

i.e., the factor of $\frac{x_3^2}{2}$ in $e_{\alpha\beta}(\boldsymbol{v})$ is indeed as announced.

(v) Let the set $\Omega = \omega \times]-\varepsilon, \varepsilon[$ and let the vector field $\boldsymbol{v} = (v_i) \in \boldsymbol{H}^1(\Omega)$ be defined as in part (iv). By part (iv), the assumption that $\gamma_{\alpha\beta}(\boldsymbol{\eta}) = \rho_{\alpha\beta}(\boldsymbol{\eta}) = 0$ in ω implies that

$$e_{ij}(\boldsymbol{v}) = 0 \text{ in } \Omega.$$

Therefore, by part (iii), there exist two vectors $\boldsymbol{a}, \boldsymbol{b} \in \mathbb{R}^3$ such that

$$v_i(y, x_3)\boldsymbol{g}^i(y, x_3) = \boldsymbol{a} + \boldsymbol{b} \wedge \{\boldsymbol{\theta}(y) + x_3\boldsymbol{a}_3(y)\} \text{ for all } (y, x_3) \in \overline{\Omega}.$$

Hence

$$\eta_i(y)\boldsymbol{a}^i(y) = v_i(y, x_3)\boldsymbol{g}^i(y, x_3)|_{x_3=0} = \boldsymbol{a} + \boldsymbol{b} \wedge \boldsymbol{\theta}(y) \text{ for all } y \in \overline{\omega},$$

and part (a) of the theorem is established.

Let next $\gamma_0 \subset \gamma$ be such that *length* $\gamma_0 > 0$. If in addition $\eta_i = \partial_\nu\eta_3 = 0$ on γ_0, the functions $\chi_\alpha = -(\partial_\alpha\eta_3 + b^\sigma_\alpha\eta_\sigma)$ vanish on γ_0, since $\eta_3 = \partial_\nu\eta_3 = 0$ on γ_0 implies $\partial_\alpha\eta_3 = 0$ on γ_0. Part (iv) then shows that

$$v_i = (v_j\boldsymbol{g}^j) \cdot \boldsymbol{g}_i = (\eta_j\boldsymbol{a}^j + x_3\mathcal{X}_\alpha\boldsymbol{a}^\alpha) \cdot \boldsymbol{g}_i = 0 \text{ on } \Gamma_0 := \gamma_0 \times [-\varepsilon, \varepsilon].$$

Since *area* $\Gamma_0 > 0$, part (iii) implies that $\boldsymbol{v} = \boldsymbol{0}$ in $\overline{\Omega}$, hence that $\boldsymbol{\eta} = \boldsymbol{0}$ on $\overline{\omega}$ and part (b) of the theorem is established. □

We are now in a position to prove the announced **Korn's inequality on a surface, "with boundary conditions"**.

This inequality was first proved by Bernadou & Ciarlet [1976]. It was later given other proofs by Ciarlet & Miara [1992] and Bernadou, Ciarlet & Miara [1994]; then by Akian [2003] and Ciarlet & S. Mardare [2001], who showed that it can be directly derived from the three-dimensional Korn inequality in curvilinear coordinates (this idea goes back to Destuynder [1985]); then by Blouza & Le Dret [1999], who showed that it still holds under a less stringent smoothness assumption on the mapping $\boldsymbol{\theta}$. We follow here the proof of Bernadou, Ciarlet & Miara [1994].

Theorem 2.5.3. *Let ω be a domain in $\mathbb{R}^2$, let $\boldsymbol{\theta} \in \mathcal{C}^3(\overline{\omega}; \boldsymbol{E}^3)$ be an injective immersion, let γ_0 be a $d\gamma$-measurable subset of $\gamma = \partial\omega$ that satisfies length $\gamma_0 > 0$, and let the space $\boldsymbol{V}(\omega)$ be defined as:*

$$\boldsymbol{V}(\omega) := \{\boldsymbol{\eta} = (\eta_i) \in H^1(\omega) \times H^1(\omega) \times H^2(\omega);\ \eta_i = \partial_\nu \eta_3 = 0 \text{ on } \gamma_0\}.$$

Given $\boldsymbol{\eta} = (\eta_i) \in H^1(\omega) \times H^1(\omega) \times H^2(\omega)$, let

$$\gamma_{\alpha\beta}(\boldsymbol{\eta}) := \Big\{\frac{1}{2}(\partial_\beta \widetilde{\boldsymbol{\eta}} \cdot \boldsymbol{a}_\alpha + \partial_\alpha \widetilde{\boldsymbol{\eta}} \cdot \boldsymbol{a}_\beta\Big\} \in L^2(\omega),$$

$$\rho_{\alpha\beta}(\boldsymbol{\eta}) := \Big\{(\partial_{\alpha\beta} \widetilde{\boldsymbol{\eta}} - \Gamma^\sigma_{\alpha\beta}\partial_\sigma \widetilde{\boldsymbol{\eta}}) \cdot \boldsymbol{a}_3\Big\} \in L^2(\omega)$$

denote the covariant components of the linearized change of metric and linearized change of curvature tensors associated with the displacement field $\widetilde{\boldsymbol{\eta}} := \eta_i \boldsymbol{a}^i$ of the surface $S = \boldsymbol{\theta}(\overline{\omega})$. Then there exists a constant $c = c(\omega, \gamma_0, \boldsymbol{\theta})$ such that

$$\Big\{\sum_\alpha \|\eta_\alpha\|^2_{1,\omega} + \|\eta_3\|^2_{2,\omega}\Big\}^{1/2} \le c\Big\{\sum_{\alpha,\beta} \|\gamma_{\alpha\beta}(\boldsymbol{\eta})\|^2_{0,\omega} + \sum_{\alpha,\beta} \|\rho_{\alpha\beta}(\boldsymbol{\eta})\|^2_{0,\omega}\Big\}^{1/2}$$

for all $\boldsymbol{\eta} = (\eta_i) \in \boldsymbol{V}(\omega)$.

Proof. Let

$$\|\boldsymbol{\eta}\|_{H^1(\omega)\times H^1(\omega)\times H^2(\omega)} := \Big\{\sum_\alpha \|\eta_\alpha\|^2_{1,\omega} + \|\eta_3\|^2_{2,\omega}\Big\}^{1/2}.$$

If the announced inequality is false, there exists a sequence $(\boldsymbol{\eta}^k)_{k=1}^\infty$ of vector fields $\boldsymbol{\eta}^k \in \boldsymbol{V}(\omega)$ such that

$$\|\boldsymbol{\eta}^k\|_{H^1(\omega)\times H^1(\omega)\times H^2(\omega)} = 1 \text{ for all } k,$$

$$\lim_{k\to\infty} \Big\{\sum_{\alpha,\beta} \|\gamma_{\alpha\beta}(\boldsymbol{\eta}^k)\|^2_{0,\omega} + \sum_{\alpha,\beta} \|\rho_{\alpha\beta}(\boldsymbol{\eta}^k)\|^2_{0,\omega}\Big\}^{1/2} = 0.$$

Since the sequence $(\boldsymbol{\eta}^k)_{k=1}^{\infty}$ is bounded in $H^1(\omega) \times H^1(\omega) \times H^2(\omega)$, a subsequence $(\boldsymbol{\eta}^\ell)_{\ell=1}^{\infty}$ converges in $L^2(\omega) \times L^2(\omega) \times H^1(\omega)$ by the *Rellich-Kondrašov theorem.* Furthermore, each sequence $(\gamma_{\alpha\beta}(\boldsymbol{\eta}^\ell))_{\ell=1}^{\infty}$ and $(\rho_{\alpha\beta}(\boldsymbol{\eta}^\ell))_{\ell=1}^{\infty}$ also converges in $L^2(\omega)$ (to 0, but this information is not used at this stage) since

$$\lim_{\ell\to\infty} \Big\{ \sum_{\alpha,\beta} \|\gamma_{\alpha\beta}(\boldsymbol{\eta}^\ell)\|_{0,\omega}^2 + \sum_{\alpha,\beta} \|\rho_{\alpha\beta}(\boldsymbol{\eta}^\ell)\|_{0,\omega}^2 \Big\}^{1/2} = 0.$$

The subsequence $(\boldsymbol{\eta}^\ell)_{\ell=1}^{\infty}$ is thus a Cauchy sequence with respect to the norm

$$\boldsymbol{\eta} \to \Big\{ \sum_{\alpha} \|\eta_\alpha\|_{0,\omega}^2 + \|\eta_3\|_{1,\omega}^2 + \sum_{\alpha,\beta} \|\gamma_{\alpha\beta}(\boldsymbol{\eta})\|_{0,\omega}^2 + \sum_{\alpha,\beta} |\rho_{\alpha\beta}(\boldsymbol{\eta})|_{0,\omega}^2 \Big\}^{1/2},$$

hence with respect to the norm $\|\cdot\|_{H^1(\omega)\times H^1(\omega)\times H^2(\omega)}$ by *Korn's inequality without boundary conditions* (Theorem 2.5.1).

The space $\boldsymbol{V}(\omega)$ being complete as a closed subspace of the space $H^1(\omega)\times H^1(\omega) \times H^2(\omega)$, there exists $\boldsymbol{\eta} \in \boldsymbol{V}(\omega)$ such that

$$\boldsymbol{\eta}^\ell \to \boldsymbol{\eta} \text{ in } H^1(\omega) \times H^1(\omega) \times H^2(\omega),$$

and the limit $\boldsymbol{\eta}$ satisfies

$$\begin{aligned}
\|\gamma_{\alpha\beta}(\boldsymbol{\eta})\|_{0,\omega} &= \lim_{\ell\to\infty} \|\gamma_{\alpha\beta}(\boldsymbol{\eta}^\ell)\|_{0,\omega} = 0,\\
\|\rho_{\alpha\beta}(\boldsymbol{\eta})\|_{0,\omega} &= \lim_{\ell\to\infty} \|\rho_{\alpha\beta}(\boldsymbol{\eta}^\ell)\|_{0,\omega} = 0.
\end{aligned}$$

Hence $\boldsymbol{\eta} = \mathbf{0}$ by Theorem 2.5.2. But this contradicts the relations $\|\boldsymbol{\eta}^\ell\|_{H^1(\omega)\times H^1(\omega)\times H^2(\omega)} = 1$ for all $\ell \geq 1$, and the proof is complete. □

If the mapping $\boldsymbol{\theta}$ is of the form $\boldsymbol{\theta}(y_1, y_2) = (y_1, y_2, 0)$ for all $(y_1, y_2) \in \overline{\omega}$, the inequality of Theorem 2.5.3 reduces to two distinct inequalities (obtained by letting first $\eta_\alpha = 0$, then $\eta_3 = 0$):

$$\|\eta_3\|_{2,\omega} \leq c \Big\{ \sum_{\alpha,\beta} \|\partial_{\alpha\beta}\eta_3\|_{0,\omega}^2 \Big\}^{1/2}$$

for all $\eta_3 \in H^2(\omega)$ satisfying $\eta_3 = \partial_\nu \eta_3 = 0$ on γ_0, and

$$\Big\{ \sum_{\alpha} \|\eta_\alpha\|_{1,\omega}^2 \Big\}^{1/2} \leq c \Big\{ \sum_{\alpha} \Big\| \frac{1}{2}(\partial_\beta\eta_\alpha + \partial_\alpha\eta_\beta) \Big\|_{0,\omega}^2 \Big\}^{1/2}$$

for all $\eta_\alpha \in \boldsymbol{H}^1(\omega)$ satisfying $\eta_\alpha = 0$ on γ_0. The first inequality is a well-known property of Sobolev spaces. The second inequality is the *two-dimensional Korn inequality in Cartesian coordinates.* Both play a central rôle in the *existence theory* for *linear two-dimensional plate equations* (see, e.g., Ciarlet [1997, Theorems 1.5-1 and 1.5-2]).

As shown by Blouza & Le Dret [1999], Le Dret [2004], and Anicic, Le Dret & Raoult [2005], the regularity assumptions made on the mapping $\boldsymbol{\theta}$ and on the field $\boldsymbol{\eta} = (\eta_i)$ in both the infinitesimal rigid displacement lemma and the Korn inequality on a surface of Theorems 2.5.2 and 2.5.3 can be substantially weakened.

It is remarkable that, *for specific geometries and boundary conditions, a Korn inequality can be established that only involves the linearized change of metric tensors.* More specifically, Ciarlet & Lods [1996] and Ciarlet & Sanchez-Palencia [1996] have established the following **Korn inequality "on an elliptic surface"**:

Let ω be a domain in $\mathbb{R}^2$ and let $\boldsymbol{\theta} \in \mathcal{C}^{2,1}(\overline{\omega}; \boldsymbol{E}^3)$ be an injective immersion with the property that *the surface* $S = \boldsymbol{\theta}(\overline{\omega})$ is **elliptic**, in the sense that all its points are elliptic (this means that the Gaussian curvature is > 0 everywhere on S; cf. Section 1.5). Then *there exists a constant* $c_M = c_M(\omega, \boldsymbol{\theta}) > 0$ *such that*

$$\Big\{ \sum_\alpha \|\eta_\alpha\|^2_{1,\omega} + \|\eta_3\|^2_{0,\omega} \Big\}^{1/2} \le c_M \Big\{ \sum_{\alpha,\beta} \|\gamma_{\alpha\beta}(\boldsymbol{\eta})\|^2_{0,\omega} \Big\}^{1/2}$$

for all $\boldsymbol{\eta} = (\eta_i) \in H^1_0(\omega) \times H^1_0(\omega) \times L^2(\omega)$.

Remarks 2.5.4. (1) The norm $\|\eta_3\|_{2,\omega}$ appearing in the left-hand side of the Korn inequality on a "general" surface (Theorem 2.5.3) is now replaced by the norm $\|\eta_3\|_{0,\omega}$. This replacement reflects that it is enough that $\boldsymbol{\eta} = (\eta_i) \in H^1(\omega) \times H^1(\omega) \times L^2(\omega)$ in order that $\gamma_{\alpha\beta}(\boldsymbol{\eta}) \in L^2(\omega)$. As a result, no boundary condition can be imposed on η_3.

(2) The Korn inequality on an elliptic surface was first established by Destuynder [1985, Theorem 6.1 and 6.5], under the additional assumption that the $\mathcal{C}^0(\overline{\omega})$-norms of the Christoffel symbols are small enough.

Only compact surfaces defined by a *single* injective immersion $\boldsymbol{\theta} \in \mathcal{C}^3(\overline{\omega})$ have been considered so far. By contrast, a *compact surface* S *"without boundary"* (such as an ellipsoid or a torus) is defined by means of a *finite number* $I \ge 2$ *of injective immersions* $\boldsymbol{\theta}_i \in \mathcal{C}^3(\overline{\omega}_i)$, $1 \le i \le I$, where the sets ω_i are domains in $\mathbb{R}^2$, in such a way that $S = \bigcup_{i \in I} \boldsymbol{\theta}_i(\omega_i)$. As shown by S. Mardare [2003a], the Korn inequality "without boundary conditions" (Theorem 2.5.1) can be both extended to such surfaces without boundary.

2.6 Existence and Uniqueness Theorems for the Linear Koiter Shell Equations

Let the space $\boldsymbol{V}(\omega)$ be defined by

$$\boldsymbol{V}(\omega) := \{\boldsymbol{\eta} = (\eta_i) \in H^1(\omega) \times H^1(\omega) \times H^2(\omega);\ \eta_i = \partial_\nu \eta_3 = 0 \text{ on } \gamma_0\},$$

where γ_0 is a $d\gamma$-measurable subset of $\gamma := \partial\omega$ that satisfies *length* $\gamma_0 > 0$. Our primary objective consists in showing that the bilinear form $B : \boldsymbol{V}(\omega) \times \boldsymbol{V}(\omega) \to \mathbb{R}$ defined by

$$B(\boldsymbol{\zeta}, \boldsymbol{\eta}) := \int_\omega \Big\{ \varepsilon a^{\alpha\beta\sigma\tau} \gamma_{\sigma\tau}(\boldsymbol{\zeta}) \gamma_{\alpha\beta}(\boldsymbol{\eta}) + \frac{\varepsilon^3}{3} a^{\alpha\beta\sigma\tau} \rho_{\sigma\tau}(\boldsymbol{\zeta}) \rho_{\alpha\beta}(\boldsymbol{\eta}) \Big\} \sqrt{a}\, dy$$

for all $(\boldsymbol{\zeta}, \boldsymbol{\eta}) \in \boldsymbol{V}(\omega) \times \boldsymbol{V}(\omega)$ is $\boldsymbol{V}(\omega)$-elliptic.

As a preliminary, we establish the *uniform positive-definiteness of the elasticity tensor of the shell*, given here by means of its *contravariant components* $a^{\alpha\beta\sigma\tau}$ (note that the assumptions on the Lamé constants, viz., $3\lambda + 2\mu > 0$ and $\mu > 0$, are those usually made for elastic materials).

Theorem 2.6.1. *Let ω be a domain in $\mathbb{R}^2$, let $\boldsymbol{\theta} \in \mathcal{C}^3(\overline{\omega}; \boldsymbol{E}^3)$ be an injective immersion, let $a^{\alpha\beta}$ denote the contravariant components of the metric tensor of the surface $\boldsymbol{\theta}(\overline{\omega})$, let the contravariant components of the two-dimensional elasticity tensor of the shell be given by*

$$a^{\alpha\beta\sigma\tau} = \frac{4\lambda\mu}{\lambda + 2\mu} a^{\alpha\beta} a^{\sigma\tau} + 2\mu (a^{\alpha\sigma} a^{\beta\tau} + a^{\alpha\tau} a^{\beta\sigma}),$$

and assume that $3\lambda + 2\mu > 0$ and $\mu > 0$. Then there exists a constant $c_e = c_e(\omega, \boldsymbol{\theta}, \lambda, \mu) > 0$ such that

$$\sum_{\alpha,\beta} |t_{\alpha\beta}|^2 \le c_e a^{\alpha\beta\sigma\tau}(y) t_{\sigma\tau} t_{\alpha\beta}$$

for all $y \in \overline{\omega}$ and all symmetric matrices $(t_{\alpha\beta})$.

Proof. In what follows, $\mathbb{M}^2$ and $\mathbb{S}^2$ respectively designate the set of all real matrices of order two and the set of all real symmetric matrices of order two. The elegant proof of part (ii) given below is due to Cristinel Mardare.

(i) To begin with, we establish a crucial inequality. *Let χ and μ be two constants satisfying $\chi + \mu > 0$ and $\mu > 0$. Then there exists a constant $\gamma = \gamma(\chi, \mu) > 0$ such that*

$$\gamma \operatorname{tr}(\boldsymbol{B}^T \boldsymbol{B}) \le \chi (\operatorname{tr} \boldsymbol{B})^2 + 2\mu \operatorname{tr}(\boldsymbol{B}^T \boldsymbol{B}) \text{ for all } \boldsymbol{B} \in \mathbb{M}^2.$$

If $\chi \geq 0$ and $\mu > 0$, this inequality holds with $\gamma = 2\mu$. It thus remains to consider the case where $-\mu < \chi < 0$ and $\mu > 0$. Given any matrix $\boldsymbol{B} \in \mathbb{M}^2$, define the matrix $\boldsymbol{C} \in \mathbb{M}^2$ by

$$\boldsymbol{C} = \mathbb{A}\boldsymbol{B} := \chi(\operatorname{tr}\boldsymbol{B})\boldsymbol{I} + 2\mu\boldsymbol{B}.$$

The linear mapping $\mathbb{A} : \mathbb{M}^2 \to \mathbb{M}^2$ defined in this fashion can be easily inverted if $\chi + \mu \neq 0$ and $\mu \neq 0$, as

$$\boldsymbol{B} = \mathbb{A}^{-1}\boldsymbol{C} = -\frac{\chi}{4\mu(\chi+\mu)}(\operatorname{tr}\boldsymbol{C})\boldsymbol{I} + \frac{1}{2\mu}\boldsymbol{C}.$$

Noting that the bilinear mapping

$$(\boldsymbol{B}, \boldsymbol{C}) \in \mathbb{M}^2 \times \mathbb{M}^2 \to \boldsymbol{B} : \boldsymbol{C} := \operatorname{tr}\boldsymbol{B}^T\boldsymbol{C}$$

defines an inner product over the space $\mathbb{M}^2$, we thus obtain

$$\begin{aligned}\chi(\operatorname{tr}\boldsymbol{B})^2 + 2\mu\operatorname{tr}(\boldsymbol{B}^T\boldsymbol{B}) &= (\mathbb{A}\boldsymbol{B}) : \boldsymbol{B} = \boldsymbol{C} : \mathbb{A}^{-1}\boldsymbol{C}\\ &= -\frac{\chi}{4\mu(\chi+\mu)}(\operatorname{tr}\boldsymbol{C})^2 + \frac{1}{2\mu}\operatorname{tr}(\boldsymbol{C}^T\boldsymbol{C}) \geq \frac{1}{2\mu}\boldsymbol{C} : \boldsymbol{C}\end{aligned}$$

for any $\boldsymbol{B} = \mathbb{A}^{-1}\boldsymbol{C} \in \mathbb{M}^2$ if $-\mu < \chi < 0$ and $\mu > 0$. Since there clearly exists a constant $\beta = \beta(\chi, \mu) > 0$ such that

$$\boldsymbol{B} : \boldsymbol{B} \leq \beta\boldsymbol{C} : \boldsymbol{C} \text{ for all } \boldsymbol{B} = \mathbb{A}^{-1}\boldsymbol{C} \in \mathbb{M}^2,$$

the announced inequality also holds if $-\mu < \chi < 0$ and $\mu > 0$, with $\gamma = (2\mu\beta)^{-1}$ in this case.

(ii) We next show that, *for any $y \in \overline{\omega}$ and any nonzero symmetric matrix $(t_{\alpha\beta})$,*

$$a^{\alpha\beta\sigma\tau}(y)t_{\sigma\tau}t_{\alpha\beta} \geq \gamma a^{\alpha\sigma}(y)a^{\beta\tau}(y)t_{\sigma\tau}t_{\alpha\beta} > 0,$$

where $\gamma = \gamma(\chi, \mu) > 0$ is the constant found in (i).

Given any $y \in \overline{\omega}$ and any symmetric matrix $(t_{\alpha\beta})$, let

$$\boldsymbol{A}(y) = (a^{\alpha\beta}(y)) \text{ and } \boldsymbol{T} = (t_{\alpha\beta}),$$

let $\boldsymbol{K}(y) \in \mathbb{S}^2$ be the unique square root of $\boldsymbol{A}(y)$ (i.e., the unique positive-definite symmetric matrix that satisfies $(\boldsymbol{K}(y))^2 = \boldsymbol{A}(y)$), and let

$$\boldsymbol{B}(y) := \boldsymbol{K}(y)\boldsymbol{T}\boldsymbol{K}(y) \in \mathbb{S}^2.$$

Then

$$\frac{1}{2}a^{\alpha\beta\sigma\tau}(y)t_{\sigma\tau}t_{\alpha\beta} = \chi\Big(\operatorname{tr}\boldsymbol{B}(y)\Big)^2 + 2\mu\operatorname{tr}\Big(\boldsymbol{B}(y)^T\boldsymbol{B}(y)\Big) \text{ with } \chi := \frac{2\lambda\mu}{\lambda+2\mu}.$$

By the inequality established in part (i), there thus exists a constant $\alpha(\lambda, \mu) > 0$ such that

$$\frac{1}{2}a^{\alpha\beta\sigma\tau}(y)t_{\sigma\tau}t_{\alpha\beta} \geq \alpha \operatorname{tr}\left(\boldsymbol{B}(y)^T \boldsymbol{B}(y)\right)$$

if $\chi + \mu > 0$ and $\mu > 0$, or equivalently, if $3\lambda + 2\mu > 0$ and $\mu > 0$.

(iii) *Conclusion:* Since the mapping

$$(y, (t_{\alpha\beta})) \in \boldsymbol{K} := \overline{\omega} \times \left\{(t_{\alpha\beta}) \in \mathbb{S}^2; \sum_{\alpha,\beta} |t_{\alpha\beta}|^2 = 1\right\} \longrightarrow a^{\alpha\sigma}(y)a^{\beta\tau}(y)t_{\sigma\tau}t_{\alpha\beta},$$

is continuous and its domain of definition is compact, we infer that

$$\delta = \delta(\omega; \boldsymbol{\theta}) := \inf_{(y,(t_{\alpha\beta}))\in\boldsymbol{K}} a^{\alpha\sigma}(y)a^{\beta\tau}(y)t_{\sigma\tau}t_{\alpha\beta} > 0.$$

Hence

$$\delta \sum_{\alpha,\beta} |t_{\alpha\beta}|^2 \leq a^{\alpha\sigma}(y)a^{\beta\tau}(y)t_{\sigma\tau}t_{\alpha\beta}$$

and thus

$$\sum_{\alpha,\beta} |t_{\alpha\beta}|^2 \leq c_e a^{\alpha\beta\sigma\tau}(y)t_{\sigma\tau}t_{\alpha\beta}$$

for all $y \in \overline{\omega}$ and all symmetric matrices $(t_{\alpha\beta})$, with $c_e := (\gamma\delta)^{-1}$. □

Combined with Korn's inequality "with boundary conditions" (Theorem 2.5.3), the positive definiteness of the elasticity tensor leads to the existence of a *weak* solution, i.e., a solution to the variational equations of the *linear Koiter shell equations.*

Theorem 2.6.2. *Let ω be a domain in $\mathbb{R}^2$, let γ_0 be a subset of $\gamma = \partial\omega$ with* length $\gamma_0 > 0$, *and let $\boldsymbol{\theta} \in \mathcal{C}^3(\overline{\omega}; \boldsymbol{E}^3)$ be an injective immersion. Finally, let there be given constants λ and μ that satisfy $3\lambda + 2\mu > 0$ and $\mu > 0$, and functions $p^\alpha \in L^r(\omega)$ for some $r > 1$ and $p^3 \in L^1(\omega)$.*

Then there is one and only one solution $\boldsymbol{\zeta} = (\zeta_i)$ to the variational problem:

$$\boldsymbol{\zeta} \in \boldsymbol{V}(\omega) = \{\boldsymbol{\eta} = (\eta_i) \in H^1(\omega) \times H^1(\omega) \times H^2(\omega);\ \eta_i = \partial_\nu \eta_3 = 0 \text{ on } \gamma_0\},$$

$$\int_\omega \left\{\varepsilon a^{\alpha\beta\sigma\tau}\gamma_{\sigma\tau}(\boldsymbol{\zeta})\gamma_{\alpha\beta}(\boldsymbol{\eta}) + \frac{\varepsilon^3}{3}a^{\alpha\beta\sigma\tau}\rho_{\sigma\tau}(\boldsymbol{\zeta})\rho_{\alpha\beta}(\boldsymbol{\eta})\right\}\sqrt{a}\,\mathrm{d}y$$
$$= \int_\omega p^i \eta_i \sqrt{a}\,\mathrm{d}y \text{ for all } \boldsymbol{\eta} = (\eta_i) \in \boldsymbol{V}(\omega),$$

where

$$a^{\alpha\beta\sigma\tau} = \frac{4\lambda\mu}{\lambda+2\mu} a^{\alpha\beta} a^{\sigma\tau} + 2\mu(a^{\alpha\sigma} a^{\beta\tau} + a^{\alpha\tau} a^{\beta\sigma}),$$

$$\gamma_{\alpha\beta}(\boldsymbol{\eta}) = \frac{1}{2}(\partial_\beta \widetilde{\boldsymbol{\eta}} \cdot \boldsymbol{a}_\alpha + \partial_\alpha \widetilde{\boldsymbol{\eta}} \cdot \boldsymbol{a}_\beta) \text{ and } \rho_{\alpha\beta}(\boldsymbol{\eta}) = (\partial_{\alpha\beta}\widetilde{\boldsymbol{\eta}} - \Gamma^\sigma_{\alpha\beta}\partial_\sigma \widetilde{\boldsymbol{\eta}}) \cdot \boldsymbol{a}_3,$$

where $\widetilde{\boldsymbol{\eta}} := \eta_i \boldsymbol{a}^i$.

The field $\boldsymbol{\zeta} \in \boldsymbol{V}(\omega)$ *is also the unique solution to the minimization problem:*

$$j(\boldsymbol{\zeta}) = \inf_{\boldsymbol{\eta} \in \boldsymbol{V}(\omega)} j(\boldsymbol{\eta}),$$

where

$$j(\boldsymbol{\eta}) := \frac{1}{2}\int_\omega \Big\{\varepsilon a^{\alpha\beta\sigma\tau}\gamma_{\sigma\tau}(\boldsymbol{\eta})\gamma_{\alpha\beta}(\boldsymbol{\eta}) + \frac{\varepsilon^3}{3} a^{\alpha\beta\sigma\tau}\rho_{\sigma\tau}(\boldsymbol{\eta})\rho_{\alpha\beta}(\boldsymbol{\eta})\Big\}\sqrt{a}\,dy - \int_\omega p^i \eta_i \sqrt{a}\,dy.$$

Proof. As a closed subspace of $H^1(\omega) \times H^1(\omega) \times H^2(\omega)$, the space $\boldsymbol{V}(\omega)$ is a Hilbert space. The assumptions made on the mapping $\boldsymbol{\theta}$ ensure in particular that the vector fields $\boldsymbol{a}_i$ and $\boldsymbol{a}^i$ belong to $\mathcal{C}^2(\overline{\omega}; \mathbb{R}^3)$ and that the functions $a^{\alpha\beta\sigma\tau}$, $\Gamma^\sigma_{\alpha\beta}$, and a are continuous on the compact set $\overline{\omega}$. Hence the bilinear form defined by the left-hand side of the variational equations is *continuous over the space* $H^1(\omega) \times H^1(\omega) \times H^2(\omega)$.

The continuous embeddings of the space $H^1(\omega)$ into the space $L^s(\omega)$ for any $s \geq 1$ and of the space $H^2(\omega)$ into the space $\mathcal{C}^1(\overline{\omega})$ show that the linear form defined by the right-hand side is continuous over the same space.

Since the symmetric matrix $(a_{\alpha\beta}(y))$ is positive-definite for all $y \in \overline{\omega}$, there exists a_0 such that $a(y) \geq a_0 > 0$ for all $y \in \overline{\omega}$.

Finally, the Korn inequality "with boundary conditions" (Theorem 2.5.3) and the uniform positive definiteness of the elasticity tensor of the shell (Theorem 2.6.1) together imply that

$$\min\Big\{\varepsilon, \frac{\varepsilon^3}{3}\Big\} c_e^{-1} c^{-2} \sqrt{a_0}\Big(\sum_\alpha \|\eta_\alpha\|^2_{1,\omega} + \|\eta_3\|^2_{2,\omega}\Big) \leq \int_\omega \Big\{\varepsilon a^{\alpha\beta\sigma\tau}\gamma_{\sigma\tau}(\boldsymbol{\eta})\gamma_{\alpha\beta}(\boldsymbol{\eta}) + \frac{\varepsilon^3}{3} a^{\alpha\beta\sigma\tau}\rho_{\sigma\tau}(\boldsymbol{\eta})\rho_{\alpha\beta}(\boldsymbol{\eta})\Big\}\sqrt{a}\,dy$$

for all $\boldsymbol{\eta} = (\eta_i) \in \boldsymbol{V}(\omega)$. Hence *the bilinear form* B *is* $\boldsymbol{V}(\omega)$*-elliptic.*

The *Lax-Milgram* lemma then shows that the variational equations have one and only one solution. Since the bilinear form is symmetric, this solution is also the unique solution of the minimization problem stated in the theorem. □

The above existence and uniqueness result applies to linearized *pure displacement* and *displacement-traction* problems, i.e., those that correspond to *length* $\gamma_0 > 0$.

An immediate consequence of Theorem 2.6.2 is the existence and uniqueness of a *displacement field* $\zeta_i \boldsymbol{a}^i$ *of the middle surface* S, whose components $\zeta_\alpha \in H^1(\omega)$ and $\zeta_3 \in H^2(\omega)$ are thus obtained by finding the solution $\boldsymbol{\zeta} = (\zeta_i)$ to the variational equations of Theorem 2.6.2. Since the vector fields $\boldsymbol{a}^i$ formed by the covariant bases belong to the space $\mathcal{C}^2(\overline{\omega}; \mathbb{R}^3)$ by assumption, the vector fields $\zeta_\alpha \boldsymbol{a}^\alpha$ and $\zeta_3 \boldsymbol{a}^3$ belong respectively to the spaces $H^1(\omega; \mathbb{R}^3)$ and $H^2(\omega; \mathbb{R}^3)$.

We next derive the *boundary value problem* that is, at least formally, equivalent to the variational equations of Theorem 2.6.2. In what follows, $\gamma_1 := \gamma - \gamma_0$, (ν_α) is the unit outer normal vector along γ, $\tau_1 := -\nu_2$, $\tau_2 := \nu_1$, and $\partial_\tau \chi := \tau_\alpha \partial_\alpha \chi$ denotes the tangential derivative of χ in the direction of the vector (τ_α).

Theorem 2.6.3. *Let ω be a domain in $\mathbb{R}^2$ and let $\boldsymbol{\theta} \in \mathcal{C}^3(\overline{\omega}; \boldsymbol{E}^3)$ be an injective immersion. Assume that the boundary γ of ω and the functions p^i are smooth enough. If the solution $\boldsymbol{\zeta} = (\zeta_i)$ to the variational equations found in* Theorem 2.6.2 *is smooth enough, then $\boldsymbol{\zeta}$ is also a solution to the following boundary value problem:*

$$
\begin{aligned}
m^{\alpha\beta}|_{\alpha\beta} - b_\alpha^\sigma b_{\sigma\beta} m^{\alpha\beta} - b_{\alpha\beta} n^{\alpha\beta} &= p^3 \text{ in } \omega, \\
-(n^{\alpha\beta} + b_\sigma^\alpha m^{\sigma\beta})|_\beta - b_\sigma^\alpha (m^{\sigma\beta}|_\beta) &= p^\alpha \text{ in } \omega, \\
\zeta_i = \partial_\nu \zeta_3 &= 0 \text{ on } \gamma_0, \\
m^{\alpha\beta} \nu_\alpha \nu_\beta &= 0 \text{ on } \gamma_1, \\
(m^{\alpha\beta}|_\alpha)\nu_\beta + \partial_\tau (m^{\alpha\beta} \nu_\alpha \tau_\beta) &= 0 \text{ on } \gamma_1, \\
(n^{\alpha\beta} + 2 b_\sigma^\alpha m^{\sigma\beta})\nu_\beta &= 0 \text{ on } \gamma_1,
\end{aligned}
$$

where

$$
n^{\alpha\beta} := \varepsilon a^{\alpha\beta\sigma\tau} \gamma_{\sigma\tau}(\boldsymbol{\zeta}) \text{ and } m^{\alpha\beta} := \frac{\varepsilon^3}{3} a^{\alpha\beta\sigma\tau} \rho_{\sigma\tau}(\boldsymbol{\zeta}),
$$

and, for an arbitrary tensor field with smooth enough covariant components $t^{\alpha\beta} : \overline{\omega} \to \mathbb{R}$,

$$
\begin{aligned}
t^{\alpha\beta}|_\beta &:= \partial_\beta t^{\alpha\beta} + \Gamma_{\beta\sigma}^\alpha t^{\beta\sigma} + \Gamma_{\beta\sigma}^\beta t^{\alpha\sigma}, \\
t^{\alpha\beta}|_{\alpha\beta} &:= \partial_\alpha (t^{\alpha\beta}|_\beta) + \Gamma_{\alpha\sigma}^\sigma (t^{\alpha\beta}|_\beta).
\end{aligned}
$$

Proof. For simplicity, we give the proof only in the case where $\gamma_0 = \gamma$, i.e., when the space $\boldsymbol{V}(\omega)$ of Theorem 2.6.2 reduces to

$$
\boldsymbol{V}(\omega) = H_0^1(\omega) \times H_0^1(\omega) \times H_0^2(\omega).
$$

The extension to the case where *length* $\gamma_1 > 0$ is straightforward.

In what follows, we assume that the solution $\boldsymbol{\zeta}$ is "smooth enough" in the sense that $n^{\alpha\beta} \in H^1(\omega)$ and $m^{\alpha\beta} \in H^2(\omega)$.

(i) We first establish the relations

$$\partial_\alpha \sqrt{a} = \sqrt{a}\Gamma^\sigma_{\sigma\alpha}.$$

Let $\boldsymbol{A}$ denote the matrix of order three with $\boldsymbol{a}_1, \boldsymbol{a}_2, \boldsymbol{a}_3$ as its column vectors, so that $\sqrt{a} = \det \boldsymbol{A}$ (see part (i) of the proof of Theorem 2.6.2). Consequently,

$$\begin{aligned}\partial_\alpha \sqrt{a} &= \det(\partial_\alpha \boldsymbol{a}_1, \boldsymbol{a}_2, \boldsymbol{a}_3) + \det(\boldsymbol{a}_1, \partial_\alpha \boldsymbol{a}_2, \boldsymbol{a}_3) + \det(\boldsymbol{a}_1, \boldsymbol{a}_2, \partial_\alpha \boldsymbol{a}_3)\\ &= (\Gamma^1_{1\alpha} + \Gamma^2_{2\alpha} + \Gamma^3_{3\alpha}) \det(\boldsymbol{a}_1, \boldsymbol{a}_2, \boldsymbol{a}_3) = \sqrt{a}\Gamma^\sigma_{\sigma\alpha}\end{aligned}$$

since $\partial_\alpha \boldsymbol{a}_\beta = \Gamma^\sigma_{\beta\alpha}\boldsymbol{a}_\sigma + b_{\alpha\beta}\boldsymbol{a}_3$ (Theorem 1.6.1).

(ii) Using the *Green formula in Sobolev spaces* (see, e.g., Nečas [1967]) and assuming that the functions $n^{\alpha\beta} = n^{\beta\alpha}$ are in $H^1(\omega)$, we first transform the first integral appearing in the left-hand side of the variational equations. This gives, for all $\boldsymbol{\eta} = (\eta_i) \in H^1_0(\omega) \times H^1_0(\omega) \times L^2(\omega)$, hence *a fortiori* for all $\boldsymbol{\eta} = (\eta_i) \in H^1_0(\omega) \times H^1_0(\omega) \times H^2_0(\omega)$,

$$\begin{aligned}&\int_\omega a^{\alpha\beta\sigma\tau}\gamma_{\sigma\tau}(\boldsymbol{\zeta})\gamma_{\alpha\beta}(\boldsymbol{\eta})\sqrt{a}\,\mathrm{d}y = \int_\omega n^{\alpha\beta}\gamma_{\alpha\beta}(\boldsymbol{\eta})\sqrt{a}\,\mathrm{d}y\\ &= \int_\omega \sqrt{a}n^{\alpha\beta}\Big(\frac{1}{2}(\partial_\beta\eta_\alpha + \partial_\alpha\eta_\beta) - \Gamma^\sigma_{\alpha\beta}\eta_\sigma - b_{\alpha\beta}\eta_3\Big)\,\mathrm{d}y\\ &= \int_\omega \sqrt{a}n^{\alpha\beta}\partial_\beta\eta_\alpha\,\mathrm{d}y - \int_\omega \sqrt{a}n^{\alpha\beta}\Gamma^\sigma_{\alpha\beta}\eta_\sigma\,\mathrm{d}y - \int_\omega \sqrt{a}n^{\alpha\beta}b_{\alpha\beta}\eta_3\,\mathrm{d}y\\ &= -\int_\omega \partial_\beta(\sqrt{a}n^{\alpha\beta})\eta_\alpha\,\mathrm{d}y - \int_\omega \sqrt{a}n^{\alpha\beta}\Gamma^\sigma_{\alpha\beta}\eta_\sigma\,\mathrm{d}y - \int_\omega \sqrt{a}n^{\alpha\beta}b_{\alpha\beta}\eta_3\,\mathrm{d}y\\ &= -\int_\omega \sqrt{a}(\partial_\beta n^{\alpha\beta} + \Gamma^\alpha_{\tau\beta}n^{\tau\beta} + \Gamma^\beta_{\beta\tau}n^{\alpha\tau})\eta_\alpha\,\mathrm{d}y - \int_\omega \sqrt{a}n^{\alpha\beta}b_{\alpha\beta}\eta_3\,\mathrm{d}y\\ &= -\int_\omega \sqrt{a}\Big\{(n^{\alpha\beta}|_\beta)\eta_\alpha + b_{\alpha\beta}n^{\alpha\beta}\eta_3\Big\}\,\mathrm{d}y.\end{aligned}$$

(iii) We then likewise transform the second integral appearing in the left-hand side of the variational equations, viz.,

$$\frac{1}{3}\int_\omega a^{\alpha\beta\sigma\tau}\rho_{\sigma\tau}(\boldsymbol{\zeta})\rho_{\alpha\beta}(\boldsymbol{\eta})\sqrt{a}\,dy = \int_\omega m^{\alpha\beta}\rho_{\alpha\beta}(\boldsymbol{\eta})\sqrt{a}\,dy$$

$$= \int_\omega \sqrt{a}\,m^{\alpha\beta}\partial_{\alpha\beta}\eta_3\,dy$$

$$+ \int_\omega \sqrt{a}\,m^{\alpha\beta}(2b^\sigma_\alpha\partial_\beta\eta_\sigma - \Gamma^\sigma_{\alpha\beta}\partial_\sigma\eta_3)\,dy$$

$$+ \int_\omega \sqrt{a}\,m^{\alpha\beta}(-2b^\tau_\beta\Gamma^\sigma_{\alpha\tau}\eta_\sigma + b^\sigma_\beta|_\alpha\eta_\sigma - b^\sigma_\alpha b_{\sigma\beta}\eta_3)\,dy,$$

for all $\boldsymbol{\eta} = (\eta_i) \in H^1_0(\omega)\times H^1_0(\omega)\times H^2_0(\omega)$. Using the symmetry $m^{\alpha\beta} = m^{\beta\alpha}$, the relation $\partial_\beta\sqrt{a} = \sqrt{a}\,\Gamma^\sigma_{\beta\sigma}$ (cf. part (i)), and the same Green formula as in part (ii), we obtain

$$\int_\omega m^{\alpha\beta}\rho_{\alpha\beta}(\boldsymbol{\eta})\sqrt{a}\,dy = -\int_\omega \sqrt{a}(\partial_\beta m^{\alpha\beta} + \Gamma^\sigma_{\beta\sigma}m^{\alpha\beta} + \Gamma^\alpha_{\sigma\beta}m^{\sigma\beta})\partial_\alpha\eta_3\,dy$$

$$+2\int_\omega \sqrt{a}m^{\alpha\beta}b^\sigma_\alpha\partial_\beta\eta_\sigma\,dy$$

$$+\int_\omega \sqrt{a}m^{\alpha\beta}(-2b^\tau_\beta\Gamma^\sigma_{\alpha\tau}\eta_\sigma + b^\sigma_\beta|_\alpha\eta_\sigma - b^\sigma_\alpha b_{\sigma\beta}\eta_3)\,dy.$$

The same Green formula further shows that

$$-\int_\omega \sqrt{a}(\partial_\beta m^{\alpha\beta} + \Gamma^\sigma_{\beta\sigma}m^{\alpha\beta} + \Gamma^\alpha_{\sigma\beta}m^{\sigma\beta})\partial_\alpha\eta_3\,dy$$

$$= -\int_\omega \sqrt{a}(m^{\alpha\beta}|_\beta)\partial_\alpha\eta_3\,dy = \int_\omega \partial_\alpha(\sqrt{a}\,m^{\alpha\beta}|_\beta)\eta_3\,dy$$

$$= \int_\omega \sqrt{a}(m^{\alpha\beta}|_{\alpha\beta})\eta_3\,dy,$$

$$2\int_\omega \sqrt{a}\,m^{\alpha\beta}b^\sigma_\alpha\partial_\beta\eta_\sigma\,dy = -2\int_\omega \sqrt{a}\{\partial_\beta(b^\sigma_\alpha m^{\alpha\beta}) + \Gamma^\tau_{\beta\tau}b^\sigma_\alpha m^{\alpha\beta}\}\eta_\sigma\,dy.$$

Consequently,

$$\int_\omega m^{\alpha\beta}\rho_{\alpha\beta}(\boldsymbol{\eta})\sqrt{a}\,dy = \int_\omega \sqrt{a}\{-2(b^\alpha_\sigma m^{\sigma\beta})|_\beta + (b^\alpha_\beta|_\sigma)m^{\sigma\beta}\}\eta_\alpha\,dy$$

$$+\int_\omega \sqrt{a}\{m^{\alpha\beta}|_{\alpha\beta} - b^\sigma_\alpha b_{\sigma\beta}m^{\alpha\beta}\}\eta_3\,dy.$$

Using in this relation the easily verified formula

$$(b^\alpha_\sigma m^{\sigma\beta})|_\beta = (b^\alpha_\beta|_\sigma)m^{\sigma\beta} + b^\alpha_\sigma(m^{\sigma\beta}|_\beta)$$

and the symmetry relations $b^\alpha_\beta|_\sigma = b^\alpha_\sigma|_\beta$ (Theorem 2.3.2), we finally obtain

$$\int_\omega m^{\alpha\beta}\rho_{\alpha\beta}(\boldsymbol{\eta})\sqrt{a}\,\mathrm{d}y = -\int_\omega \sqrt{a}\{(b^\alpha_\sigma m^{\sigma\beta})|_\beta + b^\alpha_\sigma(m^{\sigma\beta}|_\beta)\}\eta_\alpha\,\mathrm{d}y \\ - \int_\omega \sqrt{a}\{b^\sigma_\alpha b_{\sigma\beta}m^{\alpha\beta} - m^{\alpha\beta}|_{\alpha\beta}\}\eta_3\,\mathrm{d}y.$$

(iv) By parts (ii) and (iii), the variational equations

$$\int_\omega \Big\{a^{\alpha\beta\sigma\tau}\gamma_{\sigma\tau}(\boldsymbol{\zeta})\gamma_{\alpha\beta}(\boldsymbol{\eta}) + \frac{1}{3}a^{\alpha\beta\sigma\tau}\rho_{\sigma\tau}(\boldsymbol{\zeta})\rho_{\alpha\beta}(\boldsymbol{\eta}) - p^i\eta_i\Big\}\sqrt{a}\,\mathrm{d}y = 0$$

imply that

$$\int_\omega \sqrt{a}\{(n^{\alpha\beta} + b^\alpha_\sigma m^{\sigma\beta})|_\beta + b^\alpha_\sigma(m^{\sigma\beta}|_\beta) + p^\alpha\}\eta_\alpha\,\mathrm{d}y \\ + \int_\omega \sqrt{a}\{b_{\alpha\beta}n^{\alpha\beta} + b^\sigma_\alpha b_{\sigma\beta}m^{\alpha\beta} - m^{\alpha\beta}|_{\alpha\beta} + p^3\}\eta_3\,\mathrm{d}y = 0$$

for all $(\eta_i) \in H^1_0(\omega) \times H^1_0(\omega) \times H^2_0(\omega)$. The announced partial differential equations are thus satisfied in ω. □

The functions

$$n^{\alpha\beta} = \varepsilon a^{\alpha\beta\sigma\tau}\gamma_{\sigma\tau}(\boldsymbol{\zeta})$$

are the **contravariant components of the linearized stress resultant tensor field** inside the shell, and the functions

$$m^{\alpha\beta} = \frac{\varepsilon^3}{3}a^{\alpha\beta\sigma\tau}\rho_{\sigma\tau}(\boldsymbol{\zeta})$$

are the **contravariant components of the linearized stress couple**, or **linearized bending moment, tensor field** inside the shell.

The functions

$$t^{\alpha\beta}|_\beta = \partial_\beta t^{\alpha\beta} + \Gamma^\alpha_{\beta\sigma}t^{\beta\sigma} + \Gamma^\beta_{\beta\sigma}t^{\alpha\sigma},$$
$$t^{\alpha\beta}|_{\alpha\beta} = \partial_\alpha(t^{\alpha\beta}|_\beta) + \Gamma^\sigma_{\alpha\sigma}(t^{\alpha\beta}|_\beta),$$

which have naturally appeared in the course of the proof of Theorem 2.6.3, constitute examples of **first-order**, and **second order, covariant derivatives of a tensor field defined on a surface**, here by means of its contravariant components $t^{\alpha\beta} : \overline{\omega} \to \mathbb{R}$.

Finally, we state a *regularity result* that provides an instance where the *weak solution*, viz., the solution of the variational equations, is also a *classical solution*, viz., a solution of the associated boundary value problem. The proof of this result, which is due to Alexandrescu [1994], is long and delicate and for this reason is only briefly sketched here.

Theorem 2.6.4. *Let ω be a domain in $\mathbb{R}^2$ with boundary γ and let $\boldsymbol{\theta} : \overline{\omega} \to \boldsymbol{E}^3$ be an injective immersion. Assume that, for some integer $m \geq 0$ and some real number $q > 1$, γ is of class $\mathcal{C}^{m+4}$, $\boldsymbol{\theta} \in \mathcal{C}^{m+4}(\overline{\omega}; \boldsymbol{E}^3)$, $p^\alpha \in W^{m+1,q}(\omega)$, and $p^3 \in W^{m,q}(\omega)$. Finally, assume that $\gamma_0 = \gamma$. Then the weak solution found in* Theorem 2.6.2 *satisfies*

$$\boldsymbol{\zeta} = (\zeta_i) \in W^{m+3,q}(\omega) \times W^{m+3,q}(\omega) \times W^{m+4,q}(\omega).$$

Sketch of the proof. To begin with, assume that the boundary γ is of class $\mathcal{C}^4$ and the mapping $\boldsymbol{\theta}$ belongs to the space $\mathcal{C}^4(\overline{\omega}; \boldsymbol{E}^3)$.

One first verifies that the linear system of partial differential equations found in Theorem 2.6.3 (which is of the third order with respect to the unknowns ζ_α and of the fourth order with respect to the unknown ζ_3) is *uniformly elliptic* and satisfies the *supplementing condition on L* and the *complementing boundary conditions*, in the sense of Agmon, Douglis & Nirenberg [1964].

One then verifies that the same system is also *strongly elliptic* in the sense of Nečas [1967, p. 185]. A regularity result of Nečas [1967, Lemma 3.2, p. 260] then shows that the weak solution $\boldsymbol{\zeta} = (\zeta_i)$ found in Theorem 2.6.2, which belongs to the space $H_0^1(\omega) \times H_0^1(\omega) \times H_0^2(\omega)$ since $\gamma_0 = \gamma$ by assumption, satisfies

$$\boldsymbol{\zeta} = (\zeta_i) \in H^3(\omega) \times H^3(\omega) \times H^4(\omega)$$

if $p^\alpha \in H^1(\omega)$ and $p^3 \in L^2(\omega)$.

A result of Geymonat [1965, Theorem 3.5] about the *index* of the associated linear operator then implies that

$$\boldsymbol{\zeta}^\varepsilon = (\zeta_i^\varepsilon) \in W^{3,q}(\omega) \times W^{3,q}(\omega) \times W^{4,q}(\omega)$$

if $p^\alpha \in W^{1,q}(\omega)$ and $p^3 \in L^q(\omega)$ for some $q > 1$.

Assume finally that, for some integer $m \geq 1$ and some real number $q > 1$, γ is of class $\mathcal{C}^{m+4}$ and $\boldsymbol{\theta} \in \mathcal{C}^{m+4}(\overline{\omega}; \boldsymbol{E}^3)$. Then a regularity result of Agmon, Douglis & Nirenberg [1964] implies that

$$\boldsymbol{\zeta} = (\zeta_i) \in W^{m+3,q}(\omega) \times W^{m+3,q}(\omega) \times W^{m+4,q}(\omega).$$

if $p^\alpha \in W^{m+1,q}(\omega)$ and $p^3 \in W^{m,q}(\omega)$. □

Bibliography

AGMON, S.; DOUGLIS, A.; NIRENBERG, L. [1964]: Estimates near the boundary for solutions of elliptic partial differential equations satisfying general boundary conditions. II, *Comm. Pure Appl. Math.* **17**, 35–92.

AKIAN, J.L. [2003]: A simple proof of the ellipticity of Koiter's model, *Analysis and Applications* **1**, 1–16.

ALEXANDRESCU, O. [1994]: Théorème d'existence pour le modèle bidimensionnel de coque non linéaire de W.T. Koiter, *C.R. Acad. Sci. Paris, Sér. I*, **319**, 899–902.

AMROUCHE, C.; GIRAULT, V. [1994]: Decomposition of vector spaces and application to the Stokes problem in arbitrary dimension, *Czech. Math. J.* **44**, 109–140.

ANICIC, S.; LE DRET, H.; RAOULT, A. [2005]: The infinitesimal rigid displacement lemma in Lipschitz coordinates and application to shells with minimal regularity, *Math. Methods Appl. Sci.* **27**, 1283–1299.

BAMBERGER, Y. [1981]: *Mécanique de l'Ingénieur, Volume II,* Hermann, Paris.

BERGER M.; GOSTIAUX, B. [1987]: *Géométrie Différentielle: Variétés, Courbes et Surfaces*, Presses Universitaires de France, Paris.

BERNADOU, M.; CIARLET, P.G. [1976]: Sur l'ellipticité du modèle linéaire de coques de W.T. Koiter, in *Computing Methods in Applied Sciences and Engineering* (R. GLOWINSKI & J.L. LIONS, Editors), pp. 89–136, Lecture Notes in Economics and Mathematical Systems, **134**, Springer-Verlag, Heidelberg.

BERNADOU, M.; CIARLET, P.G.; MIARA, B. [1994]: Existence theorems for two-dimensional linear shell theories, *J. Elasticity* **34**, 111–138.

BLOUZA, A.; LE DRET, H. [1999]: Existence and uniqueness for the linear Koiter model for shells with little regularity, *Quart. Appl. Math.* **57**, 317–337.

BOLLEY, P.; CAMUS, J. [1976]: Régularité pour une classe de problèmes aux limites elliptiques dégénérés variationnels, *C.R. Acad. Sci. Paris, Sér. A*, **282**, 45–47.

BORCHERS, W.; SOHR, H. [1990]: On the equations $\operatorname{rot} v = g$ and $\operatorname{div} u = f$ with zero boundary conditions, *Hokkaido Math. J.* **19**, 67–87.

DO CARMO, M.P. [1976]: *Differential Geometry of Curves and Surfaces*, Prentice-Hall, Englewood Cliffs.

CIARLET, P.G. [1982]: *Introduction à l'Analyse Numérique Matricielle et à l'Optimisation*, Masson, Paris (English translation: *Introduction to Numerical Linear Algebra and Optimisation*, Cambridge University Press, Cambridge, 1989).

CIARLET, P.G. [1988]: *Mathematical Elasticity, Volume I: Three-Dimensional Elasticity*, North-Holland, Amsterdam.

CIARLET, P.G. [1997]: *Mathematical Elasticity, Volume II: Theory of Plates*, North-Holland, Amsterdam.

CIARLET, P.G. [2000a]: *Mathematical Elasticity, Volume III: Theory of Shells*, North-Holland, Amsterdam.

CIARLET, P.G. [2000b]: Un modèle bi-dimensionnel non linéaire de coque analogue à celui de W.T. Koiter, *C.R. Acad. Sci. Paris, Sér. I*, **331**, 405–410.

CIARLET, P.G. [2005]: *An Introduction to Differential Geometry with Applications to Geometry*, Springer, Dordrecht.

CIARLET, P.G.; LARSONNEUR, F. [2001]: On the recovery of a surface with prescribed first and second fundamental forms, *J. Math. Pures Appl.* **81**, 167–185.

CIARLET, P.G.; LODS, V. [1996]: On the ellipticity of linear membrane shell equations, *J. Math. Pures Appl.* **75**, 107–124.

CIARLET, P.G.; MARDARE, S. [2001]: On Korn's inequalities in curvilinear coordinates, *Math. Models Methods Appl. Sci.* **11**, 1379–1391.

CIARLET, P.G.; MIARA, B. [1992]: On the ellipticity of linear shell models, *Z. angew. Math. Phys.* **43**, 243–253.

CIARLET, P.G.; ROQUEFORT, A. [2001]: Justification of a two-dimensional nonlinear shell model of Koiter's type, *Chinese Ann. Math.* **22B**, 129–244.

CIARLET, P.G.; SANCHEZ-PALENCIA, E. [1996]: An existence and uniqueness theorem for the two-dimensional linear membrane shell equations, *J. Math. Pures Appl.* **75**, 51–67.

DACOROGNA, B. [1989]: *Direct Methods in the Calculus of Variations*, Springer, Berlin.

DESTUYNDER, P. [1985]: A classification of thin shell theories, *Acta Applicandae Mathematicae* **4**, 15–63.

DUVAUT, G.; LIONS, J.L. [1972]: *Les Inéquations en Mécanique et en Physique*, Dunod, Paris (English translation: *Inequalities in Mechanics and Physics*, Springer-Verlag, Berlin, 1976).

FRIESECKE, G.; JAMES, R.D.; MORA, M.G.; MÜLLER, S. [2003]: Derivation of nonlinear bending theory for shells from three dimensional nonlinear elasticity by Gamma-convergence, *C.R. Acad. Sci. Paris, Sér. I*, **336**, 697–702.

GAUSS, C.F. [1828]: Disquisitiones generales circas superficies curvas, *Commentationes societatis regiae scientiarum Gottingensis recentiores* **6**, Gttingen.

GEYMONAT, G. [1965]: Sui problemi ai limiti per i sistemi lineari ellittici, *Ann. Mat. Pura Appl.* **69**, 207–284.

GEYMONAT, G.; GILARDI, G. [1998]: Contre-exemples à l'inégalité de Korn et au lemme de Lions dans des domaines irréguliers, in *Equations aux Dérivées Partielles et Applications. Articles Dédiés à Jacques-Louis Lions*, pp. 541–548, Gauthier-Villars, Paris.

GEYMONAT, G.; SUQUET, P. [1986]: Functional spaces for Norton-Hoff materials, *Math. Methods Appl. Sci.* **8**, 206–222.

HARTMAN, P.; WINTNER, A. [1950]: On the embedding problem in differential geometry, *Amer. J. Math.* **72**, 553–564.

JOHN, F. [1965]: Estimates for the derivatives of the stresses in a thin shell and interior shell equations, *Comm. Pure Appl. Math.* **18**, 235–267.

JOHN, F. [1971]: Refined interior equations for thin elastic shells, *Comm. Pure Appl. Math.* **18**, 235–267.

KLINGENBERG, W. [1973]: *Eine Vorlesung über Differentialgeometrie,* Springer-Verlag, Berlin (English translation: *A Course in Differential Geometry*, Springer-Verlag, Berlin, 1978).

KOITER, W.T. [1966]: On the nonlinear theory of thin elastic shells, *Proc. Kon. Ned. Akad. Wetensch.* **B69**, 1–54.

KOITER, W.T. [1970]: On the foundations of the linear theory of thin elastic shells, *Proc. Kon. Ned. Akad. Wetensch.* **B73**, 169–195.

KÜHNEL, W. [2002]: *Differentialgeometrie,* Fried. Vieweg & Sohn, Wiesbaden (English translation: *Differential Geometry: Curves-Surfaces-Manifolds,* American Mathematical Society, Providence, 2002).

LE DRET, H. [2004]: Well-posedness for Koiter and Naghdi shells with a G_1-midsurface, *Analysis and Applications* **2**, 365–388.

LE DRET, H.; RAOULT, A. [1996]: The membrane shell model in nonlinear elasticity: A variational asymptotic derivation, *J. Nonlinear Sci.* **6**, 59–84.

LODS, V. & MIARA, B. [1998]: Nonlinearly elastic shell models. II. The flexural model, *Arch. Rational Mech. Anal.* **142**, 355–374.

MAGENES, E.; STAMPACCHIA, G. [1958]: I problemi al contorno per le equazioni differenziali di tipo ellittico, *Ann. Scuola Norm. Sup. Pisa* **12**, 247–358.

MARDARE, S. [2003a]: Inequality of Korn's type on compact surfaces without boundary, *Chinese Annals Math.* **24B**, 191–204

MARDARE, S. [2003b]: The fundamental theorem of surface theory for surfaces with little regularity, *J. Elasticity* **73**, 251–290.

MARDARE, S. [2005]: On Pfaff systems with L^p coefficients and their applications in differential geometry, *J. Math. Pures Appl.,* to appear.

MIARA, B. [1998]: Nonlinearly elastic shell models. I. The membrane model, *Arch. Rational Mech. Anal.* **142**, 331–353.

MORREY, JR., C.B. [1952]: Quasi-convexity and the lower semicontinuity of multiple integrals, *Pacific J. Math.* **2**, 25–53.

NEČAS, J. [1967]: *Les Méthodes Directes en Théorie des Equations Elliptiques*, Masson, Paris.

SCHWARTZ, L. [1966]: *Théorie des Distributions*, Hermann, Paris.

SCHWARTZ, L. [1992]: *Analyse II: Calcul Différentiel et Equations Différentielles*, Hermann, Paris.

SPIVAK, M. [1999]: *A Comprehensive Introduction to Differential Geometry, Volumes I to V, Third Edition,* Publish or Perish, Berkeley.

STEIN, E. [1970]: *Singular Integrals and Differentiability Properties of Functions*, Princeton University Press.

STOKER, J.J. [1969]: *Differential Geometry*, John Wiley, New York.

TARTAR, L. [1978]: *Topics in Nonlinear Analysis*, Publications Mathématiques d'Orsay No. 78.13, Université de Paris-Sud, Orsay.

Non-Brittle and Brittle Thin Films - Lectures I, II

Irene Fonseca

Department of Mathematical Sciences, Carnegie Mellon University

Abstract Using Γ-convergence techniques the relaxed 2-D energy is obtained for non-brittle and brittle thin films as a result of a dimension reduction analysis for thin domains.

1 Introduction

The content of this lecture is based on the paper

A. Braides and I. Fonseca, Brittle Thin Films, *Appl. Math. Optim.* **44** (2001), 299–323.

We consider a thin 3D (elastic) domain represented by

$$\Omega(\varepsilon) := \{(x_1, x_2, x_3) : (x_1, x_2) \in \omega \text{ and } |x_3| < \varepsilon f_\varepsilon(x_1, x_2)\},$$

where ω is a bounded domain of $\mathbb{R}^2$ and $f_\varepsilon(x_1, x_2)$ determines the ε-dependent profile $x_3 = \pm f_\varepsilon(x_1, x_2)$, and $\varepsilon > 0$. Define $\Omega := \omega \times \left(-\frac{1}{2}, \frac{1}{2}\right)$, $\Gamma := \partial\omega \times \left(-\frac{1}{2}, \frac{1}{2}\right)$, and $\Gamma_\varepsilon := \partial\omega \times \left(-\frac{1}{2}\varepsilon, \frac{1}{2}\varepsilon\right)$.

The elastic energy density in $\Omega(\varepsilon)$ is $\mathcal{W}(\varepsilon)(x_1, x_2, x_3; \cdot)$, and the thin body may be subject to a body load with density $F(\varepsilon)(x_1, x_2, x_3)$ (and possibly surface loads as well on $\Sigma_\varepsilon^\pm := \omega \times \{x_3 = \pm\varepsilon f_\varepsilon(x_1, x_2)\}$). In order to reach equilibrium, the deformation $u(\varepsilon)$ seeks to minimize

$$w \mapsto \int_{\Omega(\varepsilon)} \mathcal{W}(\varepsilon)(x; \nabla w)\, dx - \int_{\Omega(\varepsilon)} F(\varepsilon) \cdot w\, dx$$

among all kinematically admissible fields w.

*The research of I. Fonseca was partially funded by the National Science Foundation (NSF) under NSF Grants DMS-0103799 and DMS-0401763, and by the Center for Nonlinear Analysis under NSF Grants DMS-9803791, DMS-0405343 and DMS-0635983.

As it is usual, we reformulate this problem on a fixed domain through a change of variables,

$$\begin{aligned} \Omega_\varepsilon &:= \{(x_1,x_2,x_3) : (x_1,x_2,\varepsilon x_3) \in \Omega(\varepsilon)\}, \\ u_\varepsilon(x_1,x_2,x_3) &:= u(\varepsilon)(x_1,x_2,\varepsilon x_3), \\ W_\varepsilon(x_1,x_2,x_3;\cdot) &:= W(\varepsilon)(x_1,x_2,\varepsilon x_3;\cdot), \\ F_\varepsilon(x_1,x_2,x_3) &:= F(\varepsilon)(x_1,x_2,\varepsilon x_3), \end{aligned}$$

so, equivalently, u_ε seeks to minimize

$$v \mapsto \int_{\Omega_\varepsilon} W_\varepsilon \left(x; \nabla_\alpha v \Big| \frac{1}{\varepsilon}\nabla_3 v\right) dx - \int_{\Omega_\varepsilon} F_\varepsilon \cdot v \, dx$$

among all kinematically admissible fields v on Ω_ε. Here and in what follows we adopt the following notation: Greek letters will always run from 1 to 2 when taken as indices. Thus coordinates will be denoted by x_α, x_3, and $(F_\alpha|F_3)$ stands for the 3×3 matrix with column elements F_1, F_2, F_3 (3 vectors in $\mathbb{R}^3$). We will identify $W^{1,p}(\Omega) \cap \left\{u : \frac{\partial u}{\partial x_3} = 0\right\}$ with $W^{1,p}(\omega)$ ($\Omega := \omega \times (-1,1)$). As it is usual, $\to$ will denote strong convergence whereas $\rightharpoonup$ (resp. $\overset{*}{\rightharpoonup}$) will denote weak (resp. weak-*) convergence. Throughout we designate by C a positive constant whose value may change within one formula and from one formula to the next.

For the purpose of these lectures we consider the case where no loads are present and we direct our attention to the study of the $\Gamma(L^p)$–limit ($1 < p < +\infty$) of the functionals

$$v \mapsto E_\varepsilon(v;\omega) := \int_{\Omega_\varepsilon} W_\varepsilon \left(x; \nabla_\alpha v \Big| \frac{1}{\varepsilon}\nabla_3 v\right) dx.$$

Under appropriate growth conditions, minimizers of E_ε – if they exist – will L^p-converge to minimizers of that $\Gamma(L^p)$–limit, and thus a characterization of the latter will provide the asymptotic effective energy for equilibria states of Ω_ε.

We recall that if $\{E_n\}$ is a sequence of functions from a Banach space X into $\overline{\mathbb{R}}$ and E is a function from X into $\overline{\mathbb{R}}$, then

$\Gamma 1$. E is the $\Gamma(X)$– lim inf of E_n if, for any x in X,

$$E(x) = \inf_{\{x_n\}} \{\liminf_{n\to+\infty} E_n(x_n) : x_n \to x \text{ in } X\},$$

$\Gamma 2$. E is the $\Gamma(X)$– lim sup of E_ε if, for every x in X,

$$E(x) = \inf_{\{x_n\}} \{\limsup_{n\to+\infty} E_n(x_n) : x_n \to x \text{ in } X\}.$$

Also

Γ3. if $\Gamma(X)-\liminf E_n = \Gamma(X)-\limsup E_n$ then the common value is called the $\Gamma(X)$-limit of E_n.
Therefore, $E(u) = \Gamma-\lim E_n(u)$ if and only if
i) whenever $u_n \to u$ in X then

$$E(u) \leq \liminf_{n\to+\infty} E_n(u_n),$$

ii) there exists a (*recovery*) sequence $\{u_n\}$ such that $u_n \to u$ in X and

$$E(u) = \liminf_{n\to+\infty} E_n(u_n).$$

Moreover, given a family of maps $E_\varepsilon : X \to \overline{\mathbb{R}}$, $\varepsilon > 0$, if $u \in X$ then we say that

Γ4. $\Gamma(X)-\lim E_\varepsilon(u) = E(u)$ if $E(u) = \Gamma(X)-\lim E_{\varepsilon_n}(u)$ for every sequence $\{\varepsilon_n\}$ such that $\varepsilon_n \to 0^+$.
Hence, it can be shown that $\Gamma(X)-\lim E_\varepsilon(u) = E(u)$ if and only if
i) for every sequences $\{u_n\}$ and $\{\varepsilon_n\}$ such that $u_n \to u$ in X and $\varepsilon_n \to 0^+$ then

$$E(u) \leq \liminf_{n\to+\infty} E_{\varepsilon_n}(u_n),$$

ii) for every sequence $\{\varepsilon_n\}$ converging to 0^+ there exists a sequence $\{u_n\}$ such that $u_n \to u$ in X and

$$E(u) = \lim_{n\to+\infty} E_{\varepsilon_n}(u_n).$$

2 The Theorem of Le Dret and Raoult

Here we consider the case wheree $f_\varepsilon(x_\alpha) = 1/2$ for all $x_\alpha \in \omega$, so that we are led to the study the limit of the family of scaled energies

$$\int_\Omega W\left(\nabla_\alpha u\Big|\frac{1}{\varepsilon}\nabla_3 u\right) dx.$$

We recall the notion of *quasiconvexity* (see (23)) .

Definition 2.0.1. *Let $f : \mathbb{R}^{d\times N} \to [0, +\infty)$ be a Borel measurable function. Then f is said to be quasiconvex if for every $\varphi \in W_0^{1,\infty}(Q; \mathbb{R}^d)$ and for every $F \in \mathbb{R}^{d\times N}$*

$$f(F) \leq \int_Q f(F + \nabla\varphi)\, dx$$

where $Q := \left(-\frac{1}{2}, \frac{1}{2}\right)^N$.

Given a Borel function $f : \mathbb{R}^{d\times N} \to [0,+\infty)$ we define its *quasiconvex envelope* as

$$Qf(F) := \inf \left\{ \int_Q f(F + \nabla\varphi)\, dx : \varphi \in W_0^{1,\infty}(Q;\mathbb{R}^d) \right\}. \tag{2.0.1}$$

It can be shown that the previous definition is independent of the domain of integration, in that if f is quasiconvex and if U is any open, bounded subset of $\mathbb{R}^N$ with $\mathcal{L}^N(\partial U) = 0$, where $\mathcal{L}^N$ stands for the N-dimensional Lebesgue measure on $\mathbb{R}^N$ (often we write indistinguishably $|U|$ and $\mathcal{L}^N(U)$), then

$$f(F) \le \frac{1}{|U|} \int_U f(F + \nabla\varphi)\, dx$$

for every $F \in \mathbb{R}^{d\times N}$ and $\varphi \in W_0^{1,\infty}(U;\mathbb{R}^d)$. Moreover, Qf is quasiconvex.

If, in addition, f satisfies a growth condition of the type

$$0 \le f(F) \le C(1 + |F|^p) \tag{2.0.2}$$

for some $1 \le p < +\infty$, $C > 0$, and for all $F \in \mathbb{R}^{d\times N}$, then a simple density result shows that the test functions in (2.0.1) may be taken in $W_0^{1,p}(Q;\mathbb{R}^d)$ or in $W_\#^{1,p}(Q;\mathbb{R}^d)$, where the latter is the space of functions $\varphi \in W_{\text{loc}}^{1,p}(\mathbb{R}^N;\mathbb{R}^d)$ which are Q-periodic, i.e. $\varphi(x + k\mathrm{e}_i) = \varphi(x)$ for all $x \in \mathbb{R}^N$, all $k \in \mathbb{Z}^N$, and all $i = 1,\ldots,N$, where $\{\mathrm{e}_1,\ldots,\mathrm{e}_N\}$ is the standard orthonormal basis of $\mathbb{R}^N$. If f is quasiconvex and satisfies (2.0.2) then it also verifies the p-Lipschitz condition

$$|f(F) - f(F')| \le C(1 + |F|^{p-1} + |F'|^{p-1})|F - F'|. \tag{2.0.3}$$

The notion of quasiconvexity plays a pivotal role in the study of lowersemicontinuity and relaxation of functionals involving derivatives. In particular (see (1)):

Theorem 2.0.2. *Let $f : \mathbb{R}^{d\times N} \to [0,+\infty)$ be a continuous function, $1 \le p < +\infty$, and*

$$0 \le f(F) \le C(1 + |F|^p)$$

for some $C > 0$ and for all $F \in \mathbb{R}^{d\times N}$. Let Ω be an open, bounded subset of $\mathbb{R}^N$. Then given $u \in W^{1,p}(\Omega;\mathbb{R}^d)$

$$\inf_{\{u_n\}} \left\{ \liminf_{n\to\infty} \int_\Omega f(\nabla u_n) dx : u_n \rightharpoonup u \text{ in } W^{1,p}(\Omega;\mathbb{R}^d) \right\} = \int_\Omega Qf(\nabla u)\, dx.$$

Consequently,

$$u \in W^{1,p}(\Omega) \mapsto \int_\Omega f(\nabla u)\, dx$$

is sequential weakly lowersemicontinuous if and only if f is quasiconvex.

Going back to the thin film context, let $W : \mathbb{R}^{3\times 3} \to [0, +\infty)$ be a continuous function,

$$\frac{1}{C}|F|^p - C \leq W(F) \leq C(1 + |F|^p) \tag{2.0.4}$$

for some $C > 0$, $1 < p < +\infty$, and for all $F \in \mathbb{R}^{3\times 3}$. Define $\overline{W}(\overline{F}) := \min\{W(\overline{F}|F_3) : F_3 \in \mathbb{R}^3\}$, $\overline{F}$ is a 3×2 matrix. Then $\overline{W} : \mathbb{R}^{3\times 2} \to \mathbb{R}$ determines naturally two quasiconvex envelopes; precisely, given $\overline{F} \in \mathbb{R}^{2\times 3}$ we may define

$$Q_2\overline{W}(\overline{F}) := \inf\left\{\int_{Q'} \overline{W}(\overline{F} + \nabla_\alpha \varphi(x))\, dx_\alpha : \ \varphi \in W^{1,\infty}_{\#}(Q'; \mathbb{R}^3)\right\},$$

and also

$$Q_3\overline{W}(\overline{F}) := \inf\left\{\int_{Q} \overline{W}(\overline{F} + \nabla_\alpha \varphi(x))\, dx : \ \varphi \in W^{1,\infty}_{\#}(Q; \mathbb{R}^3)\right\},$$

where $Q' := \left(-\frac{1}{2}, \frac{1}{2}\right)^2$. It can be shown easily that $Q_2\overline{W} = Q_3\overline{W}$.

Concerning the next theorem, we refer to (31), (32) and (33).

Theorem 2.0.3. *The $\Gamma(L^1)$-limit of*

$$E_\varepsilon(u) := \begin{cases} \int_\Omega W\left(\nabla_\alpha u \middle| \frac{1}{\varepsilon}\nabla_3 u\right) dx & \textit{if} u \in W^{1,p}(\omega; \mathbb{R}^3), \\ +\infty & \textit{otherwise}, \end{cases}$$

is the functional

$$J_0(u) := \begin{cases} \int_\omega Q_2\overline{W}\left(\nabla_\alpha u\right) dx & \textit{if} u \in W^{1,p}(\omega; \mathbb{R}^3), \\ +\infty & \textit{otherwise}. \end{cases}$$

Proof. In view of (2.0.4) it is clear that if $\varepsilon_n \to 0^+$, and if $u_n \in W^{1,p}(\Omega; \mathbb{R}^3)$ are such that $u_n \to u$ in L^1 and $\sup_n E_{\varepsilon_n}(u_n) < +\infty$, then $u \in W^{1,p}(\Omega; \mathbb{R}^3)$, $\nabla_3 u = 0$, and we may identify u with a function in $W^{1,p}(\omega; \mathbb{R}^3)$.

Let now $u \in W^{1,p}(\omega; \mathbb{R}^3)$ and $\varepsilon_n \to 0^+$. By Theorem 2.0.2 we have

$$\begin{aligned} \liminf_{n\to\infty} E_{\varepsilon_n}(u_n) &\geq \liminf_{n\to\infty} \int_\Omega Q_3\overline{W}(\nabla_\alpha u_n)\, dx \\ &\geq \int_\Omega Q_3\overline{W}(\nabla_\alpha u)\, dx = \int_\omega Q_2\overline{W}(\nabla_\alpha u)\, dx, \end{aligned}$$

where we used the fact that $Q_3\overline{W}$ inherits the growth condition

$$0 \leq Q_3\overline{W}(\overline{F}) \leq C(1 + |\overline{F}|^p).$$

This shows that $\Gamma(L^1) - \liminf E_\varepsilon \geq J_0$.

Conversely, let $u \in W^{1,p}(\omega;\mathbb{R}^3)$ and again by Theorem 2.0.2 there exists a sequence $\{u_n\} \subset W^{1,p}(\omega;\mathbb{R}^3)$ such that $u_n \rightharpoonup u$ in $W^{1,p}(\omega;\mathbb{R}^3)$ and

$$\int_\omega Q_2\overline{QW}(\nabla_\alpha u)\,dx = \lim_{n\to\infty}\int_\omega \overline{QW}(\nabla_\alpha u_n)\,dx.$$

In turn, Aumann's Measurability Selection criterion guarantees the existence of measurable functions $z_n : \omega \to \mathbb{R}^3$ such that

$$+\infty > \overline{QW}(\nabla_\alpha u_n(x_\alpha)) = QW(\nabla_\alpha u_n(x_\alpha)|z_n(x_\alpha)).$$

Using the coercivity condition (2.0.4), we deduce that $\{z_n\}$ is bounded in $L^p(\omega;\mathbb{R}^3)$. A density argument and (2.0.4) allow us to assume, without loss of generality, that $z_n \in W^{1,p}(\omega;\mathbb{R}^3)$. If necessary, relabel (u_n, z_n) in such a way that

$$\lim_{n\to\infty} \varepsilon_n||z_n||_{W^{1,p}(\omega)} = 0.$$

Define

$$v_n(x) := u_n(x_\alpha) + \varepsilon_n x_3 z_n(x_\alpha).$$

Clearly $v_n \to u$ in $L^1(\Omega;\mathbb{R}^3)$ and

$$\begin{aligned}\int_\omega Q_2\overline{W}(\nabla_\alpha u)\,dx_\alpha &\geq \int_\omega Q_2\overline{QW}(\nabla_\alpha u)\,dx_\alpha \\ &\geq \lim_{n\to\infty}\int_\omega QW(\nabla_\alpha u_n(x_\alpha)|z_n(x_\alpha))\,dx_\alpha \\ &= \lim_{n\to\infty}\int_\Omega QW\left(\nabla_\alpha v_n(x)\Big|\frac{1}{\varepsilon_n}\nabla_3 v_n(x)\right)\,dx,\end{aligned}$$

where we used the fact that QW satisfies (2.0.3).

Applying Theorem 2.0.2 to the energy

$$F \mapsto W\left(F_\alpha\Big|\frac{1}{\varepsilon_n}F_3\right),$$

for every $n \in \mathbb{N}$ there exists a sequence $\{v_{n,k}\}$ converging weakly in $W^{1,p}(\Omega;\mathbb{R}^3)$ to v_n such that

$$\int_\Omega QW\left(\nabla_\alpha v_n(x)\Big|\frac{1}{\varepsilon_n}\nabla_3 v_n(x)\right)\,dx = \lim_{k\to\infty}\int_\Omega W\left(\nabla_\alpha v_{n,k}(x)\Big|\frac{1}{\varepsilon_n}\nabla_3 v_{n,k}(x)\right)\,dx.$$

Extract a diagonal subsequence $w_n = v_{n,k(n)}$ so that $w_n \to u$ in L^1 and

$$\int_\omega Q_2\overline{W}(\nabla_\alpha u)\,dx_\alpha = \lim_{n\to\infty}\int_\Omega W\left(\nabla_\alpha w_n(x)\Big|\frac{1}{\varepsilon_n}\nabla_3 w_n(x)\right)\,dx,$$

thus asserting that $\Gamma(L^1) - \limsup E_\varepsilon \leq J_0$. □

3 Fracture in Thin Films and BV Spaces

Next we take into account the possibility of fracture in a thin film, and, accordingly, we add to the formulation above a crack initiation energy. Precisely, we consider

$$\int_{\Omega_\varepsilon\setminus K} W(\nabla v)\,dx + \int_{K\cap\Omega_\varepsilon} \vartheta(v^+ - v^-, \nu(K))\,d\mathcal{H}^2,$$

where K denotes an unknown crack surface, $\mathcal{H}^2(K)$ is the 2-dimensional Hausdorff measure of K, $\nu(K)$ is the normal to the crack surface in the reference configuration, $v \in W^{1,p}(\Omega_\varepsilon \setminus K; \mathbb{R}^3)$ is the deformation, defined in the complement of the crack, and $v^\pm$ are the traces of u on both sides of K. The surface energy density ϑ is a fracture initiation term defined on $\mathbb{R}^3 \times S^2$. In Griffith's theory (30) ϑ is constant; we have added a possible dependence on $\nu(K)$ to account for anisotropy, and a dependence on the jump $u^+ - u^-$ to include Barenblatt's approach (12). Again, by performing a change of variables, and extending ϑ to $\mathbb{R}^3 \times \mathbb{R}^3$ as a homogeneous function of degree 1 in the variable ν, we arrive at the re-scaled energy

$$\int_{\Omega\setminus K_\varepsilon} W\left(\nabla_\alpha u\Big|\frac{1}{\varepsilon}\nabla_3 u\right)\,dx_\alpha dx_3 + \int_{K_\varepsilon\cap\Omega} \vartheta\left(u^+ - u^-, \nu_\alpha(K_\varepsilon), \frac{1}{\varepsilon}\nu_3(K_\varepsilon)\right)\,d\mathcal{H}^2.$$

It is well known that free discontinuity problems formulated in terms of variable domains $\Omega \setminus K_\varepsilon$ and unknown admissible field u are very difficult to handle, and often it is more convenient to set up the problem within the De Giorgi and Ambrosio framework of *special functions of bounded variation* $SBV(\Omega;\mathbb{R}^3)$ (see (8)), by replacing K_ε by the set $S(u)$ of essential discontinuity points for u and interpreting ∇u as an approximate gradient, thus reducing the simultaneous variation in the crack site and in the deformation to a variation on the discontinuous function u only.

Let $\mathcal{M}(\Omega)$ be the set of all signed Radon measures on Ω with bounded total variation. We say that $u \in L^1(\Omega;\mathbb{R}^d)$ is a *function of bounded variation*, and we write $u \in BV(\Omega;\mathbb{R}^d)$, if every its distributional first derivatives

$D_i u_j$ belong to $\mathcal{M}(\Omega)$. We denote by Du the $d \times N$-valued measure whose entries are $D_i u_j$. $S(u)$ is the complement of the Lebesgue set of u; i.e., $x \notin S(u)$ if and only if

$$\lim_{\rho \to 0^+} \rho^{-N} \int_{B_\rho(x)} |u(y) - z| \, dy = 0$$

for some $z \in \mathbb{R}^m$. If $u \in L^1(R^d)$ then the set $S(u)$ is Lebesgue-negligible, and if, in addition, $u \in BV(\Omega; \mathbb{R}^d)$ then the Hausdorff dimension of $S(u)$ is at most $N-1$; precisely, $S(u)$ is *rectifiable* in that there is a countable sequence of C^1 hypersurfaces Γ_i which covers $\mathcal{H}^{N-1}$-almost all of $S(u)$; i.e., $\mathcal{H}^{N-1}(S(u) \setminus \bigcup_{i=1}^{\infty} \Gamma_i) = 0$. Moreover, for $\mathcal{H}^{N-1}$-almost every $x \in S(u)$ it is possible to find a, $b \in \mathbb{R}^d$ and $\nu \in S^{N-1}$ such that

$$\lim_{\rho \to 0^+} \rho^{-N} \int_{B_\rho^\nu(x)} |u(y) - a| \, dy = 0, \qquad \lim_{\rho \to 0^+} \rho^{-N} \int_{B_\rho^{-\nu}(x)} |u(y) - b| \, dy = 0,$$

where $B_\rho^\nu(x) := \{y \in B_\rho(x) : \langle y - x, \nu \rangle > 0\}$. The triplet (a, b, ν) is uniquely determined up to a change of sign of ν and an interchange between a and b, and in the sequel it will be denoted by $(u^+, u^-, \nu(u))$.

The distributional derivative of a function $u \in BV(\Omega; \mathbb{R}^d)$ admits the decomposition

$$Du = \nabla u \, \mathcal{L}^N \llcorner \Omega + J_u + C_u,$$

where $\nabla u \, \mathcal{L}^N$ is the *Lebesgue part* of Du,

$$J_u := (u^+ - u^-) \otimes \nu(u) \, \mathcal{H}^{N-1} \llcorner S(u)$$

is the *Hausdorff part* or *jump* part of Du, and C_u is the *Cantor part* of Du. We recall that the measure C_u is singular with respect to the Lebesgue measure and it is 'diffuse'; i.e., $C_u(S) = 0$ for every set S with $\mathcal{H}^{N-1}(S) < +\infty$. For the general exposition of the theory of functions of bounded variation we refer the reader to Ambrosio, Fusco and Pallara (9), Evans and Gariepy (27), Federer (28), and Ziemer (35). A quick introduction can be found in (16) Section 1.

We say that a function $u \in BV(\Omega; \mathbb{R}^d)$ is a *special function of bounded variation* if $C_u \equiv 0$, or, equivalently, if

$$Du = \nabla u \, \mathcal{L}_N \llcorner \Omega + (u^+ - u^-) \otimes \nu(u) \, \mathcal{H}^{N-1} \llcorner S(u).$$

We denote the space of special functions of bounded variation by $SBV(\Omega; \mathbb{R}^d)$. The introduction of this space is due to De Giorgi and Ambrosio (25). $GSBV(\Omega)$ is defined as the space of scalar functions in $L^1(\Omega)$ such that for all $T > 0$ the truncations $u_T := (-T) \wedge (u \vee T)$ belong to $SBV(\Omega)$. We say that $u \in GSBV(\Omega; \mathbb{R}^d)$ if every component of u is in $GSBV(\Omega)$. For the properties of functions in $SBV(\Omega; \mathbb{R}^d)$ and $GSBV(\Omega; \mathbb{R}^d)$ we refer to the book by Ambrosio, Fusco and Pallara (9).

Let $p > 1$. $SBV_p(\Omega; \mathbb{R}^d)$ and $GSBV_p(\Omega; \mathbb{R}^d)$ are defined as the subspaces of functions u of $SBV(\Omega; \mathbb{R}^d)$ and $GSBV(\Omega; \mathbb{R}^d)$, respectively, such that

$$\mathcal{H}^{N-1}(S(u) \cap \Omega) < +\infty \qquad \text{and} \qquad \nabla u \in L^p(\Omega; \mathbb{R}^{d\times N}).$$

These spaces are natural domains for the treatment of energies with bulk and surface contributions in the case where the bulk energy density grows superlinearly at infinity.

A continuous function $\vartheta : \mathbb{R}^d \times S^{N-1} \to \mathbb{R}$ is said to be *BV-elliptic* if functions with an hyperplane as jump set are local minimizers for the corresponding surface integral functional among piecewise constant functions with respect to periodic perturbations; precisely, if Q_ν is a unit cube in $\mathbb{R}^N$ with center 0 and one face orthogonal to ν and $GSBV_\#(Q_\nu; \mathbb{R}^d, z)$ denotes the space of functions $u \in GSBV_\infty(Q_\nu; \mathbb{R}^d)$ which are 1-periodic in the direction orthogonal to ν, with $\nabla u = 0$ a.e., and such that $u(x) = \pm z/2$ if $\langle x, \nu\rangle = \pm 1/2$, then

$$\vartheta(z,\nu) \le \int_{S(u)\cap Q_\nu} \vartheta(u^+ - u^-, \nu(u))\, d\mathcal{H}^{N-1}$$

for all $z \in \mathbb{R}^d$, $\nu \in S^{N-1}$ and for all $u \in GSBV_\#(Q_\nu; \mathbb{R}^d, z)$. The *BV-elliptic envelope* $R\vartheta$ of a continuous function $\vartheta : \mathbb{R}^d \times S^{N-1} \to \mathbb{R}$ is the largest BV-elliptic function not larger than ϑ. We have

$$R\vartheta(z,\nu) := \inf\left\{\int_{S(u)\cap Q_\nu} \vartheta(u^+ - u^-, \nu(u))\, d\mathcal{H}^{N-1} :\ u \in GSBV_\#(Q_\nu; \mathbb{R}^d, z)\right\}$$

for all $z \in \mathbb{R}^d$ and $\nu \in S^{N-1}$ (see (7), (20)).

The following closure and lower semicontinuity theorem holds on $GSBV_p(\Omega)$ (see (2), (9)).

Theorem 3.0.4. *Let W be a quasiconvex function, let ϑ be a BV-elliptic function, with*

$$0 \le W(F) \le C(1 + |F|^p), \qquad 0 \le \vartheta(z,\nu) \le C(1 + |z|)$$

for some $C > 0$, and for all $F \in \mathbb{R}^{d\times N}$, $z \in \mathbb{R}^d$, and $\nu \in S^{N-1}$. If $\{u_n\}$ is a sequence in $GSBV_p(\Omega; \mathbb{R}^d)$ with $u_n \to u$ in $L^1(\Omega; \mathbb{R}^d)$, and if

$$\sup_n\left\{\int_\Omega |\nabla u_n|^p dx + \mathcal{H}^{N-1}(S(u_n) \cap \Omega)\right\} < +\infty, \tag{3.0.5}$$

then $u \in GSBV_p(\Omega;\mathbb{R}^d)$,

$$\int_{S(u)} \vartheta(u^+ - u^-, \nu(u))\, d\mathcal{H}^{N-1} \leq \liminf_n \int_{S(u_n)} \vartheta(u_n^+ - u_n^-, \nu(u_n))\, d\mathcal{H}^{N-1},$$

and

$$\int_\Omega W(\nabla u)dx \leq \liminf_n \int_\Omega W(\nabla u_n)dx.$$

Definition 3.0.5. *We say that a sequence* $\{u_n\}$ *in* $GSBV(\Omega;\mathbb{R}^d)$ *converges weakly in* $GSBV_p(\Omega;\mathbb{R}^d)$ *to a function* $u \in GSBV_p(\Omega;\mathbb{R}^d)$ *if* $u_n \to u$ *in* $L^1(\Omega;\mathbb{R}^d)$*, and (3.0.5) holds.*

In view of this definition, we may say that Theorem 3.0.4 provides a closure result for weak convergence in $GSBV_p(\Omega;\mathbb{R}^d)$.

The following relaxation theorem may be found in (20).

Theorem 3.0.6. *[Relaxation in* $GSBV_p(\Omega;\mathbb{R}^d)$*] If* W *and* ϑ *are Borel functions satisfying the growth conditions*

$$\frac{1}{C}|F|^p - C \leq W(F) \leq C(1+|F|^p) \qquad 1+|z| \leq \vartheta(z,\nu) \leq C(1+|z|)$$

for all $F \in \mathbb{R}^{d\times N}$, $z \in \mathbb{R}^d$, $\nu \in S^{N-1}$, *and for some* $C > 0$, *then the lower semicontinuous envelope of the functional*

$$u \mapsto \int_\Omega W(\nabla u)dx + \int_{S(u)} \vartheta(u^+ - u^-, \nu(u))\, d\mathcal{H}^{N-1}$$

on $SBV_p(\Omega;\mathbb{R}^d)$ *with respect to the* $L^1(\Omega;\mathbb{R}^d)$ *convergence is given by*

$$u \mapsto \int_\Omega QW(\nabla u)dx + \int_{S(u)} R\vartheta(u^+ - u^-, \nu(u))\, d\mathcal{H}^{N-1}.$$

4 Brittle Thin Films

Let $W : \mathbb{R}^{3\times 3} \to [0,+\infty)$ and $\vartheta : (\mathbb{R}^3\setminus\{0\}) \times S^2 \to [0,+\infty)$ denote bulk and surface energy density, respectively. The '3D-energy' of a deformation $v \in GSBV_p(\Omega_\varepsilon;\mathbb{R}^3)$ of the thin film occupying Ω_ε as a reference configuration is given by

$$\int_{\Omega_\varepsilon} W(\nabla v)dx + \int_{S(v)\cap\Omega_\varepsilon} \vartheta(v^+ - v^-, \nu(v))\, d\mathcal{H}^2$$

for functions $v \in GSBV_p(\Omega_\varepsilon;\mathbb{R}^3)$. Note that there is a 1-1 correspondence between rank-1-matrices in $\mathbb{R}^{3\times 3}$ and $(\mathbb{R}^3 \setminus \{0\}) \times S^2$, so that we could as well consider the surface energy as defined on matrices by $\theta(z \otimes \nu) = \vartheta(z,\nu)$, so as to have a symmetric notation with that of the bulk part. However, we prefer to adopt the above notation, as it better describes the decoupled dependence of surface energy densities with respect to the amplitude of the jump, z, and the normal to the jump set, ν.

We extend ϑ by positive 1-homogeneity as follows:

$$\vartheta(z,\eta) = \begin{cases} |\eta|\vartheta\Big(z, \frac{\eta}{|\eta|}\Big) & \text{if } \eta \neq 0, \\ 0 & \text{if } \eta = 0. \end{cases}$$

Changing variables and setting

$$u(x_\alpha, x_3) := v(x_\alpha, \varepsilon x_3), \quad (\nu_\alpha, \nu_3) := \nu,$$

clearly $u \in GSBV_p(\Omega;\mathbb{R}^3)$ and the integral above becomes

$$\begin{aligned}
&\int_{\Omega_\varepsilon} W\Big(\nabla_\alpha u\left(x_\alpha, \frac{x_3}{\varepsilon}\right)\Big|\frac{1}{\varepsilon}\nabla_3 u\left(x_\alpha, \frac{x_3}{\varepsilon}\right)\Big)dx \\
&\qquad + \int_{S(v)\cap\Omega_\varepsilon} \vartheta\Big(u^+\left(x_\alpha, \frac{x_3}{\varepsilon}\right) - u^-\left(x_\alpha, \frac{x_3}{\varepsilon}\right), \\
&\qquad\qquad \nu_\alpha(u)\left(x_\alpha, \frac{x_3}{\varepsilon}\right), \frac{1}{\varepsilon}\nu_3(u)\left(x_\alpha, \frac{x_3}{\varepsilon}\right)\Big)\frac{\varepsilon}{\sqrt{\varepsilon^2|\nu_\alpha|^2(v) + \nu_3^2(v)}}\, d\mathcal{H}^2 \\
&= \varepsilon\left[\int_\Omega W\Big(\nabla_\alpha u(x)\Big|\frac{1}{\varepsilon}\nabla_3 u(x)\Big)dx\right. \\
&\qquad \left. + \int_{S(u)\cap\Omega} \vartheta\Big(u^+(x) - u^-(x), \nu_\alpha(u)(x), \frac{1}{\varepsilon}\nu_3(u)(x)\Big)\, d\mathcal{H}^2\right].
\end{aligned}$$

Consider the ε-scaled $3D$-energies $J_\varepsilon : L^1(\Omega;\mathbb{R}^3) \to [0,+\infty]$ defined by

$$J_\varepsilon(u) := \begin{cases} \int_\Omega W\Big(\nabla_\alpha u\Big|\frac{1}{\varepsilon}\nabla_3 u\Big)dx & + \int_{S(u)} \vartheta\Big(u^+ - u^-, \nu_\alpha(u), \frac{1}{\varepsilon}\nu_3(u)\Big)\, d\mathcal{H}^2 \\ & \text{if } u \in GSBV_p(\Omega;\mathbb{R}^3) \\ +\infty & \text{otherwise.} \end{cases} \tag{4.0.6}$$

We introduce the space

$$\mathcal{V} := \{u \in SBV_p(\Omega;\mathbb{R}^3) : \ \nabla_3 u = 0 \text{ a.e. and } \nu_3(u) = 0 \ \mathcal{H}^2\text{-a.e.}\}.$$

Note that if $u \in \mathcal{V}$ then $D_3 u = 0$ in the sense of distributions, so that $\mathcal{V}$ can be identified with $SBV_p(\omega;\mathbb{R}^3)$.

Associated to a function $\widehat{\vartheta} : (\mathbb{R}^3 \setminus \{0\}) \times \mathbb{R}^2 \to \mathbb{R}$ we may introduce two BV-elliptic envelopes, namely

$$R_2\widehat{\vartheta}(z,\eta) :=$$
$$\inf \left\{ \int_{S(u)\cap Q'_\eta} \widehat{\vartheta}(u^+ - u^-, \nu(u))\, d\mathcal{H}^1 : \; u \in GSBV_{\#}(Q'_\eta; \mathbb{R}^3, z) \right\}$$

for all $z \in \mathbb{R}^3$ and $\eta \in S^1$, where Q'_η is a unit square on the plane, centered at zero and with two faces orthogonal to η, and

$$R_3\widehat{\vartheta}(z,\eta) :=$$
$$\inf \left\{ \int_{S(u)\cap Q_{(\eta,0)}} \widehat{\vartheta}(u^+ - u^-, \nu_\alpha(u))\, d\mathcal{H}^2 : \; u \in GSBV_{\#}(Q_{(\eta,0)}; \mathbb{R}^3, z) \right\}.$$

In the latter envelope $Q_{(\eta,0)}$ stands for a unit cube in $\mathbb{R}^3$ centered at zero and with two faces orthogonal to $(\eta, 0)$.

The main result now is the following.

Theorem 4.0.7. *Let* $W : \mathbb{R}^{3\times 3} \to [0,+\infty)$ *and* $\vartheta : (\mathbb{R}^3 \setminus \{0\}) \times \mathbb{R}^3 \to [0,+\infty)$ *be continuous functions,* ϑ *is symmetric and positively homogeneous of degree* 1 *in the second variable; i.e.,*

$$\vartheta(z,\nu) = \vartheta(-z,-\nu), \qquad \vartheta(z,t\nu) = t\vartheta(z,\nu) \text{ for all } t > 0,\; z,\nu \in \mathbb{R}^3.$$

Let W *and* ϑ *satisfy the growth and coercivity conditions*

$$\frac{1}{C}|F|^p - C \le W(F) \le C(1+|F|^p), \qquad 1+|z| \le \vartheta(z,\nu) \le C(1+|z|)$$

for all $F \in \mathbb{R}^{3\times 3}$, $z \in \mathbb{R}^3$, $\nu \in S^2$, *and for some* $C > 0$ *and* $1 < p < +\infty$. *Suppose, in addition, that* ϑ *satisfies the Lipschitz condition*

$$|\vartheta(z,\nu) - \vartheta(z',\nu)| \le L\,|z - z'|$$

for all $F, F' \in \mathbb{R}^{3\times 3}$, $z, z' \in \mathbb{R}^3$, $\nu \in S^2$, *and for some* $C > 0$. *Then the functionals* J_ε *defined in (4.0.6)* Γ*-converge with respect to the* $L^1(\Omega;\mathbb{R}^3)$ *convergence as* $\varepsilon \to 0^+$ *to the functional* $J_0 : L^1(\Omega;\mathbb{R}^3) \to [0,+\infty]$ *defined by*

$$J_0(u) := \begin{cases} \displaystyle\int_\omega Q_2\overline{W}(\nabla_\alpha u)\, dx_\alpha + \int_{S(u)\cap\omega} R_2\overline{\vartheta}(u^+ - u^-, \nu_\alpha(u))\, d\mathcal{H}^1 \text{ if } u \in \text{\textbackslash} \\ +\infty \qquad \textit{otherwise,} \end{cases}$$

where

$$\overline{W}(\overline{F}) := \inf\{W(F) : \ F = (\overline{F}|F_3), \ F_3 \in \mathbb{R}^3\}$$

for $F \in \mathbb{R}^{3\times 3}$, *and*

$$\overline{\vartheta}(z,\eta) := \inf\{\vartheta(z,\eta,\xi) : \ \xi \in \mathbb{R}\}$$

for $z \in \mathbb{R}^3$ *and* $\eta \in \mathbb{R}^2 \setminus \{0\}$.

5 The case $p = 1$

In the case where $p = 1$ it is well known that the analog of Theorem 3.0.4 and 3.0.6 do not hold. Γ-limit and relaxation problems must be described in the whole space of functions of bounded variation, and in general there is an interaction phenomenon between the bulk and interfacial energy densities. In this section we show that we can still deal with the thin-film limit in this case by combining the description given for $p = 1$ and a relaxation result for functionals in BV, established by Bouchitté, Fonseca and Mascarenhas (15) Section 4.2 (see also (13), (18)).

Relaxation in BV

Let $W : \mathbb{R}^{d\times N} \to [0,+\infty)$ and $\vartheta : \mathbb{R}^d \times S^{N-1} \to [0,+\infty)$ be continuous functions. We recall that the *recession function of* f is defined by

$$W^\infty(F) := \limsup_{t\to+\infty} \frac{W(tF)}{t}.$$

If W is Lipschitz (respectively, quasiconvex) and has linear growth (see (H1) below), then W^∞ is a Lipschitz (respectively, quasiconvex), positively 1-homogeneous function (see (29)). Moreover, we define the positively 1-homogeneous function

$$\vartheta_0(z,\nu) := \limsup_{t\to 0^+} \frac{\vartheta(tz,\nu)}{t}.$$

We suppose that W and ϑ satisfy the following hypotheses:
(H1) there exists $C > 0$ such that

$$\frac{1}{C}|F| - C \le W(x,F) \le C(1+|F|)$$

for all $F \in \mathbb{R}^{d\times N}$;

(H2) there exist constants $C, L > 0,\ 0 < r < 1$, such that

$$\left| W^\infty(F) - \frac{W(tF)}{t} \right| \le C \frac{1}{t^r}$$

for every $F \in \mathbb{R}^{d \times N}$ with $|F| = 1$ and for all $t > 0$ such that $t > L$;

(H3) there exists $C > 0$ such that

$$\frac{1}{C}|z| \le \vartheta(z, \nu) \le C|z|$$

for all $(z, \nu) \in \mathbb{R}^d S^{N-1}$;

(H4) there exist constants $C, l, s > 0$ such that

$$\left| \vartheta_0(z, \nu) - \frac{\vartheta(tz, \nu)}{t} \right| \le Ct^s$$

for every $(z, \nu) \in \mathbb{R}^d \times S^{N-1}$ with $|z| = 1$ and for all t such that $0 < t < l$.

Let $(z, \nu) \in \mathbb{R}^d \times S^{N-1}$, let $\{\nu_1, \dots, \nu_{N-1}, \nu\}$ form an orthonormal basis of $\mathbb{R}^N$, and define the class of admissible functions

$$\mathcal{A}(z, \nu) := \Big\{ v \in SBV_{\text{loc}}(S_\nu; \mathbb{R}^d) : v(y) = 0 \text{ if } \langle y, \nu \rangle = -\frac{1}{2},$$
$$v(y) = z \text{ if } \langle y, \nu \rangle = \frac{1}{2},\ v \text{ is 1-periodic in the directions of } \nu_1, \dots, \nu_{N-1} \Big\},$$

where S_ν is the strip

$$S_\nu := \left\{ y \in \mathbb{R}^N : |\langle y, \nu \rangle| < \frac{1}{2} \right\},$$

and where the boundary values of v are understood in the sense of traces.

We use the notation

$$\int_\Omega W(dC_u) := \int_\Omega W\Big(\frac{dC_u}{d|C_u|}\Big) d|C_u|,$$

where $|C_u|$ is the total variation measure of C_u.

Theorem 5.0.8. *Under hypotheses (H1)–(H4), with W Lipschitz continuous, if $u \in BV(\Omega; \mathbb{R}^d)$ then the lower semicontinuous envelope $\mathcal{F}$ with respect to the L^1 convergence of the functional*

$$u \in SBV(\Omega; \mathbb{R}^d) \mapsto \int_\Omega W(\nabla u)\, dx + \int_{S(u) \cap \Omega} \vartheta(u^+ - u^-, \nu) d\mathcal{H}^{N-1},$$

is given by

$$\mathcal{F}(u) = \int_\Omega g(\nabla u)dx + \int_{S(u)\cap\Omega} h(u^+ - u^-, \nu)d\mathcal{H}^{N-1} + \int_\Omega g^\infty(dC_u),$$

where

$$\begin{aligned} g(A) \quad := \quad & \inf\Big\{\textstyle\int_Q W(\nabla u)dx + \int_{S(u)\cap(Q} \vartheta_0(u^+ - u^-, \nu)d\mathcal{H}^{N-1} : \\ & u \in SBV(Q;\mathbb{R}^d), u\,|_{\partial Q}\,(x) = Ax\Big\}, \end{aligned}$$

and

$$h(z,\nu) :=$$
$$\inf\left\{\int_{Q_\nu} W^\infty(\nabla u(x))dx + \int_{S(u)\cap Q_\nu} \vartheta(u^+ - u^-, \nu)d\mathcal{H}^{N-1} : u \in \mathcal{A}(z,\nu)\right\}.$$

We will now apply this general result to the energies J_ε introduced in (4.0.6). Let

$$\begin{aligned} g_2(\overline{F}) \quad := \quad & \inf\left\{\int_{(0,1)^2} \overline{W}(\nabla_\alpha u)\; dx_\alpha + \int_{S(u)\cap(0,1)^2} (\overline{\vartheta})_0(u^+ - u^-, \nu_\alpha)d\mathcal{H}^1 : \right. \\ & \left. u \in SBV((0,1)^2;\mathbb{R}^3), u\,|_{\partial(0,1)^2}\,(x) = \overline{F}x\right\}, \end{aligned}$$

$$\begin{aligned} g_3(\overline{F}) \quad := \quad & \inf\left\{\int_{(0,1)^3} \overline{W}(\nabla_\alpha u)\; dx + \int_{S(u)\cap(0,1)^3} (\overline{\vartheta})_0(u^+ - u^-, \nu_\alpha)d\mathcal{H}^2 : \right. \\ & \left. u \in SBV((0,1)^3;\mathbb{R}^3), u\,|_{\partial(0,1)^3}\,(x) = \overline{F}x\right\}, \end{aligned}$$

and, similarly,

$$\begin{aligned} h_2(z,\eta) \quad := \quad & \inf\left\{\int_{Q'_\eta} \overline{W}^\infty(\nabla_\alpha u(x))dx_\alpha \right. \\ & + \left. \int_{S(u)\cap Q'_\eta} \overline{\vartheta}(u^+ - u^-, \eta)d\mathcal{H}^1 : u \in \mathcal{A}'(z,\eta)\right\}, \end{aligned}$$

$$\begin{aligned} h_3(z,\eta) \quad := \quad & \inf\left\{\int_{Q_{(\eta,0)}} \overline{W}^\infty(\nabla_\alpha u(x))dx \right. \\ & + \left. \int_{S(u)\cap Q_{(\eta,0)}} \overline{\vartheta}(u^+ - u^-, \eta)d\mathcal{H}^2 : u \in \mathcal{A}(z,(\eta,0))\right\}, \end{aligned}$$

where

$$\mathcal{A}'(z,\eta) := \Big\{ v \in SBV_{\mathrm{loc}}(S_\nu \cap \omega; \mathbb{R}^m) : v(y) = 0 \text{ if } \langle y, \eta\rangle = -\frac{1}{2},$$

$$v(y) = z \text{ if } \langle y, \eta\rangle = \frac{1}{2}, v \text{ is 1-periodic in the direction of } \rho \Big\},$$

with ρ orthogonal to η on the plane.

Theorem 5.0.9. *Let $W : \mathbb{R}^{3\times 3} \to [0,+\infty)$ and $\vartheta : (\mathbb{R}^3 \setminus \{0\}) \times \mathbb{R}^3 \to [0,+\infty)$ be continuous functions, and let ϑ be symmetric and positively homogeneous of degree 1 in the second variable; i.e.,*

$$\vartheta(z,\nu) = \vartheta(-z,-\nu), \qquad \vartheta(z,t\nu) = t\vartheta(z,\nu) \text{ for all } t > 0,\ z,\nu \in \mathbb{R}^3.$$

Suppose that W and ϑ satisfy the growth and coercivity conditions

$$\frac{1}{C}|F| - C \le W(F) \le C(1+|F|), \qquad |z| \le \vartheta(z,\nu) \le C|z|$$

for all $F \in \mathbb{R}^{3\times 3}$, $z \in \mathbb{R}^3$, $\nu \in S^2$, and for some $C > 0$. Let W and ϑ satisfy (H2) and (H4), respectively. The functionals J_ε defined in (4.0.6), with $GSBV_p(\Omega;\mathbb{R}^3)$ replaced by $SBV(\Omega;\mathbb{R}^3)$, Γ-converge with respect to the $L^1(\Omega;\mathbb{R}^3)$ convergence as $\varepsilon \to 0^+$ to the functional $J_0 : L^1(\Omega;\mathbb{R}^3) \to [0,+\infty]$ defined by

$$J_0(u) = \begin{cases} \int_\omega g_2(\nabla_\alpha u)dx_\alpha + \int_{S(u)\cap\omega} h_2(u^+ - u^-, \nu_\alpha(u))d\mathcal{H}^1 + \int_\Omega g_2^\infty(dC_u) \\ \qquad\qquad \textit{if } u \in BV(\Omega;\mathbb{R}^3), D_3 u = 0, \\ +\infty \qquad\qquad \textit{otherwise.} \end{cases}$$

Bibliography

[1] Acerbi E., Fusco, N., Semicontinuity problems in the calculus of variations, *Arch. Rat. Mech. Anal.* **86** (1984), 125–145.

[2] Ambrosio L., A compactness theorem for a new class of functions of bounded variation. *Boll. Un. Mat. Ital.* **3-B** (1989), 857–881.

[3] Ambrosio, L., Existence theory for a new class of variational problems. *Arch. Rat. Mech. Anal.* **111** (1990), 291–322.

[4] Ambrosio, L., Variational problems in SBV and image segmentation. *Acta Appl. Math.* **17** (1989), 1–40.

[5] Ambrosio, L., On the lower semicontinuity of quasi-convex integrals defined in $SBV(\Omega;\mathbb{R}^k)$. *Nonlinear Anal.* **23** (1994), 405–425.

[6] Ambrosio, L., Braides, A., Functionals defined on partitions of sets of finite perimeter, I: integral representation and Γ-convergence. *J. Math. Pures. Appl.* **69** (1990), 285–305.

[7] Ambrosio, L., Braides, A., Functionals defined on partitions of sets of finite perimeter, II: semicontinuity, relaxation and homogenization. *J. Math. Pures. Appl.* **69** (1990), 307–333.

[8] Ambrosio, L, Braides, A., Energies in SBV and variational models in fracture mechanics. *Homogenization and Applications to Material Sciences*, (D. Cioranescu, A. Damlamian, P. Donato eds.), GAKUTO, Gakkōtosho, Tokyo, Japan, (1997), 1–22.

[9] Ambrosio, L., Fusco, N., Pallara, D., *Special Functions of Bounded Variation and Free Discontinuity Problems*, Mathematical Monographs, Oxford University Press, (2000).

[10] Anzellotti, G., Baldo, S., Percivale, D., Dimension-reduction in variational problems, asymptotic development in Γ-convergence and thin structures in elasticity. *Asymptotic Anal.* **9** (1994), 61–100.

[11] Ball, J. M., Murat, F., $W^{1,p}$ quasiconvexity and variational problems for multiple integrals. *J. Funct. Anal.* **58** (1984), 225-253.

[12] Barenblatt, G. I., The mathematical theory of equilibrium cracks in brittle fracture. *Adv. Appl. Mech.* **7** (1962), 55–129.

[13] Barroso, A. C., Bouchitté, G., Buttazzo, G., Fonseca, I., Relaxation in $BV(\Omega;\mathbb{R}^p)$ of energies involving bulk and surface energy contributions. *Arch. Rat. Mech. Anal.* **135** (1996), 107–173.

[14] Bouchitté, G., Braides, A., Buttazzo, G., Relaxation results for some free discontinuity problems. *J. Reine Angew. Math.* **458** (1995), 1–18.

[15] Bouchitté, G., Fonseca, I., Mascarenhas, L., A global method for relaxation. *Arch. Rat. Mech. Anal.* **145** (1998), 51–98.

[16] Braides, A., *Approximation of Free-Discontinuity Problems. Lecture Notes in Mathematics,* Springer Verlag, Berlin, (1998).

[17] Braides, A., Chiadò, Piat V., Integral representation results for functionals defined on $SBV(\Omega;\mathbb{R}^m)$. *J. Math. Pures Appl.* **75** (1996), 595–626.

[18] Braides, A., Coscia, A., The interaction between bulk energy and surface energy in multiple integrals. *Proc. Royal Soc. Edin.* **124A** (1994), 737–756.

[19] Braides A., Defranceschi, A., *Homogenization of Multiple Integral.* Oxford University Press, Oxford, (1998).

[20] Braides, A., Defranceschi A, Vitali E Homogenization of free-discontinuity problems. *Arch. Rat. Mech. Anal.* **135** (1996), 297–356.

[21] Braides, A., Fonseca, I., Francfort, G., 3D-2D Asymptotic analysis for inhomogeneous thin films. *Indiana U. Math. J.* (2000), 1367– 1404.

[22] Caillerie, D., Thin elastic and periodic plates. *Math. Meth. Appl. Sci.* **6** (1984), 159–191.

[23] Dacorogna, B., *Direct Methods in the Calculus of Variations.* Springer-Verlag, Berlin, (1989).

[24] Dal Maso, G., *An Introduction to* Γ*-convergence.* Birkhäuser, Boston, (1993).

[25] De Giorgi, E., Ambrosio, L., Un nuovo funzionale del calcolo delle variazioni. *tti Accad. Naz. Lincei Rend. Cl. Sci. Fis. Mat. Natur.* **82** (1988), 199–210.

[26] De Giorgi, E., Franzoni, T., Su un tipo di convergenza variazionale. *Atti Accad. Naz. Lincei Rend. Cl. Sci. Mat.* **58** (1975), 842–850.

[27] Evans, L. C., Gariepy, R. F., *Measure theory and fine properties of functions.* CRC Press, Boca Raton, (1992).

[28] Federer, H., *Geometric Measure Theory.* Springer Verlag, New York, (1969).

[29] Fonseca I, Müller, S., Relaxation of quasiconvex functionals in $BV(\Omega;\mathbb{R}^p)$ for integrands $f(x,u,\nabla u)$. *Arch. Rat. Mech. Anal.* **123** (1993), 1–49.

[30] Griffith, A. A., The phenomenon of rupture and flow in solids. *Phil Trans. Royal Soc. London A* **221** (1920), 163–198.

[31] Le Dret, H., Raoult, A., The nonlinear membrane model as variational limit of nonlinear three-dimensional elasticity. *J. Math. Pures Appl.* **74** (1995), 549–578.

[32] Le Dret, H., Raoult, A., The membrane shell model in nonlinear elasticity: a variational asymptotic derivation. *J. Nonlinear Sc.*, **6** (1996), 59–84.

[33] Le Dret, H., Raoult, A., Variational convergence for nonlinear shell models with directors and related semicontinuity and relaxation results. *Arch. Rat. Mech. Anal.*, **154** (2000), 101–134.

[34] Mumford, D., Shah, J., Optimal approximation by piecewise smooth functions and associated variational problems. *Comm. Pure Appl. Math.* **17** (1989), 577–685.

[35] Ziemer, W. *Differentiable Functions.* Springer, Berlin, (1989).

A Young measure approach to a nonlinear membrane model - Lecture III

Irene Fonseca

Department of Mathematical Sciences, Carnegie Mellon University

Abstract An integral representation of a relaxed functional deduced via 3D-2D dimension reduction is obtained in terms of a special class of parameterized probability measures (bending Young measures).

1 Introduction

The content of this lecture is based on the paper

M. Bocea and I. Fonseca, A Young Measure Approach to a Nonlinear Membrane Model Involving the Bending Moment, *Proc. R. Soc. Edin.* ***134A*** *(2004), 845–883.*

In the sequel we assume that the elastic energy density $W : \Omega \times \mathbb{R}^{3\times 3} \to \mathbb{R}$ is a Carathéodory function satisfying the p-growth and coercivity condition

$$\frac{1}{C}|F|^p - C \leq W(x; F) \leq C(1 + |F|^p) \tag{1.0.1}$$

$\mathcal{L}^3$-a.e. $x \in \Omega$ and for all $F \in \mathbb{R}^{3\times 3}$, where $C > 0$, and

$$\tilde{E}_\varepsilon(u) := \int_\Omega W\left(x; \nabla_\alpha v \Big| \frac{1}{\varepsilon}\nabla_3 v\right) dx.$$

By (1.0.1), any minimizing sequence is scaled-bounded, i.e.

$$\sup_\varepsilon \int_\Omega \left|\left(\nabla_\alpha v_\varepsilon \Big| \frac{1}{\varepsilon}\nabla_3 v_\varepsilon\right)\right|^p dx < +\infty.$$

*The research of I. Fonseca was partially funded by the National Science Foundation (NSF) under NSF Grants DMS-0103799 and DMS-0401763, and by the Center for Nonlinear Analysis under NSF Grants DMS-9803791, DMS-0405343 and DMS-0635983.

and thus, up to a subsequence (not relabeled), $\{v_\varepsilon\}$ and $\{\frac{1}{\varepsilon}\nabla_3 v_\varepsilon\}$ converge weakly to some functions v and b in $W^{1,p}(\Omega;\mathbb{R}^3)$ and $L^p(\Omega;\mathbb{R}^3)$, respectively (with v depending only on x_α). In particular, $\{\frac{1}{\varepsilon}\int_I \nabla_3 v_\varepsilon(\cdot,x_3)\,dx_3\}$ converges weakly to $\bar{b}$ in $L^p(\omega;\mathbb{R}^3)$, where $I := (-1/2,1/2)$ and $\bar{b}(x_\alpha) := \int_I b(x_\alpha,x_3)\,dx_3$.

We obtain an integral representation of the relaxed energy functional $E : W^{1,p}(\omega;\mathbb{R}^3)\times L^p(\omega,\mathbb{R}^3)\to\mathbb{R}\cup\{+\infty\}$ defined by

$$E(v,b) := \inf\left\{ \liminf_{n\to+\infty}\int_\Omega W\left(x;\nabla_\alpha v_n\Big|\frac{1}{\varepsilon_n}\nabla_3 v_n\right)dx : \varepsilon_n\to 0^+,\right.$$

$$\left. v_n \rightharpoonup v \text{ in } W^{1,p}(\Omega;\mathbb{R}^3),\ \frac{1}{\varepsilon_n}\int_I \nabla_3 v_n(\cdot,x_3)dx_3 \rightharpoonup b \text{ in } L^p(\omega,\mathbb{R}^3)\right\},$$

in terms of a suitable class of Young measures, defined as follows

Definition 1.0.1. *Let $\Omega := \omega\times I$ where $\omega\subset\mathbb{R}^2$ is an open domain and let $1<p<+\infty$. A Young measure $\mu := \{\mu_x\}_{x\in\Omega}$ is called a bending Young measure if there exist sequences $\{\varepsilon_n\}\subset(0,+\infty)$ and $\{v_n\}\subset W^{1,p}(\Omega;\mathbb{R}^3)$, with $\varepsilon_n\to 0^+$, and functions $v\in W^{1,p}(\omega;\mathbb{R}^3)$, $b\in L^p(\omega;\mathbb{R}^3)$, such that*

(i) the sequence $\left\{\left(\nabla_\alpha v_n\big|\frac{1}{\varepsilon_n}\nabla_3 v_n\right)\right\}$ is bounded in $L^p(\Omega,\mathbb{M}^{3\times 3})$ and it generates μ;

(ii) $v_n\rightharpoonup v$ in $W^{1,p}(\Omega;\mathbb{R}^3)$;

(iii) $\displaystyle\frac{1}{\varepsilon_n}\int_I \nabla_3 v_n(\cdot,x_3)dx_3 \rightharpoonup b$ in $L^p(\omega,\mathbb{R}^3)$.

The weak limit v of v_n in $W^{1,p}(\Omega;\mathbb{R}^3)$ is called an underlying deformation for μ while the weak limit b of $\displaystyle\frac{1}{\varepsilon_n}\int_I \nabla_3 v_n(\cdot,x_3)dx_3$ in $L^p(\omega;\mathbb{R}^3)$ is called the bending moment density associated to μ. We denote by $Y(v,b)$ the set of bending Young measures with underlying deformation v and associated bending moment density b.

The relaxation result is

Theorem 1.0.2. *Let $1<p<+\infty$, $\Omega := \omega\times I$ where $\omega\subset\mathbb{R}^2$ is an open bounded Lipschitz domain, and let $W:\Omega\times\mathbb{R}^{3\times 3}\to\mathbb{R}$ be a Carathéodory function satisfying the p-growth and coercivity condition (1.0.1) for $\mathcal{L}^3$-a.e $x\in\Omega$ and for all $A\in\mathbb{R}^{3\times 3}$. Then, for every $v\in W^{1,p}(\omega;\mathbb{R}^3)$ and $b\in$*

$L^p(\omega, \mathbb{R}^3)$,

$$E(v,b) = \inf_{\nu \in Y(v,b)} \int_\Omega \int_{\mathbb{R}^{3\times 3}} W(x;A)d\nu_x(A)dx. \tag{1.0.2}$$

Moreover, the infimum on the right hand side is attained, i.e. there exists a bending Young measure $\mu \in Y(v,b)$ such that

$$\inf_{\nu \in Y(v,b)} \int_\Omega \int_{\mathbb{R}^{3\times 3}} W(x;A)d\nu_x(A)dx = \int_\Omega \int_{\mathbb{R}^{3\times 3}} W(x;A)d\mu_x(A)dx. \tag{1.0.3}$$

Remark 1.0.3.

a) *If we replace (iii) in Definition 1.0.1 by*
(iii′) $\frac{1}{\varepsilon_n}\nabla_3 v_n \rightharpoonup c \in L^p(\Omega;\mathbb{R}^3)$ in $L^p(\Omega,\mathbb{R}^3)$,
we recover the definition of α-gradient p-Young measures (see (10)) when $1 < p < +\infty$. In this case the weak limit c of $\frac{1}{\varepsilon_n}\nabla_3 v_n$ in $L^p(\Omega;\mathbb{R}^3)$ is called a Cosserat vector associated to μ. The set of α-gradient p- Young measures with underlying deformation v and associated Cosserat vector c is denoted by $Y^{\alpha-p}_{v,c}$. Note that if we set $b := \int_I c(\cdot, x_3)dx_3$, then $Y^{\alpha-p}_{v,c} \subset Y(v,b)$. The corresponding integral representation result (see Theorem 1.1 in (10)) for the relaxed functional $\mathcal{W} : W^{1,p}(\omega;\mathbb{R}^3) \times L^p(\Omega,\mathbb{R}^3) \to \mathbb{R}$ defined by

$$\mathcal{W}(v,c) := \inf \left\{ \liminf_{n\to+\infty} \int_\Omega W\left(x; \nabla_\alpha v_n \middle| \frac{1}{\varepsilon_n}\nabla_3 v_n\right) dx : \varepsilon_n \to 0^+, \right.$$
$$\left. v_n \rightharpoonup v \quad in \quad W^{1,p}(\Omega;\mathbb{R}^3), \frac{1}{\varepsilon_n}\nabla_3 v_n \rightharpoonup c \quad in \quad L^p(\Omega,\mathbb{R}^3) \right\},$$

in terms of α- gradient p- Young measures states that there exists $\mu^0 \in Y^{\alpha-p}_{v,c}$ such that

$$\begin{aligned} \mathcal{W}(v,c) &= \inf_{\nu \in Y^{\alpha-p}_{v,c}} \int_\Omega \int_{\mathbb{R}^{3\times 3}} W(x;A)d\nu_x(A)dx \\ &= \int_\Omega \int_{\mathbb{R}^{3\times 3}} W(x;A)d\mu^0_x(A)dx. \end{aligned}$$

b) *We refer to Freddi and Paroni (18) for another study involving Young measures in the membrane theory.*

The main effort in this work is to provide a complete characterization of the class of bending Young measures involved in our relaxation result (Theorem 1.0.2); in the case of gradient Young measures, as in (18), such a characterization already exists (20). The core of the paper is the identification of algebraic and analytical conditions on parameterized probability measures both necessary and sufficient to guarantee that they belong to $Y(v,b)$ (Theorem 1.0.4). The characterization of Young measures generated by gradients bounded in L^p is due to Kinderlehrer and Pedregal (see (19), (20)), and it was subsequently generalized to the realm of A- quasiconvexity by Fonseca and Müller (see (16)).

Before making our main result precise, we need to introduce some notation. Let $p-$ $\mathrm{Lip}(I\times\mathbb{R}^{3\times 3})$ be the space of functions $f: I\times\mathbb{R}^{3\times 3}\to\mathbb{R}$ such that

$$|f(x,X)-f(y,Y)|\leq c(1+|X|^{p-1}+|Y|^{p-1})(|x-y|+|X-Y|)$$

for all $(x,X),(y,Y)\in I\times\mathbb{R}^{3\times 3}$ and for $c>0$, and for every $f\in p-$ $\mathrm{Lip}(I\times\mathbb{R}^{3\times 3})$ given, let $\mathrm{Lip}(f)$ be defined by

$$\begin{aligned}\mathrm{Lip}(f):=\inf\{c>0:\ & |f(x,X)-f(y,Y)|\leq c(1+|X|^{p-1}\\ & +|Y|^{p-1})(|x-y|+|X-Y|),\\ & \text{forall }\ (x,X),(y,Y)\in I\times\mathbb{R}^{3\times 3}\}.\end{aligned}$$

Then $p-$ $\mathrm{Lip}(I\times\mathbb{R}^{3\times 3})$, endowed with the norm

$$\|f\|:=|f(0,0)|+\ \mathrm{Lip}(f),$$

is a Banach space. Define $\mathcal{E}_p$ as the set of function $f\in p-$ $\mathrm{Lip}(I\times\mathbb{R}^{3\times 3})$ for which $I\times\mathbb{R}^{3\times 3}\ni(x,X)\longmapsto\frac{f(x,X)}{1+|X|^p}$ admits a continuous extension to $[-\frac{1}{2},\frac{1}{2}]\times(\mathbb{R}^{3\times 3}\cup\{+\infty\})$, where $\mathbb{R}^{3\times 3}\cup\{+\infty\}$ stands for the one point compactification of $\mathbb{R}^{3\times 3}$ (function spaces of this type have been previously considered (with $p=1$) by Kristensen (21)). Note that $(\mathcal{E}_p,\|\cdot\|)$ is a (non-separable) Banach space. As an example, the function $g: I\times\mathbb{R}^{3\times 3}\to\mathbb{R}$ defined by $g(x,X):=|X|^p$ belongs to $\mathcal{E}_p$.

Set $Q':=(0,1)^2$. For $f\in p$-$\mathrm{Lip}(I\times\mathbb{R}^{3\times 3})$, $A\in\mathbb{R}^{3\times 2}$, and $a\in\mathbb{R}^3$, define the potential

$$\begin{aligned}\widehat{Q}f(A|a):=\ & \inf_{(\varphi,\lambda)}\Bigg\{\int_{Q'\times I} f\left(x_3,\left(A+\nabla_\alpha\varphi\Big|\lambda\nabla_3\varphi\right)\right)dx:\lambda\in\mathbb{R},\\ & \varphi\in W^{1,p}(\mathbb{R}^2\times I;\mathbb{R}^3),\varphi(\cdot,x_3)\ \text{ is }\ Q'-\ \text{periodic}\\ & \text{for }\ \mathcal{L}^1-\text{ a.e. }x_3\in I,\lambda\int_{Q'\times I}\nabla_3\varphi\,dx=a\Bigg\}.\end{aligned}$$

We are now ready to state

Theorem 1.0.4. *Let $1 < p < +\infty$, $\Omega := \omega \times I$ and let $u \in W^{1,p}(\omega;\mathbb{R}^3)$ and $b \in L^p(\omega,\mathbb{R}^3)$ be given. A weak* measurable family $\nu = \{\nu_x\}_{x\in\Omega}$ of probability measures belongs to $Y(u,b)$ if and only if*

$$\int_I \int_{\mathbb{R}^{3\times 3}} \Lambda d\nu_x(\Lambda)dx_3 = \big(\nabla u(x_\alpha)\big|b(x_\alpha)\big) \text{ for } \mathcal{L}^2 - a.e. \ x_\alpha \in \omega, \quad (1.0.4)$$

$$\int_\Omega \int_{\mathbb{R}^{3\times 3}} |\Lambda|^p d\nu_x(\Lambda)dx < +\infty, \quad (1.0.5)$$

and

$$\int_I \int_{\mathbb{M}^{3\times 2}\times\mathbb{R}^3} f\left(x_3, \left(\Lambda - \nabla u(x_\alpha)\Big|\lambda - b(x_\alpha)\right)\right) d\nu_x(\Lambda|\lambda)dx_3 \geq \widehat{Q}f(0|0), \quad (1.0.6)$$

$$\text{for } \mathcal{L}^2 - a.e. \ x_\alpha \in \omega \text{ and for all } f \in \mathcal{E}_p.$$

2 Preliminaries

Let Ω be an open subset of $\mathbb{R}^N$ $(N \geq 1)$. The Lebesgue measure in $\mathbb{R}^N$ is designated by $\mathcal{L}^N$, and H^{N-1} stands for the $(N-1)$-dimensional Hausdorff measure in $\mathbb{R}^N$. Let $C(\Omega;\mathbb{R}^d)$ be the space of $\mathbb{R}^d$- valued continuous functions on Ω and define

$$C_0(\Omega;\mathbb{R}^d) := \Big\{\varphi \in C(\Omega;\mathbb{R}^d) : \text{for every } \varepsilon > 0 \text{ there exists a compact set } K \subset \Omega \text{ such that } |\varphi(x)| \leq \varepsilon \text{ if } x \in \Omega \setminus K\Big\},$$

where $|\cdot|$ stands for the Euclidian norm in $\mathbb{R}^d$. Endowed with the supremum norm, $C_0(\Omega;\mathbb{R}^d)$ is a separable Banach space. In view of Riesz's Theorem, the dual space $C_0(\Omega;\mathbb{R}^d)'$ can be identified with the space $\mathcal{M}(\Omega;\mathbb{R}^d)$ of bounded $\mathbb{R}^d$- valued Radon measures on Ω with the norm $\|\mu\| := |\mu|(\Omega)$ via the duality pairing

$$\langle \mu, \varphi\rangle = \int_\Omega \varphi(x)\frac{d\mu}{d|\mu|}(x)d|\mu|(x),$$

where $|\mu|$ stands for the total variation of μ and is a non-negative, finite Radon measure on Ω.

A map $\nu : \Omega \to \mathcal{M}(\mathbb{R}^d)$ is said to be *weak * measurable* if $x \mapsto \langle \nu(x), \varphi\rangle$ is measurable for every $\varphi \in C_0(\Omega)$. In the sequel we will denote $\nu(x)$ by ν_x.

Definition 2.0.5. *(i) A function* $f : \Omega \times \mathbb{R}^d \to \mathbb{R} \cup \{+\infty\}$ *is called a normal integrand if* f *is Borel measurable and* $f(x, \cdot) : \mathbb{R}^d \to \mathbb{R} \cup \{+\infty\}$ *is lower semicontinuous for every* $x \in \Omega$.

(ii) A real-valued function $f : \Omega \times \mathbb{R}^d \to \mathbb{R}$ *is called a Carathéodory integrand if both* f *and* $-f$ *are normal integrands.*

Theorem 2.0.6. *(The Fundamental Theorem on Young measures (5), (6), (27)) Let* $\Omega \subset \mathbb{R}^N$ *be a measurable set of finite measure and let* $\{v_n\}$ *be a sequence of measurable functions,* $v_n : \Omega \to \mathbb{R}^d$. *Then there exist a subsequence* $\{v_{n_k}\} \subset \{v_n\}$ *and a weak * measurable map* $\nu : \Omega \to \mathcal{M}(\mathbb{R}^d)$ *such that the following hold:*

(i) $\nu_x \geq 0$, $\|\nu_x\|_{\mathcal{M}(\mathbb{R}^d)} := \int_{\mathbb{R}^d} d\nu_x \leq 1$, *for* $\mathcal{L}^N$*-a.e.* $x \in \Omega$;

(ii) $\varphi(v_{n_k}) \rightharpoonup < \nu_x, \varphi >= \int_{\mathbb{R}^d} \varphi(z)\, d\nu_x(z)$ *weakly-* in* $L^\infty(\Omega)$ *for all* $\varphi \in C_0(\mathbb{R}^d)$;

(iii) $\|\nu_x\|_{\mathcal{M}(\mathbb{R}^d)} = 1$ *if and only if*

$$\lim_{R\to\infty} \sup_{k\in\mathbb{N}} \mathcal{L}^N \left(\{x \in \Omega : |v_{n_k}(x)| \geq R\}\right) = 0.$$

In this case we say that ν *is a Young measure generated by the sequence* $\{v_{n_k}\}$;

(iv) if $K \subset \mathbb{R}^d$ *is compact and* $dist(v_{n_k}, K) \to 0$ *in measure, then* $supp\ \nu_x \subset K$ *for* $\mathcal{L}^N$*-a.e.* $x \in \Omega$. *The converse also holds provided that* $\|\nu_x\|_{\mathcal{M}(\mathbb{R}^d)} = 1$;

(v) if ν *is generated by* $\{v_{n_k}\}$, $f : \Omega \times \mathbb{R}^d \to \mathbb{R} \cup \{+\infty\}$ *is a normal integrand, and if* $\{f^-(\cdot, v_{n_k})\}$ *is equi-integrable, then*

$$\int_\Omega \int_{\mathbb{R}^d} f(x, \Lambda) d\nu_x(\Lambda) dx \leq \liminf_{k\to+\infty} \int_\Omega f(x, v_{n_k}(x)) dx,$$

where $f^- := -\min\{f, 0\}$;

(vi) if ν *is generated by* $\{v_{n_k}\}$ *and if* f *is a Carathéodory integrand then* $\{f(\cdot, v_{n_k})\}$ *is equi-integrable if and only if*

$$\int_\Omega \int_{\mathbb{R}^d} f(x, \Lambda) d\nu_x(\Lambda) dx = \lim_{k\to+\infty} \int_\Omega f(x, v_{n_k}(x)) dx.$$

Young measures have been originally introduced in Optimal Control Theory by L.C. Young in connection to nonconvex problems, thus providing the appropriate framework for the description of generalized minimizers in the Calculus of Variations (see (30), (31)). Later, Tartar developed the use of Young measures in the PDE framework (see (27), (28), (29)). For more details regarding the study of Young measures we refer the reader to (4), (5), (6), (17), (19), (20), (21), (26), among others.

Proposition 2.0.7. *If a sequence* $\{v_n\}$ *of measurable functions* $v_n : \Omega \to \mathbb{R}^d$ *generates a Young measure* ν *and if* $w_n \to w$ *in measure, then the sequence* $\{v_n + w_n\}$ *generates the translated Young measure* $\overline{\nu} = \{\nu_x\}_{x\in\Omega}$ *defined by* $\overline{\nu}_x := \Gamma_{w(x)}\nu_x$, *where, for* $a \in \mathbb{R}^d$, $\varphi \in C_0(\mathbb{R}^d)$ *and* $\mu \in \mathcal{M}(\mathbb{R}^d)$, *we define* $\langle \Gamma_a \mu, \varphi\rangle := \langle \mu, \varphi(\cdot + a)\rangle$. *In particular, if* $\{u_n\}$ *is a sequence of measurable mappings from* Ω *into* $\mathbb{R}^d$ *such that* $v_n - u_n \to 0$ *in measure then* $\{u_n\}$ *also generates* ν.

A Young measure ν is called a $W^{1,p}$*-gradient Young measure* if it is generated by $\{\nabla u_n\}$, with $u_n \rightharpoonup u$ in $W^{1,p}(\Omega;\mathbb{R}^d)$. The following characterization of $W^{1,p}$-gradient Young measures is due to Kinderlehrer and Pedregal (19), (20) (see also (16), (17), (21)).

Theorem 2.0.8. *Let* $1 \le p \le +\infty$. *A weak* measurable map* $\nu : \Omega \to \mathcal{M}(\mathbb{M}^{d\times N})$ *is a* $W^{1,p}$*-gradient Young measure if and only if* $\nu_x \ge 0$ *for* $\mathcal{L}^N$*-a.e.* $x \in \Omega$, *and*

(i) *there exists* $u \in W^{1,p}(\Omega;\mathbb{R}^d)$ *such that* $\langle \nu_x, id\rangle = \nabla u(x)$ *for* $\mathcal{L}^N$*-a.e.* $x \in \Omega$;

(ii) $\displaystyle\int_\Omega \int_{\mathbb{R}^{d\times N}} |\Lambda|^p d\nu_x(\Lambda)dx < +\infty$ *(supp* $\nu_x \subset K$ *for* $\mathcal{L}^N$*- a.e.* $x \in \Omega$, *where* $K \subset \mathbb{R}^{d\times N}$ *is a compact set, if* $p = \infty$ *);*

(iii) $\langle \nu_x, f\rangle \ge f(\langle \nu_x, id\rangle)$ *for* $\mathcal{L}^N$*-a.e.* $x \in \Omega$ *and for all quasiconvex functions* $f : \mathbb{R}^{d\times N} \to \mathbb{R}$, *with* $|f(A)| \le C(1 + |A|^p)$ *for some* $C > 0$ *and all* $A \in \mathbb{R}^{d\times N}$ *if* $1 \le p < +\infty$.

The following decomposition result for scaled gradients (see (9) and (14) for a proof) will play an important role in the sequel.

Theorem 2.0.9. *Let* $\Omega := \omega \times I$, *where* $\omega \subset \mathbb{R}^2$ *is an open, bounded Lipschitz domain, let* $\{\varepsilon_n\}$ *be a sequence of positive real numbers converging to zero, and let* $\{v_n\}$ *be a bounded sequence in* $W^{1,p}(\Omega,\mathbb{R}^3)$ $(1 < p < +\infty)$ *satisfying*

$$\sup_{n\in\mathbb{N}} \int_\Omega \left| \left(\nabla_\alpha v_n \middle| \frac{1}{\varepsilon_n}\nabla_3 v_n \right) \right|^p dx < +\infty.$$

Suppose further that $v_n \rightharpoonup v$ *in* $W^{1,p}(\Omega,\mathbb{R}^3)$ *and* $\frac{1}{\varepsilon_n}\nabla_3 v_n \rightharpoonup c$ *in* $L^p(\Omega,\mathbb{R}^3)$. *Then there exists a subsequence* $\{v_{n_k}\}$ *of* $\{v_n\}$ *and a sequence* $\{w_k\} \subset W^{1,\infty}(\Omega,\mathbb{R}^3)$ *such that (i)* $\lim_{k\to\infty} \mathcal{L}^3\left(\{x \in \Omega : w_k(x) \neq v_{n_k}(x)\}\right) = 0$;

(ii) $\left\{\left(\nabla_\alpha w_k \middle| \frac{1}{\varepsilon_{n_k}} \nabla_3 w_k\right)\right\}$ *is p-equi-integrable;*

(iii) $w_k \rightharpoonup v$ *in* $W^{1,p}(\Omega, \mathbb{R}^3)$;

(iv) $\frac{1}{\varepsilon_{n_k}} \nabla_3 w_k \rightharpoonup c$ *in* $L^p(\Omega, \mathbb{R}^3)$;

(v) $w_k = v$ *on* $\partial\omega \times I$.

Remark 2.0.10.

a) (v) is not explicitly presented in (9) but it easily follows by means of a standard cut-off argument;

b) if in place of $\frac{1}{\varepsilon_n} \nabla_3 v_n \rightharpoonup c$ *in* $L^p(\Omega, \mathbb{R}^3)$ *we assume that*

$$\frac{1}{\varepsilon_n} \int_I \nabla_3 v_n(\cdot, x_3)dx_3 \rightharpoonup b \in L^p(\omega; \mathbb{R}^3) \quad \text{in } L^p(\omega, \mathbb{R}^3)$$

then (i)-(iii), (v) still hold, and (iv) would need to be replaced by

$$\frac{1}{\varepsilon_{n_k}} \int_I \nabla_3 w_k(\cdot, x_3)dx_3 \rightharpoonup b \quad \text{in} \quad L^p(\omega, \mathbb{R}^3).$$

Indeed, for each $j \in \mathbb{N}$ define $\omega_j := \left\{x \in \omega : \text{ dist } (x, \partial\omega) < \frac{1}{j}\right\}$, and consider cut-off functions $\varphi_j \in C_c^\infty(\omega; [0,1])$ such that φ_j vanishes in a neighborhood of $\partial\omega$, $\varphi_j \equiv 1$ in $\omega \setminus \omega_j$, and $\|\nabla\varphi_j\|_{L^\infty(\omega;\mathbb{R}^2)} \leq Cj$, for some constant $C > 1$. Extract a subsequence $\{k(j)\}$ of $\{k\}$ such that, for each $j \in \mathbb{N}$,

$$\|w_{k(j)} - v\|_{L^p(\Omega;\mathbb{R}^3)} \leq \frac{1}{j^2}, \tag{2.0.7}$$

and define

$$\overline{w}_j(x_\alpha, x_3) := v(x_\alpha) + \varphi_j(x_\alpha)\left(w_{k(j)}(x_\alpha, x_3) - v(x_\alpha)\right).$$

It is clear that $\overline{w}_j = v$ on $\partial\omega \times I$. Since

$$\{x \in \Omega : \overline{w}_j(x) \neq v_{n_{k(j)}}(x)\} \subset (\omega_j \times I) \cup \{x \in \Omega : w_{k(j)}(x) \neq v_{n_{k(j)}}(x)\},$$

we infer that

$$\lim_{j\to\infty} \mathcal{L}^3\left(\{x \in \Omega : \overline{w}_j(x) \neq v_{n_{k(j)}}(x)\}\right) = 0.$$

Let A be a Borel subset of Ω. In view of (2.0.7) we have

$$\begin{aligned}
&\int_A \left| \left(\nabla_\alpha \overline{w}_j \Big| \frac{1}{\varepsilon_{n_{k(j)}}} \nabla_3 \overline{w}_j \right) \right|^p dx \\
&\leqq C \left(\int_A |\nabla_\alpha v|^p dx + j^p \int_A |w_{k(j)} - v|^p dx \right. \\
&\qquad \left. + \int_A \left| \left(\nabla_\alpha w_{k(j)} \Big| \frac{1}{\varepsilon_{n_{k(j)}}} \nabla_3 w_{k(j)} \right) \right|^p dx \right) \\
&\leqq C \left(\int_A |\nabla_\alpha v|^p dx + \left(\int_A |w_{k(j)} - v|^p dx \right)^{\frac{1}{2}} \right. \\
&\qquad \left. + \int_A \left| \left(\nabla_\alpha w_{k(j)} \Big| \frac{1}{\varepsilon_{n_{k(j)}}} \nabla_3 w_{k(j)} \right) \right|^p dx \right),
\end{aligned}$$

which implies that the sequence $\left\{ \left(\nabla_\alpha \overline{w}_j \Big| \frac{1}{\varepsilon_{n_{k(j)}}} \nabla_3 \overline{w}_j \right) \right\}$ is p-equi-integrable. Next, relabel $\{k(j)\}$ by $\{k\}$ and $\{\overline{w}_j\}$ by $\{w_k\}$ and we have obtained the sequence $\{w_k\}$ with the desired properties.

Note that we also have

$$\begin{aligned}
\limsup_{k\to\infty} \int_\Omega \left| \left(\nabla_\alpha w_k \Big| \frac{1}{\varepsilon_{n_k}} \nabla_3 w_k \right) \right|^p dx &\leq C \left(\int_\Omega |\nabla_\alpha v|^p dx \right. \\
&\left. + \limsup_{n\to\infty} \int_\Omega \left| \left(\nabla_\alpha v_n \Big| \frac{1}{\varepsilon_n} \nabla_3 v_n \right) \right|^p dx \right) < +\infty, \qquad (2.0.8)
\end{aligned}$$

where C is a positive constant that only depends on p.

3 Proof of Theorem 1.0.2

We will first prove that

$$E(v,b) \geq \inf_{\nu \in Y(v,b)} \int_\Omega \int_{\mathbb{R}^{3\times 3}} W(x,A) d\nu_x(A) dx. \qquad (3.0.9)$$

If $E(v,b) = +\infty$ then there is nothing to prove. Assume now that $E(v,b) < +\infty$, and for every $k \in \mathbb{N}$ let $\{\varepsilon_n^k\} \subset (0,+\infty)$ and $\{v_n^k\} \subset W^{1,p}(\Omega;\mathbb{R}^3)$ be such that $\varepsilon_n^k \to 0^+$, $v_n^k \rightharpoonup v$ in $W^{1,p}(\Omega;\mathbb{R}^3)$, $\displaystyle\frac{1}{\varepsilon_n^k}\int_I \nabla_3 v_n^k(\cdot,x_3)dx_3 \rightharpoonup b$ in $L^p(\omega;\mathbb{R}^3)$ (as $n\to\infty$), and

$$E(v,b) \le \liminf_{n\to+\infty}\int_\Omega W\left(x;\nabla_\alpha v_n^k\Big|\frac{1}{\varepsilon_n^k}\nabla_3 v_n^k\right)dx < E(v,b)+\frac{1}{k}. \quad (3.0.10)$$

Extract an increasing subsequence $\{n(j,k)\}_j$ of $\{n\}$ so that

$$\lim_{j\to+\infty}\int_\Omega W\left(x;\nabla_\alpha v_{n(j,k)}^k\Big|\frac{1}{\varepsilon_{n(j,k)}^k}\nabla_3 v_{n(j,k)}^k\right)dx =$$

$$\liminf_{n\to+\infty}\int_\Omega W\left(x;\nabla_\alpha v_n^k\Big|\frac{1}{\varepsilon_n^k}\nabla_3 v_n^k\right)dx < E(v,b)+\frac{1}{k} < +\infty. \quad (3.0.11)$$

The coercivity condition in (1.0.1) guarantees that the sequence $\left\{\left(\nabla_\alpha v_{n(j,k)}^k\Big|\frac{1}{\varepsilon_{n(j,k)}^k}\nabla_3 v_{n(j,k)}^k\right)\right\}_j$ is bounded in $L^p(\Omega;\mathbb{M}^{3\times 3})$ and, consequently, we can extract a subsequence (not relabeled) generating a Young measure $\nu^{(k)}=\left\{\nu_x^{(k)}\right\}_{x\in\Omega}\in Y(v,b)$. By Theorem 2.0.6 (iv) and taking into account (3.0.11), we have

$$\int_\Omega\int_{\mathbb{R}^{3\times 3}} W(x;A)d\nu_x^{(k)}(A)dx \le \lim_{j\to+\infty}\int_\Omega W\left(x;\nabla_\alpha v_{n(j,k)}^k\Big|\frac{1}{\varepsilon_{n(j,k)}^k}\nabla_3 v_{n(j,k)}^k\right)dx.$$

Since $\nu^{(k)}\in Y(v,b)$, and in view of (3.0.10), (3.0.11), we obtain that

$$\inf_{\nu\in Y(v,b)}\int_\Omega\int_{\mathbb{R}^{3\times 3}} W(x;A)d\nu_x(A)dx \le E(v,b)+\frac{1}{k}.$$

Letting $k\to\infty$ we deduce (3.0.9).

Conversely, let $\nu\in Y(v,b)$. There exist sequences $\{\varepsilon_n\}\subset(0,+\infty)$ and $\{v_n\}\subset W^{1,p}(\Omega;\mathbb{R}^3)$ such that $\varepsilon_n\to 0^+$, $v_n\rightharpoonup v$ in $W^{1,p}(\Omega;\mathbb{R}^3)$,

$$\frac{1}{\varepsilon_n}\int_I \nabla_3 v_n(\cdot,x_3)dx_3 \rightharpoonup b \quad \text{in } L^p(\Omega;\mathbb{R}^3),$$

and ν is generated by the sequence $\left\{\left(\nabla_\alpha v_n\Big|\frac{1}{\varepsilon_n}\nabla_3 v_n\right)\right\}$, bounded in $L^p(\Omega;\mathbb{M}^{3\times 3})$. By Theorem 2.0.9 and in view of Remark 2.0.10, there exists a subsequence $\{n_k\}$ of $\{n\}$ and a sequence $\{w_k\} \subset W^{1,\infty}(\Omega, \mathbb{R}^3)$ such that

$$\lim_{k\to\infty} \mathcal{L}^3\left(\{x \in \Omega : w_k(x) \neq v_{n_k}(x)\}\right) = 0, \tag{3.0.12}$$

$$\left\{\left(\nabla_\alpha w_k \Big| \frac{1}{\varepsilon_{n_k}} \nabla_3 w_k\right)\right\} \text{ is } p- \text{ equi-integrable}, \tag{3.0.13}$$

$$w_k \rightharpoonup v \quad \text{in} \quad W^{1,p}(\Omega, \mathbb{R}^3), \tag{3.0.14}$$

and

$$\frac{1}{\varepsilon_{n_k}} \int_I \nabla_3 w_k(\cdot, x_3) dx_3 \rightharpoonup b \quad \text{in} \quad L^p(\omega, \mathbb{R}^3). \tag{3.0.15}$$

Taking into account (3.0.12) and Proposition 2.0.7, we deduce that the sequence $\left\{\left(\nabla_\alpha w_k \Big| \frac{1}{\varepsilon_{n_k}} \nabla_3 w_k\right)\right\}$ generates ν. Theorem 2.0.6 (v), together with (1.0.1) and (3.0.13), yields

$$\int_\Omega \int_{\mathbb{R}^{3\times 3}} W(x; A) d\nu_x(A) dx = \lim_{k\to+\infty} \int_\Omega W\left(x; \nabla_\alpha w_k \Big| \frac{1}{\varepsilon_{n_k}} \nabla_3 w_k\right) dx \geq E(v, b),$$

where the last inequality follows by (3.0.14) and (3.0.15). Passing to the infimum over all $\nu \in Y(v, b)$ we obtain that

$$E(v, b) \leq \inf_{\nu \in Y(v,b)} \int_\Omega \int_{\mathbb{R}^{3\times 3}} W(x; A) d\nu_x(A) dx,$$

which, together with (3.0.9), asserts (1.0.2). It remains to prove (1.0.3).

For every $n \in \mathbb{N}$, let $\{\varepsilon_{k,n}\} \subset (0, +\infty)$ and $\{v_{k,n}\} \subset W^{1,p}(\Omega; \mathbb{R}^3)$ be such that $\varepsilon_{k,n} \to 0^+$, $v_{k,n} \rightharpoonup v$ in $W^{1,p}(\Omega; \mathbb{R}^3)$, $\displaystyle\frac{1}{\varepsilon_{k,n}} \int_I \nabla_3 v_{k,n}(\cdot, x_3) dx_3 \rightharpoonup b$ in $L^p(\omega; \mathbb{R}^3)$ (as $k \to +\infty$), and

$$E(v, b) \leq \liminf_{k\to+\infty} \int_\Omega W\left(x; \nabla_\alpha v_{k,n} \Big| \frac{1}{\varepsilon_{k,n}} \nabla_3 v_{k,n}\right) dx \leq E(v, b) + \frac{1}{n}.$$

Extract an increasing subsequence $\{k(j, n)\}_j$ of $\{k\}$ so that

$$\liminf_{k\to+\infty} \int_\Omega W\left(x; v_{k,n} \Big| \frac{1}{\varepsilon_{k,n}} \nabla_3 v_{k,n}\right) dx =$$

$$\lim_{j\to+\infty} \int_\Omega W\left(x; \nabla_\alpha v_{k(j,n),n} \Big| \frac{1}{\varepsilon_{k(j,n),n}} \nabla_3 v_{k(j,n),n}\right) dx.$$

A diagonalization procedure (see (10) for details in a similar context) yields sequences $\varepsilon_n \to 0^+$ and $\{v_n\} \subset W^{1,p}(\Omega;\mathbb{R}^3)$ such that $v_n \rightharpoonup v$ in $W^{1,p}(\Omega;\mathbb{R}^3)$ $\frac{1}{\varepsilon_n}\int_I \nabla_3 v_n(\cdot, x_3)dx_3 \rightharpoonup b$ in $L^p(\omega;\mathbb{R}^3)$, and

$$\lim_{n\to+\infty}\int_\Omega W\left(x;\nabla_\alpha v_n\Big|\frac{1}{\varepsilon_n}\nabla_3 v_n\right)dx = E(v,b). \tag{3.0.16}$$

Since we may assume without loss of generality that $E(v,b) < +\infty$, the coercivity condition in (1.0.1) implies that the sequence $\left\{\left(\nabla_\alpha v_n\Big|\frac{1}{\varepsilon_n}\nabla_3 v_n\right)\right\}$ is bounded in $L^p(\Omega;\mathbb{R}^{3\times3})$. Let $\mu \in Y(v,b)$ be the Young measure generated by a subsequence (not relabeled) of $\left\{\left(\nabla_\alpha v_n\Big|\frac{1}{\varepsilon_n}\nabla_3 v_n\right)\right\}$ and let $\{n_k\} \subset \{n\}$ and $\{w_k\} \subset W^{1,\infty}(\Omega,\mathbb{R}^3)$ be the sequences provided by Theorem 2.0.9 (again, we need to take into account Remark 2.0.10), so that (3.0.12)-(3.0.15) hold. In view of Theorem 2.0.6 we have that

$$\begin{aligned} E(v,b) &\le \liminf_{k\to+\infty}\int_\Omega W\left(x;\nabla_\alpha w_k\Big|\frac{1}{\varepsilon_{n_k}}\nabla_3 w_k\right)dx \\ &= \int_\Omega\int_{\mathbb{R}^{3\times3}} W(x;A)d\mu_x(A)dx \\ &\leqq \lim_{n\to+\infty}\int_\Omega W\left(x;\nabla_\alpha v_n\Big|\frac{1}{\varepsilon_n}\nabla_3 v_n\right)dx \\ &= E(v,b), \end{aligned}$$

where we have used (1.0.1), (3.0.12)-(3.0.15) and (3.0.16). Thus,

$$\int_\Omega\int_{\mathbb{R}^{3\times3}} W(x;A)d\mu_x(A)dx = E(v,b),$$

and, in view of (1.0.2), we infer (1.0.3).

Bibliography

[1] Acerbi, E., Fusco, N., Semicontinuity problems in the calculus of variations. *Arch. Rat. Mech. Anal.* **86** (1984), 125-145.

[2] Anzelotti, E., Baldo, S., Percivale, D., Dimensional reduction in variational problems, asymptotic developments in Γ- convergence, and thin structures in elasticity. *Asymptotic Anal.* **9** (1994), 61-100.

[3] Babadjian, J-F., Francfort, G., Spatial heterogeneity in 3D-2D dimensional reduction. *ESAIM: Cont. Opt. Calc. Var.* **11** (2005), 139-160.
[4] Balder, E. J., A general approach to lower semicontinuity and lower closure in optimal control theory. *SIAM J. Control Opt.* **22** (1984), 570-598.
[5] Ball, J. M., *A version of the fundamental theorem for Young mesures,* in PDE's and Continuum Models of Phase Transitions, M. Rascle, D. Serre, and M. Slemrod, eds., Lecture Notes in Phys. 344, Springer-Verlag, Berlin, 1989, 207-215.
[6] Berliocchi, H., Lasry, J.-M., Intégrands normales et mesures paramétrées en calcul des variations. *Bull. Soc. Math. France.* **101** (1973), 129-184.
[7] Bhattacharya, K., Fonseca, I., Francfort, G., An asymptotic study of the debonding of thin films. *Arch. Rat. Mech. Anal.* **161** (2002), 205-229.
[8] Bhattacharya, K., James, R. D., A theory of thin films of martensitic materials with applications to microactuators. *J. Mech. Phys. Solids* **47** (1999), 531-576.
[9] Bocea, M., Fonseca, I., Equi-integrability results for 3D-2D dimension reduction problems. *ESAIM: Control, Optimisation and Calculus of Variations* **7** (2002), 443-470.
[10] Bocea, M., Young measure minimizers in the asymptotic analysis of thin films. *Scientific Report 03-CNA-002, Center for Nonlinear Analysis, Carnegie Mellon University* (2003).
[11] Bouchitté, G., Fonseca, I., Mascarenhas, M.L., Bending Moment in Membrane Theory. *Scientific Report 02-CNA-016, Center for Nonlinear Analysis, Carnegie Mellon University* (2002).
[12] Braides, A., Fonseca, I., Francfort, G., 3D-2D asymptotic analysis for inhomogeneous thin films. *Indiana Univ. Math. J.* **49** (2000), 1367-1404.
[13] Braides, A., Fonseca, I., Brittle thin films. *Applied Math. and Optimization* **44** (2001), 299-323.
[14] Dacorogna, B., *Direct Methods in the Calculus of Variations,* Springer-Verlag, 1989.
[15] Fonseca, I., Francfort, G., On the inadequacy of scaling of linear elasticity for 3D-2D asymptotics in a nonlinear setting. *J. Math. Pures Appl.* **80** (2001), 547-562.
[16] Fonseca, I., Müller, S.: *A*-quasiconvexity, lower semicontinuity, and Young measures. *SIAM J. Math. Anal.* **30** (1999), 1355–1390.
[17] Fonseca, I., Müller, S., Pedregal, P., Analysis of concentration and oscillation effects generated by gradients. *SIAM J. Math. Anal.* **29** (1998), 736-756.

[18] Freddi, L., Paroni, R., The energy density of martensitic thin films via dimension reduction. *Interfaces and Free Boundaries* **6** (2004), 439-459.

[19] Kinderlehrer, D., Pedregal, P., Characterizations of Young mesures generated by gradients. *Arch. Rat. Mech. Anal.* **115** (1991), 329-365.

[20] Kinderlehrer, D., Pedregal, P., Gradient Young mesures generated by sequences in Sobolev spaces. *J. Geom. Anal.* **4** (1994), 59-90.

[21] Kristensen, J., Lower semicontinuity in spaces of weakly differentiable functions. *Math. Ann.* **313** (1999), 653-710.

[22] Le Dret, H., Raoult, A., The nonlinear membrane model as variational limit of nonlinear three-dimensional elasticity. *J. Math. Pures Appl.* **74** (1995), 549-578.

[23] Le Dret, H., Raoult, A., Variational convergence for nonlinear shell models with directors and related semicontinuity and relaxation results. *Arch. Rat. Mech. Anal.* **154** (2000), 101-134.

[24] Morrey, C. B., Quasiconvexity and the semicontinuity of multiple integrals. *Pacific J. Math.* **2** (1952), 25-53.

[25] Morrey, C. B., *Multiple Integrals in the Calculus of Variations,* Springer-Verlag, Berlin, 1966.

[26] Pedregal, P., *Parametrized mesures and Variational Principles,* Birkhäuser, Boston, 1997.

[27] Tartar, L., *Compensated compactness and applications to partial differential equations,* in Nonlinear Analysis and Mechanics: Heriot-Watt Symposium, R. Knops, ed., Vol. IV, Pitman Res. Notes Math., 39, Longman, Harlow, U.K., 1979, 136-212.

[28] Tartar, L., *The compensated compactness method applied to systems of conservation laws,* in Systems of Nonlinear Partial Differential Equations, J. M. Ball, ed., D. Riebel, Dordrecht, 1983

[29] Tartar, L., *Étude des oscillations dans les équations aux dérivées partielles nonlinéaires,* in Trends and Applications of Pure Mathematics to Mechanics, Lecture Notes in Phys. 195, Springer-Verlag, Berlin, New York, 1984, 384-412.

[30] Young, L. C., Generalized curves and the existence of an attained absolute minimum in the calculus of variations. *Comptes Rendus de la Société des Sciences et des Lettres de Varsovie, classe III.* **30** (1937), 212-234.

[31] Young, L.C., *Lectures on the calculus of variations and optimal control theory.* Saunders, 1969 (reprinted by Chelsea, 1980).

Bending Moment and Higher Order Membrane Models-Lecture IV

Irene Fonseca

Department of Mathematical Sciences, Carnegie Mellon University

Abstract A model of nonlinear membrane involving second order derivatives is used to obtain a 3D-2D dimension reduction via Γ-convergence taking into account the Cosserat vector field.

1 Introduction

The content of this lecture is based on the paper

I. Fonseca, G. Francfort and G. Leoni, Thin Elastic Films: The Impact of Higher Order Perturbations, *Quart. Appl. Math.* **65** (2007), 69-98, and I. Fonseca, Erratum to Thin Elastic Films; The Impact of Higher Order Perturbations, CNA Preprint 07-CNA-003.

Here we add to the elastic energy

$$\int_{\omega\times(-\frac{\varepsilon}{2},\frac{\varepsilon}{2})} W(\nabla u)\,dx$$

an interfacial energy of the form

$$\varepsilon^{\gamma}\int_{\omega\times(-\frac{\varepsilon}{2},\frac{\varepsilon}{2})} |\nabla^2 u|^2\,dx$$

where $\nabla^2 u$ is the Hessian matrix associated to the transformation vector $u:\omega\times(-\frac{\varepsilon}{2},\frac{\varepsilon}{2})\to\mathbb{R}^3$ and $\gamma>0$. It proves to be more convenient to re-scale the resulting energy minimization problem onto a fixed domain of thickness 2 through a $\frac{1}{\varepsilon}$-dilation of the transverse variable. The problem, cast in the

*The research of I. Fonseca was partially funded by the National Science Foundation (NSF) under NSF Grants DMS-0103799 and DMS-0401763, and by the Center for Nonlinear Analysis under NSF Grants DMS-9803791, DMS-0405343 and DMS-0635983.

variational framework, consists in studying the Γ-limit, in an appropriate topology, of

$$\int_{\omega\times(-\frac{1}{2},\frac{1}{2})}\left[W\left(\nabla_\alpha u|\frac{1}{\varepsilon}\nabla_3 u\right)\right.$$
$$\left.+\varepsilon^\gamma\left(|\nabla_{\alpha\beta}u|^2+\frac{1}{\varepsilon^2}|\nabla_{\alpha 3}u|^2+\frac{1}{\varepsilon^4}|\nabla_{33}u|^2\right)\right]\,dx \qquad (1.0.1)$$

where ∇_α stands for the gradient in the plane of the film, i.e. the differential operators in (1.0.1) are defined as

$$\nabla_\alpha := (\nabla_1,\nabla_2)\,,\quad \nabla_{\alpha\beta} := (\nabla_{11},\nabla_{22},\nabla_{12})\,,\quad \nabla_{\alpha 3} := (\nabla_{13},\nabla_{23})\,.$$

In all cases, provided that the elastic density has polynomial growth – which of course excludes the setting of hyperelasticity for which W blows up as $\det\nabla u$ vanishes – the limit kinematics is identical as far as the limit of u_ε is concerned: the limit field u is independent of the transverse variable x_3; it represents the transformation of the mid-surface of the membrane. In the case $\gamma < 2$, the interfacial energy is strong enough to rigidify the cross section (the Cosserat vector is b independent of x_3) and to decouple the mid-plane deformation from the bending moment (the average of b over the thickness). Thus, the resulting energy is merely the lower semi-continuous envelope of $\int_{\omega\times(-\frac{1}{2},\frac{1}{2})} W(\nabla_\alpha u|b)\,dx$.

If $\gamma = 2$, then the behavior drastically depends on the form of the Cosserat vector b. If it is independent of x_3, then the resulting energy once again treats $\nabla_\alpha u$ and b as independent fields and the result is that obtained for $\gamma < 2$. If b does depend on x_3, then the resulting energy is more involved and we strongly suspect that it is non local.

When $\gamma > 2$ the higher order terms in the energy do not affect the Γ-limit which coincides with that obtained by Bouchitté, Fonseca and Mascarenhas in (5).

For ε, $\gamma > 0$ and $1 < p < \infty$ consider the functional

$$E_\varepsilon^\gamma : W^{1,p}\left(\Omega;\mathbb{R}^3\right)\times\mathcal{A}\left(\omega\right)\to[0,\infty]$$

defined by

$$E_\varepsilon^\gamma(u;A) := \int_{A\times I} W\left(\nabla_\alpha u|\frac{1}{\varepsilon}\nabla_3 u\right)\,dx$$
$$+\,\varepsilon^\gamma\int_{A\times I}\left(|\nabla_{\alpha\beta}u|^2+\frac{1}{\varepsilon^2}|\nabla_{\alpha 3}u|^2+\frac{1}{\varepsilon^4}|\nabla_{33}u|^2\right)\,dx$$

if $u \in W^{2,2}(\Omega;\mathbb{R}^3)$, and $E_\varepsilon^\gamma(u;A) := \infty$ otherwise. Here

$$\Omega := \omega \times I, \quad I := (-1/2, 1/2),$$

and the elastic energy density $W : \mathbb{R}^{3\times3} \to [0,\infty)$ satisfies the following hypothesis:

(H_1) W is continuous and there exists $C > 0$ such that

$$W(F) \geq \frac{1}{C}|F|^p - C$$

for all $F \in \mathbb{R}^{3\times3}$.

Theorem 1.0.1 (Compactness). *Assume that W satisfies condition (H_1). Let $\varepsilon_n \to 0^+$ and let $\{u_n\} \subset W^{2,2}(\Omega;\mathbb{R}^3)$ be such that*

$$\sup_n E_{\varepsilon_n}^\gamma(u_n;\Omega) < \infty.$$

Then there exist a subsequence $\{u_{n_k}\}$, $u \in W^{1,p}(\Omega;\mathbb{R}^3)$, with $\nabla_3 u = 0$ $\mathcal{L}^3$ a.e. in Ω, and $b \in L^p(\Omega;\mathbb{R}^3)$ such that

$$u_{n_k} \rightharpoonup u \text{ in } W^{1,p}(\Omega;\mathbb{R}^3), \quad \frac{1}{\varepsilon_{n_k}}\nabla_3 u_{n_k} \rightharpoonup b \text{ in } L^p(\Omega;\mathbb{R}^3). \tag{1.0.2}$$

Moreover if $\gamma < 2$ then $\nabla_3 b = 0$ $\mathcal{L}^3$ a.e. in Ω, while if $\gamma = 2$ then $\nabla_3 b \in L^2(\Omega;\mathbb{R}^3)$.

As a consequence of the the previous theorem, for every $\gamma > 0$ the natural ambient space for the limit energy is given by

$$\begin{aligned}\mathcal{V}^\gamma := \{&(u,b) \in W^{1,p}(\Omega;\mathbb{R}^3) \times L^p(\Omega;\mathbb{R}^3) : \nabla_3 u = 0\ \mathcal{L}^3 \text{ a.e. in } \Omega, \\ &\nabla_3 b = 0\ \mathcal{L}^3 \text{ a.e. in } \Omega \text{ if } \gamma < 2,\ \nabla_3 b \in L^2(\Omega;\mathbb{R}^3) \text{ if } \gamma = 2\}.\end{aligned}$$

In what follows we identify functions $u \in W^{1,p}(\Omega;\mathbb{R}^3)$ such that $\nabla_3 u = 0$ $\mathcal{L}^3$ a.e. in Ω with functions in $W^{1,p}(\omega;\mathbb{R}^3)$, and, similarly, functions $b \in L^p(\Omega;\mathbb{R}^3)$ such that $\nabla_3 b = 0$ $\mathcal{L}^3$ a.e. in Ω with functions in $L^p(\omega;\mathbb{R}^3)$.

In view of (1.0.2) the appropriate notion of Γ-convergence in our setting is the following:

$$E_-^\gamma = \Gamma - \liminf_{n\to\infty} E_{\varepsilon_n}^\gamma.$$

Since $E_{\varepsilon_n}^\gamma(u_n;A) = \infty$ if $u_n \notin W^{2,2}(\Omega;\mathbb{R}^3)$, it is clear that we may write

$$\begin{aligned}E_-^\gamma(u,b;A) = \inf\Big\{&\liminf_{n\to\infty} E_{\varepsilon_n}^\gamma(u_n;A) : u_n \in W^{2,2}(\Omega;\mathbb{R}^3), \\ &u_n \rightharpoonup u \text{ in } W^{1,p}(\Omega;\mathbb{R}^3), \quad \frac{1}{\varepsilon_n}\nabla_3 u_n \rightharpoonup b \text{ in } L^p(\Omega;\mathbb{R}^3)\Big\}.\end{aligned}$$

A standard diagonalization argument, together with the coercivity hypothesis (H_1), yields the following result whose proof we omit. We remark that this argument holds only for $q > 1$.

Proposition 1.0.2. *Assume that W satisfies condition (H_1). Then for any open subset $A \subset \omega$, the functional $E_-^\gamma(\cdot,\cdot;A)$ is sequentially lower semicontinuous with respect to the weak convergence in $W^{1,p}(\Omega;\mathbb{R}^3)\times L^p(\Omega;\mathbb{R}^3)$.*

Similarly, we say that a functional

$$E_+^\gamma : \mathcal{V}^\gamma \times \mathcal{A}(\omega) \to [0,\infty]$$

is the $\Gamma-\limsup$ of the sequence of functionals $\{E_{\varepsilon_n}^\gamma\}$ with respect to the weak-weak convergence in $W^{1,p}(\Omega;\mathbb{R}^3)\times L^p(\Omega;\mathbb{R}^3)$ if for every $(u,b)\in\mathcal{V}^\gamma$ and $A\in\mathcal{A}(\omega)$

$$\begin{aligned} E_+^\gamma(u,b;A) &= \inf\Big\{\limsup_{n\to\infty} E_{\varepsilon_n}^\gamma(u_n;A) : u_n \in W^{2,2}(\Omega;\mathbb{R}^3), \\ &\qquad u_n \rightharpoonup u \text{ in } W^{1,p}(\Omega;\mathbb{R}^3), \quad \tfrac{1}{\varepsilon_n}\nabla_3 u_n \rightharpoonup b \text{ in } L^p(\Omega;\mathbb{R}^3)\Big\} \end{aligned}$$

and we write

$$E_+^\gamma = \Gamma - \limsup_{n\to\infty} E_{\varepsilon_n}^\gamma .$$

We say that the sequence $\{E_{\varepsilon_n}^\gamma\}$ Γ-converges to a functional E^γ if the $\Gamma-\liminf$ and $\Gamma-\limsup$ coincide, and we write

$$E^\gamma = \Gamma - \lim_{n\to\infty} E_{\varepsilon_n}^\gamma .$$

We conclude this section with two results which will be useful in the sequel. For a proof we refer to (15), (16).

Lemma 1.0.3 (Decomposition)**.** *Let $E\subset\mathbb{R}^N$ be a bounded Lebesgue measurable set and let $\{u_n\}$ be a sequence of functions uniformly bounded in $L^p(E;\mathbb{R}^d)$, $1\le p<\infty$. For $r>0$ consider the truncation $\tau_r:\mathbb{R}^d\to\mathbb{R}^d$ defined by*

$$\tau_r(z) := \begin{cases} z & \text{if } |z|\le r, \\ \dfrac{z}{|z|}r & \text{if } |z|>r. \end{cases}$$

Then there exists a subsequence of $\{u_n\}$ (not relabeled) and an increasing sequence of numbers $r_n\to\infty$ such that the truncated sequence $\{\tau_{r_n}\circ u_n\}$ is q-equi-integrable, and

$$|\{x\in E : u_n(x)\neq(\tau_{r_n}\circ u_n)(x)\}| \to 0.$$

Theorem 1.0.4. *Let $\Omega \subset \mathbb{R}^N$ be an open bounded set, let $1 < p < \infty$, and let $\{u_n\} \subset W^{1,p}(\Omega;\mathbb{R}^d)$ be a sequence of functions converging weakly in $W^{1,p}(\Omega;\mathbb{R}^d)$ to some function $u_0 \in W^{1,p}(\Omega;\mathbb{R}^d)$. Then there exists a subsequence $\{u_{n_k}\}$ and a sequence $\{v_k\} \subset W^{1,p}(\mathbb{R}^N;\mathbb{R}^d)$ such that $\{v_k\}$ converges to u_0 weakly in $W^{1,p}(\Omega;\mathbb{R}^d)$, $v_k = u_0$ in a neighborhood of $\partial\Omega$,*

$$|\{x \in \Omega : v_k(x) \neq u_{n_k}(x)\}| \to 0 \quad \text{as } k \to \infty$$

and $\{|\nabla v_k|^p\}$ is equi-integrable.

2 Γ-convergence

For $(u,b) \in \mathcal{V}^\gamma$ and $A \in \mathcal{A}(\omega)$ define

$$H_{3,3}^\gamma(u,b;A) := \inf\left\{\liminf_{n\to\infty}\int_{A\times I} W(\nabla_\alpha u_n\,|b_n)\,dx : \{u_n\} \subset W^{1,p}(\Omega;\mathbb{R}^3),\right.$$
$$\left.\{b_n\} \subset L^p(\Omega;\mathbb{R}^3),\ u_n \rightharpoonup u \text{ in } W^{1,p}(\Omega;\mathbb{R}3),\ b_n \rightharpoonup b \text{ in } L^p(\Omega;\mathbb{R}3)\right\}$$

if $\gamma \neq 2$, and

$$H_{3,3}^2(u,b;A) := \inf\left\{\liminf_{n\to\infty}\int_{A\times I}\left[W(\nabla_\alpha u_n\,|b_n) + |\nabla_3 b_n|^2\right]dx :\right.$$
$$\{u_n\} \subset W^{1,p}(\Omega;\mathbb{R}^3), \{b_n\} \subset L^p(\Omega;\mathbb{R}^3) \text{ with } \nabla_3 b_n \in L^2(\Omega;\mathbb{R}^3),$$
$$\left.u_n \rightharpoonup u \text{ in } W^{1,p}(\Omega;\mathbb{R}^3),\ b_n \rightharpoonup b \text{ in } L^p(\Omega;\mathbb{R}^3)\right\}.$$

The energies $H_{2,3}^\gamma$ and $H_{2,3}^2$ are defined analogously except for the fact that the approximating sequences $\{u_n\}$ belong to $W^{1,p}(\omega;\mathbb{R}^3)$. In addition, if $D_3 b = 0$ then we introduce $H_{2,2}^\gamma$ and $H_{2,2}^2$ where now both $\{u_n\}$ and $\{b_n\}$ are defined in ω. In the remainder of the paper we assume that condition (H_1) is strengthened as follows:

$(H_1)'$ $W : \mathbb{R}^{3\times 3} \to [0,\infty)$ is continuous and there exists $C > 0$ such that

$$\frac{1}{C}|F|^p - C \leq W(F) \leq C(1 + |F|^q)$$

for all $F \in \mathbb{R}^{3\times 3}$.

Theorem 2.0.5 (Γ-convergence). *Assume that condition $(H_1)'$ is satisfied. If the Γ-limit E_-^γ of the family $\{E_\varepsilon^\gamma\}$ exists then it satisfies*

$$H_{3,3}^\gamma(u,b;A) \leq E_-^\gamma(u,b,A) \leq H_{2,3}^\gamma(u,b;A)$$

for all $(u,b) \in \mathcal{V}^\gamma$ and $A \in \mathcal{A}(\omega)$.

Proof. Fix a sequence $\varepsilon_n \to 0^+$ and let

$$E_-^\gamma := \Gamma - \liminf_{n\to\infty} E_{\varepsilon_n}^\gamma .$$

Lower bound. We claim that

$$E_-^\gamma (u, b; A) \geq H_{3,3}^\gamma (u, b; A) \tag{2.0.3}$$

for all $(u, b) \in \mathcal{V}^\gamma$ and $A \in \mathcal{A}(\omega)$.

Fix $(u, b) \in \mathcal{V}^\gamma$ and $A \in \mathcal{A}(\omega)$, and consider any sequence $\{u_n\} \subset W^{2,2}(\Omega; \mathbb{R}^3)$ such that $u_n \rightharpoonup u$ in $W^{1,q}(\Omega; \mathbb{R}^3)$, $\frac{1}{\varepsilon_n}\nabla_3 u_n \rightharpoonup b$ in $L^p(\Omega; \mathbb{R}^3)$. If $\gamma \neq 2$ then

$$\begin{aligned}\liminf_{n\to\infty} E_{\varepsilon_n}^\gamma (u_n; A) &\geq \liminf_{n\to\infty} \int_{A\times I} W\left(\nabla_\alpha u_n \left| \frac{1}{\varepsilon_n}\nabla_3 u_n\right.\right) dx \\ &\geqq H_{3,3}^\gamma (u, b; A),\end{aligned}$$

while if $\gamma = 2$

$$\liminf_{n\to\infty} E_{\varepsilon_n}^2 (u_n; A) \geq$$
$$\liminf_{n\to\infty} \int_{A\times I} \left[W\left(\nabla_\alpha u_n \left| \frac{1}{\varepsilon_n}\nabla_3 u_n\right.\right) + \left|\frac{1}{\varepsilon_n}\nabla_{33} u_n\right|^2 \right] dx \geq H_{3,3}^2 (u, b; A) .$$

By taking the infimum over all such sequences $\{u_n\}$ in both cases we obtain (2.0.3).

Upper bound. We claim that for every $(u, b) \in \mathcal{V}^\gamma$ and for all $A \in \mathcal{A}(\omega)$

$$E_-^\gamma (u, b; A) \leq \int_{A\times I} W(\nabla_\alpha u \,|b)\, dx \tag{2.0.4}$$

if $\gamma \neq 2$, while

$$E_-^2 (u, b; A) \leq \int_{A\times I} \left(W(\nabla_\alpha u\,|b) + |\nabla_3 b|^2 \right) dx. \tag{2.0.5}$$

Extend u as a $W^{1,p}(\mathbb{R}^3; \mathbb{R}^3)$ function with

$$\|u\|_{W^{1,p}(\mathbb{R}^3;\mathbb{R}^3)} \leq C\|u\|_{W^{1,p}(\Omega;\mathbb{R}^3)}$$

and $\nabla_3 u = 0$ $\mathcal{L}^3$ a.e. in $\mathbb{R}^3$. Extend also b as $L^p(\mathbb{R}^3; \mathbb{R}^3)$ function with

$$\|b\|_{L^p(\mathbb{R}^3;\mathbb{R}^3)} \leq C\|b\|_{L^p(\Omega;\mathbb{R}^3)},$$

and $\nabla_3 b = 0$ $\mathcal{L}^3$ a.e. in $\mathbb{R}^3$ if $\gamma < 2$, and $\nabla_3 b \in L^2\left(\mathbb{R}^3;\mathbb{R}^3\right)$ with

$$\|\nabla_3 b\|_{L^2(\mathbb{R}^3;\mathbb{R}^3)} \leq C\|\nabla_3 b\|_{L^2(\Omega;\mathbb{R}^3)},$$

if $\gamma = 2$.

Consider a mollifier of the type

$$\varphi_j(x) := \frac{1}{\delta_j^3}\varphi\left(\frac{x}{\delta_j}\right)$$

where $\delta_j \to 0^+$, and $\varphi \in C_c^\infty\left(\mathbb{R}^3\right)$ with $\int_{\mathbb{R}^3}\varphi(x)\,dx = 1$. Set

$$u_{nj}(x) := (u * \varphi_j)(x) + \varepsilon_n \int_0^{x_3} (b * \varphi_j)(x_\alpha, s)\,ds \quad \text{for } x \in \mathbb{R}^3.$$

Since $\nabla_3(u * \varphi_j) = \nabla_3 u * \varphi_j = 0$ (recall that $\nabla_3 u = 0$ $\mathcal{L}^3$ a.e. in $\mathbb{R}^3$) it follows that $\nabla_3 u_{n,j} = \varepsilon_n b * \varphi_j$, and so for every fixed j

$$\frac{1}{\varepsilon_n}\nabla_3 u_{n,j} = b * \varphi_j.$$

Moreover, for $i = 1, 2$,

$$D_i u_{n,j} = D_i(u * \varphi_j) + \varepsilon_n \int_0^{x_3} (b * D_i\varphi_j)(x_\alpha, s)\,ds,$$

and so

$$|D_i u_{n,j} - D_i(u * \varphi_j)| \leq \frac{C\varepsilon_n}{\delta_j}\|b\|_{L^p(\mathbb{R}^3;\mathbb{R}^3)},$$

where we have used the fact that, by Hölder inequality, for all $x \in \Omega$ we have

$$\begin{aligned} |(b * D_i\varphi_j)(x)| &= \left|\int_{\mathbb{R}^3} b(y)\, D_i\varphi_j(x-y)\,dy\right| \\ &\leqq \|b\|_{L^p(\mathbb{R}^3;\mathbb{R}^3)}\|D_i\varphi_j\|_{L^{p'}(\mathbb{R}^3;\mathbb{R}^2)} \leq \frac{C}{\delta_j}\|b\|_{L^p(\Omega;\mathbb{R}^3)}. \end{aligned} \tag{2.0.6}$$

For every fixed j the sequence $\{u_{n,j}\}$ is admissible for $E_-^\gamma(u * \varphi_j, b * \varphi_j; A)$, and we have

$$\begin{aligned}
E_-^\gamma(u*\varphi_j, b*\varphi_j; A) &\le \liminf_{n\to\infty} E_{\varepsilon_n}^\gamma(u_{nj}; A)\\
&\leqq \lim_{n\to\infty}\int_{A\times I} W\left(\nabla_\alpha u*\varphi_j + \frac{C\varepsilon_n}{\delta_j} O(1)\,|b*\varphi_j\right) dx\\
&+ \liminf_{n\to\infty}\int_{A\times I}\varepsilon_n^\gamma\left(\left|\nabla_{\alpha\beta}u_{n,j}\right|^2 + \frac{1}{\varepsilon_n^2}\left|\nabla_{\alpha 3}u_{n,j}\right|^2 + \frac{1}{\varepsilon_n^4}\left|\nabla_{33}u_{n,j}\right|^2\right) dx\\
&= \int_{A\times I} W\left(\nabla_\alpha u*\varphi_j\,|b*\varphi_j\right) dx \qquad (2.0.7)\\
&+ \liminf_{n\to\infty}\int_{A\times I}\varepsilon_n^\gamma\left(\left|\nabla_{\alpha\beta}u_{n,j}\right|^2 + \frac{1}{\varepsilon_n^2}\left|\nabla_{\alpha 3}u_{n,j}\right|^2 + \frac{1}{\varepsilon_n^4}\left|\nabla_{33}u_{n,j}\right|^2\right) dx,
\end{aligned}$$

where in the last equality we have used Lebesgue Dominated Convergence Theorem together with the continuity of W. We claim that

$$\lim_{n\to\infty}\int_{A\times I}\varepsilon_n^\gamma\left(\left|\nabla_{\alpha\beta}u_{n,j}\right|^2 + \frac{1}{\varepsilon_n^2}\left|\nabla_{\alpha 3}u_{n,j}\right|^2 + \frac{1}{\varepsilon_n^4}\left|\nabla_{33}u_{n,j}\right|^2\right) dx = 0 \qquad (2.0.8)$$

if $\gamma \neq 2$, while

$$\lim_{n\to\infty}\int_{A\times I}\varepsilon_n^2\left(\left|\nabla_{\alpha\beta}u_{n,j}\right|^2 + \frac{1}{\varepsilon_n^2}\left|\nabla_{\alpha 3}u_{n,j}\right|^2 + \frac{1}{\varepsilon_n^4}\left|\nabla_{33}u_{n,j}\right|^2\right) dx = \int_{A\times I}\left|\nabla_3 b*\varphi_j\right|^2 dx \qquad (2.0.9)$$

if $\gamma = 2$.

For i, $k = 1, 2$

$$\begin{aligned}
D_{ik}u_{n,j}(x) &= (D_i u * D_k\varphi_j)(x) + \varepsilon_n\int_0^{x_3}(b*D_{ik}\varphi_j)(x_\alpha, s)\,ds,\\
D_{k3}u_{n,j}(x) &= \varepsilon_n(b*D_k\varphi_j)(x),
\end{aligned}$$

and so, reasoning as in (2.0.6),

$$\begin{aligned}
\left\|\nabla_{\alpha\beta}u_{n,j}\right\|_\infty &\le \frac{C}{\delta_j}\|\nabla_\alpha u\|_{L^q(\Omega;\mathbb{R}^{3\times 2})} + \frac{C\varepsilon_n}{\delta_j^2}\|b\|_{L^q(\Omega;\mathbb{R}^3)},\\
\left\|\nabla_{\alpha 3}u_{n,j}\right\|_\infty &\le \frac{C\varepsilon_n}{\delta_j}\|b\|_{L^q(\Omega;\mathbb{R}^3)},
\end{aligned}$$

which clearly imply that

$$\lim_{n\to\infty}\int_{A\times I}\varepsilon_n^\gamma\left(\left|\nabla_{\alpha\beta}u_{n,j}\right|^2 + \frac{1}{\varepsilon_n^2}\left|\nabla_{\alpha 3}u_{n,j}\right|^2\right) dx = 0.$$

To estimate the last term in (2.0.8) note that if $\gamma < 2$ then

$$\nabla_{33} u_{n,j} = \varepsilon_n \nabla_3 b * \varphi_j \equiv 0$$

since $\nabla_3 b = 0$ $\mathcal{L}^3$ a.e. in $\mathbb{R}^3$, while if $\gamma > 2$ then

$$\nabla_{33} u_{n,j}(x) = \varepsilon_n (b * \nabla_3 \varphi_j)(x),$$

and so

$$\|\nabla_{33} u_{n,j}\|_\infty \leq \frac{C\varepsilon_n}{\delta_j} \|b\|_{L^p(\Omega;\mathbb{R}^3)}.$$

In turn

$$\varepsilon_n^{\gamma-4} \|\nabla_{33} u_{n,j}\|_\infty^2 \leq \frac{C\varepsilon_n^{\gamma-2}}{\delta_j^2} \|b\|^2_{L^p(\Omega;\mathbb{R}^3)},$$

which yields

$$\lim_{n\to\infty} \int_{A\times I} \varepsilon_n^{\gamma-4} |\nabla_{33} u_{n,j}|^2 \, dx = 0.$$

Finally, if $\gamma = 2$ then

$$\nabla_{33} u_{n,j} = \varepsilon_n \nabla_3 b * \varphi_j$$

and thus

$$\lim_{n\to\infty} \int_{A\times I} \frac{1}{\varepsilon_n^2} |\nabla_{33} u_{n,j}|^2 \, dx = \int_{A\times I} |\nabla_3 b * \varphi_j|^2 \, dx.$$

Hence (2.0.8) and (2.0.9) hold, and by (2.0.7) we deduce that

$$E_-^\gamma(u * \varphi_j, b * \varphi_j; A) \leq \begin{cases} \int_{A\times I} W(\nabla_\alpha u * \varphi_j | b * \varphi_j) \, dx & \text{if } \gamma \neq 2, \\ \int_{A\times I} \left[W(\nabla_\alpha u * \varphi_j | b * \varphi_j) + |\nabla_3 b * \varphi_j|^2 \right] dx & \text{if } \gamma = 2. \end{cases}$$

Since $u * \varphi_j \to u$ in $W^{1,p}(\Omega;\mathbb{R}^3)$, $b * \varphi_j \to b$ in $L^p(\Omega;\mathbb{R}^3)$, and when $\gamma = 2$ also $\nabla_3 b * \varphi_j \to \nabla_3 b$ in $L^p(\Omega;\mathbb{R}^3)$, letting $j \to \infty$ in the previous inequality and using Proposition 1.0.2 and $(H_1)'$ we obtain (2.0.4) and (2.0.5).

Fix $(u,b) \in \mathcal{V}^\gamma$ and $A \in \mathcal{A}(\omega)$ and let $\{u_j\} \subset W^{1,p}(\omega;\mathbb{R}^3)$ converge weakly to u in $W^{1,p}(\omega;\mathbb{R}^3)$, and $\{b_j\} \subset L^p(\Omega;\mathbb{R}^3)$ converge weakly to b in $L^p(\Omega;\mathbb{R}^3)$. Using Proposition 1.0.2, (2.0.4) and (2.0.5) we have that

$$E_-^\gamma(u,b;A) \leq \liminf_{j\to\infty} E_-^\gamma(u_j,b_j;A) \leq \liminf_{j\to\infty} \int_{A\times I} W(\nabla_\alpha u_j | b_j) \, dx$$

if $\gamma \neq 2$, and

$$E_{-}^{\gamma}(u,b;A) \leq \liminf_{j\to\infty} E_{-}^{\gamma}(u_j,b_j;A)$$
$$\leq \liminf_{j\to\infty} \int_{A\times I} \left[W\left(\nabla_\alpha u_j \,|b_j\right) + \left|\nabla_3 b_j\right|^2\right] dx.$$

Taking the infimum over all such sequences $\{u_j\}$ and $\{b_j\}$ yields

$$E_{-}^{\gamma}(u,b;A) \leq H_{2,3}^{\gamma}(u,b,A)$$

for all $(u,b) \in \mathcal{V}^\gamma$ and $A \in \mathcal{A}(\omega)$. □

We now study the energies $H_{3,3}^{\gamma}$ and $H_{3,3}^{2}$.

Theorem 2.0.6 ($\gamma \neq 2$). *Assume that $\gamma \neq 2$ and that condition $(H_1)'$ is satisfied. Then for all $(u,b) \in \mathcal{V}^\gamma$, and $A \in \mathcal{A}(\omega)$*

$$H_{3,3}^{\gamma}(u,b;A) = \int_{A\times I} (Q_3 \times C_3)[W]\left(\nabla_\alpha u(x_\alpha)\,|b(x)\right) dx,$$

where $(Q_3 \times C_3)[W]$ is the cross quasiconvexification-convexification of W,

$$(Q_3 \times C_3)[W]\left(\overline{F}\,|z\right) := \inf\left\{\int_Q W\left(\overline{F} + \nabla_\alpha \varphi(x)\,|z + \phi(x)\right) dx\right.$$
$$\left.\varphi \in W_0^{1,p}\left(Q;\mathbb{R}^3\right), \phi \in L^p\left(Q;\mathbb{R}^3\right), \int_Q \phi(x)\, dx = 0\right\}$$

for $\overline{F} \in \mathbb{R}^{2\times 3}$, $z \in \mathbb{R}^3$.

Proof. The proof of this result is standard and so we omit it (see e.g. (13)). □

Remark 2.0.7. *Note that if $\gamma \neq 2$ then*

$$H_{3,3}^{\gamma}(u,b;A) = \int_{A\times I} (Q_2 \times C_2)[W]\left(\nabla_\alpha u(x_\alpha)\,|b(x)\right) dx, \tag{2.0.10}$$

where $(Q_2 \times C_2)[W]$ is defined as

$$(Q_2 \times C_2)[W]\left(\overline{F}\,|z\right) := \inf\left\{\int_{Q'} W\left(\overline{F} + \nabla_\alpha \varphi(x_\alpha)\,|z + \phi(x_\alpha)\right) dx_\alpha\right.$$
$$\varphi \in W_0^{1,p}\left(Q';\mathbb{R}^3\right), \phi \in L^p\left(Q';\mathbb{R}^3\right), \int_{Q'} \phi(x_\alpha)\, dx_\alpha =$$

for $\overline{F} \in \mathbb{R}^{2\times 3}$, $z \in \mathbb{R}^3$.

To see this we begin by observing that in the definition of $(Q_3 \times C_3)[W]$ *(resp. of* $(Q_2 \times C_2)[W]$*) Q-periodic functions in* $W^{1,p}(Q;\mathbb{R}^3)$ *(resp. Q'-periodic functions in* $W^{1,p}(Q';\mathbb{R}^3)$ *) can also be used in lieu of* $W_0^{1,p}(Q;\mathbb{R}^3)$ *-functions (resp.* $W_0^{1,p}(Q';\mathbb{R}^3)$*); (see (2), Conjecture 3.7 and Theorem 3.1). Hence*

$$(Q_3 \times C_3)[W](\overline{F}|z) \leq (Q_2 \times C_2)[W](\overline{F}|z).$$

Conversely, for $\varepsilon > 0$ *find* φ, ϕ *admissible in the definition of* $(Q_3 \times C_3)[W]$ $(\overline{F}, z)$ *and such that*

$$(Q_3 \times C_3)[W](\overline{F}, z) \geq \int_Q W\left(\overline{F} + \nabla_\alpha \varphi(x)\,|z + \phi(x)\right) dx - \varepsilon.$$

Then

$$\begin{aligned}
&(Q_3 \times C_3)[W](\overline{F}, z) \geq \\
&\int_{-\frac{1}{2}}^{\frac{1}{2}} \int_{Q'} W\left(\overline{F} + \nabla_\alpha\varphi(x)|\overline{z}(x_3) + \phi(x) - \int_{Q'} \phi(y_\alpha, x_3)\, dy_\alpha\right) dx_\alpha\, dx_3 - \varepsilon \\
&\geqq \int_{-\frac{1}{2}}^{\frac{1}{2}} (Q_2 \times C_2)[W](F|\overline{z}(x_3))\, dx_3 - \varepsilon \\
&\geqq (Q_2 \times C_2)[W]\left(\overline{F}| \int_{-\frac{1}{2}}^{\frac{1}{2}} \overline{z}(x_3)\, dx_3\right) - \varepsilon \\
&= (Q_2 \times C_2)[W](\overline{F}|z) - \varepsilon,
\end{aligned}$$

where

$$\overline{z}(x_3) := z + \int_{Q'} \phi(y_\alpha, x_3)\, dy_\alpha,$$

and where the last inequality holds by Jensen's inequality, because $(Q_2 \times C_2)[W](F|\cdot)$ *is convex and* ϕ *has zero average over Q. Given the arbitrariness of* ε *we conclude that*

$$(Q_3 \times C_3)[W](\overline{F}|z) = (Q_2 \times C_2)[W](\overline{F}|z), \tag{2.0.11}$$

which, together with previous theorem, yields (2.0.10). *In particular if W is independent of z then we recover the well known identity*

$$Q_3 W(\overline{F}) = Q_2 W(\overline{F}).$$

3 The critical case $\gamma = 2$

In this section we study the critical case $\gamma = 2$. We recall that

$$H^2_{3,3}(u,b;A) := \inf\Big\{\liminf_{n\to\infty}\int_{A\times I}\left(W(\nabla_\alpha u_n|b_n) + |\nabla_3 b_n|^2\right)dx :$$
$$\{u_n\}\subset W^{1,p}(\Omega;\mathbb{R}^3),\ \{b_n\}\subset L^p(\Omega;\mathbb{R}^3) \text{ with } \nabla_3 b_n \in L^2(\Omega;\mathbb{R}^3),$$
$$u_n \rightharpoonup u \text{ in } W^{1,p}(\Omega;\mathbb{R}^3),\ b_n \rightharpoonup b \text{ in } L^p(\Omega;\mathbb{R}^3)\Big\}$$

for all $(u,b)\in\mathcal{V}^2$ and $A\in\mathcal{A}(\omega)$, where

$$\mathcal{V}^2 := \Big\{(u,b)\in W^{1,p}(\Omega;\mathbb{R}^3)\times L^p(\Omega;\mathbb{R}^3) : \nabla_3 u = 0\ \mathcal{L}^3 \text{ a.e. in } \Omega,$$
$$\nabla_3 b \in L^2(\Omega;\mathbb{R}^3)\Big\}.$$

We begin by characterizing $H^2(u,b,A)$ for functions b which do not depend on x_3. Note that in this case $(u,b)\in\mathcal{V}^2$ may be identified with a function in $W^{1,q}(\omega;\mathbb{R}^3)\times L^p(\omega;\mathbb{R}^3)$.

Theorem 3.0.8. *Assume that condition $(H_1)'$ is satisfied. Then for all $(u,b)\in\mathcal{V}^2$ with $\nabla_3 b = 0$ $\mathcal{L}^3$ a.e. in Ω, and for all $A\in\mathcal{A}(\omega)$*

$$H^2(u,b;A) = \int_A (Q_2\times C_2)[W](\nabla_\alpha u(x_\alpha)|b(x_\alpha))\,dx_\alpha.$$

Proof. It is clear

$$H^2_{3,3}(u,b;A) \le \inf\Big\{\liminf_{n\to\infty}\int_A W(\nabla_\alpha u_n(x_\alpha)|b_n(x_\alpha))\,dx_\alpha :$$
$$\{u_n\}\subset W^{1,q}(\omega;\mathbb{R}^3),\ \{b_n\}\subset L^p(\omega;\mathbb{R}^3),$$
$$u_n \rightharpoonup u \text{ in } W^{1,p}(\omega;\mathbb{R}^3),\ b_n\rightharpoonup b \text{ in } L^p(\omega;\mathbb{R}^3)\Big\}$$

since the infimum on the right hand side is taken over a smaller class of sequences (i.e. those independent of x_3). By standard results (see e.g (13)) the right hand side coincides with

$$\int_A (Q_2\times C_2)[W](\nabla_\alpha u(x_\alpha)|b(x_\alpha))\,dx_\alpha,$$

and so we have

$$H^2_{3,3}(u,b;A) \le \int_A (Q_2\times C_2)[W](\nabla_\alpha u(x_\alpha)|b(x_\alpha))\,dx_\alpha.$$

On the other hand, since $W \geq (Q_3 \times C_3)[W]$, by classical lower semicontinuity results we deduce that

$$H^2_{3,3}(u, b; A) \geq \int_A (Q_3 \times C_3)[W] (\nabla_\alpha u(x_\alpha) | b(x_\alpha)) \, dx_\alpha,$$

and now it suffices to recall (2.0.11). □

When the function b depends on the x_3 variable the situation is significantly more involved, and the representation obtained in Theorem 3.0.9 below is considerably less explicit.

We introduce the functional

$$\overline{\mathcal{W}} : \mathbb{R}^{2\times 3} \times W^{1,2}(I; \mathbb{R}^3) \to [0, \infty)$$

defined for $\overline{F} \in \mathbb{R}^{2\times 3}$, $b \in W^{1,2}(I; \mathbb{R}^3)$, as follows:

$$\begin{aligned} &\overline{\mathcal{W}}(\overline{F}|b) := \int_{-\frac{1}{2}}^{\frac{1}{2}} |\nabla_3 b(x_3)|^2 \, dx_3 \\ &+ \inf_{\varphi, \psi} \left\{ \int_Q \left(W\left(\overline{F} + \nabla_\alpha \varphi(x) | b(x_3) + \psi(x)\right) + |\nabla_3 \psi(x)|^2 \right) dx : \right. \\ &\varphi \in W^{1,p}(Q; \mathbb{R}^3),\ \varphi(\cdot, x_3)\ Q'\text{-periodic for } \mathcal{L}^1 \text{ a.e. } x_3, \qquad (3.0.12) \\ &\left. \psi \in L^p(Q; \mathbb{R}^3),\ \nabla_3 \psi \in L^2(Q; \mathbb{R}^3),\ \int_{Q'} \psi(x_\alpha, x_3)\, dx_\alpha = 0 \text{ for } \mathcal{L}^1 \text{ a.e. } x_3 \right\}. \end{aligned}$$

Note that

$$\begin{aligned} \overline{\mathcal{W}}(\overline{F}|b) := \inf_{\varphi, \psi} &\left\{ \int_Q \left(W\overline{F} + \nabla_\alpha \varphi(x) | b(x_3) + \psi(x) \right) \right. \\ &+ |\nabla_3 b(x_3) + \nabla_3 \psi(x)|^2 \big) \, dx : \\ &\varphi \in W^{1,q}(Q; \mathbb{R}^3),\ \varphi(\cdot, x_3)\ Q'\text{-periodic for } \mathcal{L}^1 \text{ a.e. } x_3, \\ &\left. \psi \in L^q(Q; \mathbb{R}^3),\ \nabla_3 \psi \in L^2(Q; \mathbb{R}^3),\ \int_{Q'} \psi(x_\alpha, x_3)\, dx_\alpha = 0 \text{ for } \mathcal{L}^1 \text{ a.e. } x_3 \right\}. \end{aligned}$$

Indeed the equality holds because the admissible test functions ψ satisfy

$$\int_{Q'} \nabla_3 \psi(x_\alpha, x_3) \, dx_\alpha = 0 \text{ for } \mathcal{L}^1 \text{ a.e. } x_3.$$

Theorem 3.0.9. *Assume that condition $(H_1)'$ is satisfied. Then for all $(u,b) \in \mathcal{V}^2$ and $A \in \mathcal{A}(\omega)$*

$$H^2_{3,3}(u,b;A) = \int_A \overline{\mathcal{W}}(\nabla_\alpha u(x_\alpha)|b(x_\alpha,\cdot))\,dx_\alpha,$$

where $\overline{\mathcal{W}}$ is defined in (3.0.12).

Bibliography

[1] Acerbi, E., Fusco, N., Semicontinuity problems in the calculus of variations. *Arch. Rat. Mech. Anal.*, **86**, (1984), 125–145.

[2] Ball, J. M., Murat, F., $W^{1,p}$ quasiconvexity and variational problems for multiple integrals. *J. Func. Anal.* **58**, (1984), 225–253.

[3] Bhattacharya, K., James, R. D., A theory of thin films of martensitic materials with applications to microactuators. *J. Mech. Phys. Solids* **47**, (1999), 531–576.

[4] Bouchitté, G., Fonseca, I., Mascarenhas, M. L., Bending moment in membrane theory. *J. Elasticity* **73**, (2004), 75–99.

[5] Bouchitté, G., Fonseca, I., Mascarenhas, M. L., The Cosserat Vector in Membrane Theory: A variational Approach. To appear.

[6] Braides, A., Defranceschi, A., *Homogenization of multiple integrals.* Oxford Lecture Series in Mathematics and its Applications, **12**, Oxford, (1998).

[7] Braides, A., Fonseca, I., Francfort, G., 3D–2D asymptotic analysis for inhomogeneous thin films. *Ind. U. Math. J.* **49**, 2000, 1367–1404.

[8] Dacorogna, B., *Direct methods in the calculus of variations.* Applied Mathematical Sciences, 78. Springer-Verlag, Berlin, 1989.

[9] De Giorgi, E., Letta, G., Une notion générale de convergence faible pour des fonctions croissantes d'ensemble. *Ann. Sc. Norm. Sup. Pisa Cl. Sci.*,**4**, (1977), 61–99.

[10] Evans, L. C., *Weak convergence methods for nonlinear partial differential equations.* CBMS Regional Conference Series in Mathematics, 74. Published for the Conference Board of the Mathematical Sciences, Washington, DC; by the American Mathematical Society, Providence, RI, (1990).

[11] Evans, L. C., Gariepy, R. F., *Measure theory and fine properties of functions.* Studies in Advanced Mathematics, CRC Press, Boca Raton, (1992).

[12] Fonseca, I., Francfort, G.A., 3D-2D asymptotic analysis of an optimal design problem for thin films. *J. Reine Angew. Math.* **505**, (1998), 173–202.

[13] Fonseca, I., Kinderlehrer, D., Pedregal, P., Energy functionals depending on elastic strain and chemical composition. *Calc. Var. Partial Differential Equations* **2** 3, 1994, 283–313.
[14] Fonseca, I., Leoni, G., Malý, J., Weak continuity and lower semicontinuity results for determinants. *Arch. Rat. Mech. Anal.* **178** (2005), no. 3, 411– 448.
[15] Fonseca, I., Leoni, G., Paroni, R., On lower semicontinuity in BH^p and 2-quasiconvexification. *Calc. Var. Partial Differential Equations* **17** (2003), no. 3, 283–309.
[16] Fonseca, I., Müller, S., Pedregal, P., Analysis of concentration and oscillation effects generated by gradients. *SIAM J. Math. Anal.* **29** (1998), no. 3, 736–756.
[17] Le Dret, H., Raoult, A., The nonlinear membrane model as variational limit of nonlinear three dimensional elasticity. *J. Math. Pures. Appl.* **74**, 1995, 549–578.
[18] Loeb, P.A., Talvila, E., Lusin's theorem and Bochner integration. *Sci. Math. Jpn.* **60**-1, (2004), 113–120.
[19] Shu, Y.C., Heterogeneous thin films of martensitic materials. *Arch. Rat. Mech. Anal* **153**, (2000), 39–90.

3D-2D Asymptotic Analysis for Inhomogeneous Thin Films - Lectures V, VI

Irene Fonseca

Department of Mathematical Sciences, Carnegie Mellon University

Abstract Using Γ-convergence techniques the relaxed 2-D energy is obtained as a result of a dimension reduction analysis for thin domains of the form

$$\Omega_\varepsilon := \{(x_1, x_2, x_3) : (x_1, x_2) \in \omega, |x_3| < \varepsilon f_\varepsilon(x_1, x_2)\},$$

where ω is a bounded domain of $\mathbb{R}^2$ and f_ε is an ε-dependent profile. In particular, explicit characterizations of the effective energy densities are obtained within the context of optimal design for thin films.

1 Introduction

The contents of these two lectures are based on the paper

A. Braides, I. Fonseca, G. Francfort. 3D-2D Asymptotic analysis for Inhomogeneous Thin Films, *Indiana U. Math. J.*, **49** (2000) 1367–1403.

For the purpose of these two lectures we consider the case where no loads are present and we direct our attention to the study of the $\Gamma(L^p)$–limit ($1 < p < +\infty$) of the functionals

$$v \mapsto E_\varepsilon(v; \omega) := \int_{\Omega_\varepsilon} W_\varepsilon\left(x_1, x_2, x_3; \nabla_\alpha \Big| \frac{1}{\varepsilon}\nabla_3 v\right) dx.$$

Under appropriate growth conditions, minimizers of E_ε – if they exist – will L^p-converge to minimizers of that $\Gamma(L^p)$–limit, and thus a characterization of the latter will provide the asymptotic effective energy for equilibria states of Ω_ε.

*The research of I. Fonseca was partially funded by the National Science Foundation (NSF) under NSF Grants DMS-0103799 and DMS-0401763, and by the Center for Nonlinear Analysis under NSF Grants DMS-9803791, DMS-0405343 and DMS-0635983.

2 Compactness and Γ-Limit Integral Representation

Assume that $\{W_\varepsilon(x;F)\}_\varepsilon$ is a sequence of Carathéodory functions in $\Omega \times \mathbb{R}^{3\times 3}$ such that for a.e. x in Ω and any F in $\mathbb{R}^{3\times 3}$,

$$\frac{1}{C}|F|^p - C \leq W_\varepsilon(x;F) \leq C(1+|F|^p), C>0,\ 1\leq p<\infty, \tag{2.0.1}$$

$f_\varepsilon(x_\alpha)$ are continuous functions on ω such that

$$0<\gamma\leq f_\varepsilon(x_\alpha)\leq \frac{1}{2} \quad \text{for all } x_\alpha\in\omega, \tag{2.0.2}$$

for some $\gamma>0$, and for any open subset A of ω set

$$A_\varepsilon := \{(x_\alpha,x_3) : x_\alpha\in A, |x_3|<f_\varepsilon(x_\alpha)\},$$

and

$$\partial_t A_\varepsilon := \{(x_\alpha,x_3) :\ |x_3|<f_\varepsilon(x_\alpha),\ x_\alpha\in\partial A\}.$$

Note that $\omega_\varepsilon=\Omega_\varepsilon$. For any u in $L^p(\Omega;\mathbb{R}^3)$ define

$$E_\varepsilon(v;A) := \begin{cases} \int_{A_\varepsilon} W_\varepsilon\left(x_\alpha,x_3;\nabla_\alpha v\middle|\frac{1}{\varepsilon}\nabla_3 v\right)dx & \text{if } v\in W^{1,p}(A_\varepsilon;\mathbb{R}^3),\\ +\infty, & \text{otherwise,}\end{cases}$$

and

$$J_{\{\varepsilon\}}(u;A) := \inf_{\{v_\varepsilon\}}\Big\{\liminf_{\varepsilon\to 0^+} E_\varepsilon(v_\varepsilon;A)\ :\ v_\varepsilon\in W^{1,p}(A_\varepsilon;\mathbb{R}^3) \text{ and } (v_\varepsilon-u)\chi_{A_\varepsilon}\to 0 \text{ in } L^p(\Omega;\mathbb{R}^3)\Big\}. \tag{2.0.3}$$

Note that $J_{\{\varepsilon\}}(u;\cdot)$ is an increasing function on open subsets of ω.

In the compactness theorem below we assume that f_ε is such that the trace of v_ε is meaningful on $\partial_t\omega_\varepsilon$.

Theorem 2.0.1. *Assume that $p>1$. Energy bounded sequences are compact in L^p. Precisely, if $\{v_\varepsilon\}$ is a sequence in $W^{1,p}(\Omega_\varepsilon;\mathbb{R}^3)$ with $v_\varepsilon=0$ on $\partial_t\omega_\varepsilon$, and if*

$$\sup_\varepsilon \int_{\Omega_\varepsilon} W_\varepsilon\left(x_\alpha,x_3;\nabla_\alpha v_\varepsilon\middle|\frac{1}{\varepsilon}\nabla_3 v_\varepsilon\right)dx<\infty,$$

then there exist $u\in W_0^{1,p}(\omega;\mathbb{R}^3)$ and a subsequence $\{\varepsilon_k\}$ of $\{\varepsilon\}$ such that

$$(v_{\varepsilon_k}-u)\chi_{\Omega_{\varepsilon_k}}\to 0 \textit{ in } L^p(\Omega;\mathbb{R}^3). \tag{2.0.4}$$

Proof. In view of (2.0.1), (2.0.2),

$$\int_{\omega\times(-\gamma,\gamma)}\left(|\nabla_\alpha v_\varepsilon|^p+\frac{1}{\varepsilon^p}|\nabla_3 v_\varepsilon|^p\right)dx \le \int_{\omega_\varepsilon}\left(|\nabla_\alpha v_\varepsilon|^p+\frac{1}{\varepsilon^p}|\nabla_3 v_\varepsilon|^p\right)dx<\infty, \tag{2.0.5}$$

so that by Poincaré's inequality and Rellich's Theorem there exist $u\in W^{1,p}(\omega\times(-\gamma,\gamma);\mathbb{R}^3)$ and a subsequence $\{\varepsilon_k\}$ of $\{\varepsilon\}$ such that

$$\begin{cases} v_{\varepsilon_k}\rightharpoonup u & \text{in } W^{1,p}(\omega\times(-\gamma,\gamma);\mathbb{R}^3),\\ v_{\varepsilon_k}(x_\alpha,\pm\gamma)\to u(x_\alpha,\pm\gamma) & \text{in } L^p(\omega;\mathbb{R}^3). \end{cases} \tag{2.0.6}$$

Now (2.0.5) implies that $\nabla_3 u=0$, i.e., that $u\in W_0^{1,p}(\omega;\mathbb{R}^3)$, and we have

$$\begin{aligned}
&\int_{\omega_\varepsilon}|v_{\varepsilon_k}-u|^p dx=\int_{\omega\times(-\gamma,\gamma)}|v_{\varepsilon_k}-u|^p\,dx+\\
&\quad\int_\omega\int_\gamma^{f_\varepsilon(x_\alpha)}|v_{\varepsilon_k}-u|^p\,dx+\int_\omega\int_{-f_\varepsilon(x_\alpha)}^{-\gamma}|v_{\varepsilon_k}-u|^p dx=\\
&\int_{\omega\times(-\gamma,\gamma)}|v_{\varepsilon_k}-u|^p dx+\\
&\quad\int_\omega\int_\gamma^{f_\varepsilon(x_\alpha)}\left|\int_\gamma^{x_3}\nabla_3 v_{\varepsilon_k}(x_\alpha,s)\,ds+v_{\varepsilon_k}(x_\alpha,\gamma)-u(x_\alpha)\right|^p dx+\\
&\quad\int_\omega\int_{-f_\varepsilon(x_\alpha)}^{-\gamma}\left|\int_{-\gamma}^{x_3}\nabla_3 v_{\varepsilon_k}(x_\alpha,s)\,ds+v_{\varepsilon_k}(x_\alpha,-\gamma)-u(x_\alpha)\right|^p dx\le\\
&\int_{\omega\times(-\gamma,\gamma)}|v_{\varepsilon_k}-u|^p dx+\\
&\quad C\left\{\int_\omega|v_{\varepsilon_k}(x_\alpha,\pm\gamma)-u(x_\alpha)|^p\,dx_\alpha+\int_{\omega_\varepsilon}|\nabla_3 v_{\varepsilon_k}|^p dx\right\},
\end{aligned}$$

so that (2.0.5) and (2.0.6) imply (2.0.4). □

Remark 2.0.2. *If $u\in W^{1,p}(\omega;\mathbb{R}^3)$, i.e., u does not depend upon x_3, then (2.0.1) implies that $J_{\{\varepsilon\}}(u;\omega)<\infty$, as immediately seen by taking $v_\varepsilon=u$ in the definition (2.0.3) of $J_{\{\varepsilon\}}$. Conversely, and in view of the previous theorem, if $(v_\varepsilon-u)\chi_{\Omega_\varepsilon}\to 0$ in L^1 and $\{v_\varepsilon\}$ is an energy-bounded sequence, then we may assume that $u\in W^{1,p}(\omega;\mathbb{R}^3)$.*

Let $\mathcal{C}$ be a countable collection of subsets of ω such that for any $\delta > 0$ and any open subset A of ω, there exists a finite union C_A of disjoint elements of $\mathcal{C}$ satisfying

$$\begin{cases} \overline{C}_A \subset A, \\ \mathcal{L}^2(A) \leq \mathcal{L}^2(C_A) + \delta. \end{cases}$$

For example, $\mathcal{C}$ may be taken to be the family of all open cubes contained in ω centered at points with rational coordinates and with rational side length. Denote by $\mathcal{R}$ the countable collection of all finite unions of elements of $\mathcal{C}$, i.e.,

$$\mathcal{R} := \left\{ \cup_{i=1}^{k} C_i : \ k \in \mathbb{N}, \ C_i \in \mathcal{C} \right\}.$$

We recall Proposition 7.9 in (16).

Proposition 2.0.3. *Let (X, d) be a separable metric space and let $f_n : X \to \overline{\mathbb{R}}$. There exists and increasing sequence $\{j_k\} \subset \mathbb{N}$ such that $\Gamma(d) - \lim_{k\to\infty} f_{j_k}(x)$ exists for every $x \in X$.*

In view of the separable and metrizable character of $L^p(\Omega; \mathbb{R}^3)$ and of the previous proposition, for any sequence $\{\varepsilon\} \searrow 0^+$ a diagonalization argument yields the existence of a subsequence $\{\varepsilon^{\mathcal{R}}\}$ such that, setting

$$J_{\{\varepsilon^{\mathcal{R}}\}}(u; A) := \inf_{\{v_{\varepsilon^{\mathcal{R}}}\}} \left\{ \liminf_{\varepsilon^{\mathcal{R}} \to 0^+} E_{\varepsilon^{\mathcal{R}}}(v_{\varepsilon^{\mathcal{R}}}; A) : \ v_{\varepsilon^{\mathcal{R}}} \in W^{1,p}(A_{\varepsilon^{\mathcal{R}}}; \mathbb{R}^3) \text{ and } (v_{\varepsilon^{\mathcal{R}}} - u)\chi_{A_{\varepsilon^{\mathcal{R}}}} \to 0 \text{ in } L^p(\Omega; \mathbb{R}^3) \right\},$$

then $J_{\{\varepsilon^{\mathcal{R}}\}}(\cdot; C)$ is the $\Gamma(L^p)$–limit of $E_{\varepsilon^{\mathcal{R}}}(\cdot; C)$ for every $C \in \mathcal{R}$. In particular, if $u \in W^{1,p}(\Omega; \mathbb{R}^3)$ and if C in $\mathcal{R}$, then there exists a sequence $\{v^C_{\varepsilon^{\mathcal{R}}}\}$ in $W^{1,p}(C_{\varepsilon^{\mathcal{R}}}; \mathbb{R}^3)$ such that

$$\begin{cases} (v_{\varepsilon^{\mathcal{R}}} - u)\chi_{\varepsilon^{\mathcal{R}}} \to 0 \text{ in } L^p(\Omega; \mathbb{R}^3) \\ J_{\{\varepsilon^{\mathcal{R}}\}}(u; C) = \lim_{\varepsilon^{\mathcal{R}} \to 0+} E_{\varepsilon^{\mathcal{R}}}(v_{\varepsilon^{\mathcal{R}}}; C). \end{cases} \tag{2.0.7}$$

Theorem 2.0.4. *Let $u \in W^{1,p}(\omega; \mathbb{R}^3)$, let A be an open subset of ω, and let $\{\varepsilon\} \searrow 0^+$. Then $J_{\{\varepsilon^{\mathcal{R}}\}}(\cdot; A)$ is the $\Gamma(L^p)$–limit of $E_{\varepsilon^{\mathcal{R}}}(\cdot; A)$.*

Furthermore, there exists a Carathéodory function $W_{\{\varepsilon^{\mathcal{R}}\}} : \mathbb{R}^2 \times \mathbb{R}^{3\times 2} \to \mathbb{R}$ such that

$$J_{\{\varepsilon^{\mathcal{R}}\}}(u; A) = 2 \int_A W_{\{\varepsilon^{\mathcal{R}}\}}(x_\alpha; \nabla_\alpha u)\, dx_\alpha. \tag{2.0.8}$$

We will use the following integral representation theorem (see Theorem 4.3.2 in (19)). Here $\mathcal{A}(\Omega)$ stands for the collection of all open subsets of Ω.

Theorem 2.0.5. *Let Ω be an open, bounded subset of $\mathbb{R}^N$, let $1 < p \leq +\infty$, and let $F : W^{1,p}(\Omega; \mathbb{R}^d) \times \mathcal{A}(\Omega) \to [0, +\infty)$ be such that*
(i)(locality) F is local, i.e. $F(u, U) = F(v, U)$ if $v = u$ a.e. in $U \in \mathcal{A}(\Omega)$;
(ii)(measure property) if $u \in W^{1,p}(\Omega; \mathbb{R}^d)$ then $F(u, \cdot)$ is a finite, nonnegative Radon measure;
(iii)(growth condition) $F(u, U) \leq C \int_U (a(x) + |\nabla u|^p)\, dx$ for some $C > 0$, $a \in L^1(\Omega)$, and all $(u, U) \in W^{1,p}(\Omega; \mathbb{R}^d) \times \mathcal{A}(\Omega)$;
(iv)(translation invariance) $F(u + c, U) = F(u, U)$ for all $c \in \mathbb{R}^d$ and $(u, U) \in W^{1,p}(\Omega; \mathbb{R}^d) \times \mathcal{A}(\Omega)$;
(v)(lowersemicontinuity) $F(\cdot, U)$ is sequentially weakly lowersemicontinuous in $W^{1,p}$ for all $U \in \mathcal{A}(\Omega)$.

Then there exists a Carathéodory function $f : \Omega \times \mathbb{R}^{d \times N} \to [0, +\infty)$, such that

$$0 \leq f(x, \xi) \leq C(a(x) + |\xi|^p)$$

for all $(x, \xi) \in \Omega \times \mathbb{R}^{d \times N}$ such that

$$F(u, U) = \int_U f(x, \nabla u(x))\, dx$$

for every $(u, U) \in W^{1,p}(\Omega; \mathbb{R}^d) \times \mathcal{A}(\Omega)$.

To prove (iii) above we often use the criterion below that provides sufficient conditions for a mapping $\pi : \mathcal{A}(\Omega) \to [0, +\infty)$ to be the trace of a Radon measure. It is close in spirit to De Giorgi-Letta's criterion (26).

Lemma 2.0.6. *Let π be a mapping from $\mathcal{A}(\Omega)$ into $\mathbb{R}^+$ and let μ be a finite nonnegative Radon measure on $\mathbb{R}^N$. If for any $A, B, C \in \mathcal{A}(\Omega)$,*
(i) $\pi(A) \leq \pi(A \setminus \overline{C}) + \pi(B)$, if $\overline{C} \subset B \subset A$,
(ii) for any $\varepsilon > 0$ there exists $C_\varepsilon \in \mathcal{A}(\Omega)$ with $\overline{C}_\varepsilon \subset A$ and $\pi(A \setminus \overline{C}_\varepsilon) \leq \varepsilon$,
(iii) $\pi(\Omega) \geq \mu(\mathbb{R}^N)$,
(iv) $\pi(A) \leq \mu(\overline{A})$,
then, π is the restriction of μ to $\mathcal{A}(\Omega)$.

Proof. Recalling (ii), consider $d_\varepsilon := \mathrm{dist}(\overline{C}_\varepsilon, \partial A) > 0$. Then $\overline{C}_\varepsilon \subset B_\varepsilon := \{x \in A : \ \mathrm{dist}(x, \partial A) > d_\varepsilon/2\}$, while $\overline{B}_\varepsilon \subset A$. Thus, by (i, iv), and since $\overline{C}_\varepsilon \subset B_\varepsilon \subset A$,

$$\pi(A) \leq \varepsilon + \pi(B_\varepsilon) \leq \varepsilon + \mu(\overline{B}_\varepsilon) \leq \varepsilon + \mu(A).$$

Hence, letting ε tend to 0, we have

$$\pi(A) \leq \mu(A). \tag{2.0.9}$$

Conversely, since μ is a Radon measure then it is inner regular, so that for any $\varepsilon > 0$ there exists $C_\varepsilon \in \mathcal{A}(\Omega)$ with $\overline{C}_\varepsilon \subset A$ and $\mu(A) \leq \varepsilon + \mu(C_\varepsilon)$. Hence, by (iii) and (2.0.9)

$$\begin{aligned} \mu(A) &\leq \varepsilon + \mu(\overline{C}_\varepsilon) = \varepsilon + \mu(\Omega) - \mu(\Omega \setminus \overline{C}_\varepsilon) \\ &\leq \varepsilon + \mu(\mathbb{R}^N) - \pi(\Omega \setminus \overline{C}_\varepsilon) \\ &\leq \varepsilon + \pi(\Omega) - \pi(\Omega \setminus \overline{C}_\varepsilon), \end{aligned}$$

and, since $\overline{C}_\varepsilon \subset A \subset \Omega$, (i) implies that $\mu(A) \leq \varepsilon + \pi(A)$, so that the result is obtained upon letting ε tend to 0. □

Our proofs often use diagonalization arguments, and for this we recall

Lemma 2.0.7. *Let $\{a_{k,j}\}$ be a doubly indexed sequence of real numbers $(k, j \nearrow \infty)$. If one of the following holds:*
(i) $\lim_k \lim_j a_{k,j} = L$,
(ii) $\lim_k \limsup_j a_{k,j} = L = \lim_k \liminf_j a_{k,j}$,
then there exist subsequences $\{j(k)\}_k$ and $\{k(j)\}_j$ converging to ∞ such that

$$\lim_k a_{j(k)} = \lim_j a_{k(j),j} = L.$$

The next lemma establishes that approximating sequences may as well take the value u on the lateral boundary of A_ε.

Lemma 2.0.8. *Let $u \in W^{1,p}(A;\mathbb{R}^3)$ where A is an open subset of ω. If $\{\overline{\varepsilon}\} \subset \{\varepsilon^{\mathcal{R}}\}$ and $v_{\overline{\varepsilon}} \in W^{1,p}(A_{\overline{\varepsilon}};\mathbb{R}^3)$ are such that*

$$\begin{cases} (v_{\overline{\varepsilon}} - u)\chi_{A_{\overline{\varepsilon}}} \to 0 \quad \text{in } L^p(\Omega;\mathbb{R}^3), \\ J_{\{\varepsilon^{\mathcal{R}}\}}(u;A) = \lim\limits_{\overline{\varepsilon}\to 0^+} E_{\overline{\varepsilon}}(v_{\overline{\varepsilon}};A), \end{cases}$$

then there exists a sequence $\{w_{\overline{\varepsilon}}\} \subset W^{1,p}(A_{\overline{\varepsilon}};\mathbb{R}^3)$ which satisfies (2.0.10) and is such that

$$w_{\overline{\varepsilon}} = u \text{ in}\{(x_\alpha, x_3) : x_\alpha \in A \setminus K^{\overline{\varepsilon}} \text{ and} |x_3| < f_{\overline{\varepsilon}}(x_\alpha)\}$$

for some compact set $K^{\overline{\varepsilon}} \subset A$.

Proof. The proof relies on De Giorgi's slicing argument, and on the fact that the cut-off functions considered below are independent of the variable x_3. Set

$$C := \sup_{\overline{\varepsilon}} \int_{A_{\overline{\varepsilon}}} \left(1 + |\nabla_\alpha v_{\overline{\varepsilon}}|^p + \frac{1}{\overline{\varepsilon}^p}|\nabla_3 v_{\overline{\varepsilon}}|^p\right) dx,$$

and note that $C < \infty$ by virtue of (2.0.1). Define

$$K(\overline{\varepsilon}) := \left[\!\!\left[\frac{1}{||v_{\overline{\varepsilon}} - u||^{1/2}_{L^p(A_{\overline{\varepsilon}})}} \right]\!\!\right], \tag{2.0.10}$$

where $[\![a]\!]$ stands for the integer part of the number a, and $M(\overline{\varepsilon}) := [\![\sqrt{K(\overline{\varepsilon})}]\!]$; set also

$$A(\overline{\varepsilon}) := \left\{ x_\alpha \in A : \operatorname{dist}(x_\alpha, \partial A) < \frac{M(\overline{\varepsilon})}{K(\overline{\varepsilon})} \right\}.$$

Note that, in view of (2.0.10), $K(\overline{\varepsilon}) \nearrow \infty$ while $\mathcal{L}^2(A(\overline{\varepsilon})) \searrow 0^+$ as $\overline{\varepsilon} \searrow 0^+$. Subdivide $A(\overline{\varepsilon})$ into $M(\overline{\varepsilon})$ disjoint subsets,

$$A_i^{\overline{\varepsilon}} := \left\{ x_\alpha \in A : \operatorname{dist}(x_\alpha, \partial A) \in \left[\frac{i}{K(\overline{\varepsilon})}, \frac{i+1}{K(\overline{\varepsilon})} \right) \right\}, \quad i = 0, \dots, M(\overline{\varepsilon}) - 1.$$

Then, there exists $i(\overline{\varepsilon}) \in \{0, ..., M(\overline{\varepsilon}) - 1\}$ such that

$$\int_{(A_i^{\overline{\varepsilon}})_{\overline{\varepsilon}}} \left(1 + |\nabla_\alpha v_{\overline{\varepsilon}}|^p + \frac{1}{\overline{\varepsilon}^p} |\nabla_3 v_{\overline{\varepsilon}}|^p \right) dx \le \frac{C}{M(\overline{\varepsilon})}, \tag{2.0.11}$$

where $(A^{\overline{\varepsilon}}_{i(\overline{\varepsilon})})_{\overline{\varepsilon}} := \{(x_\alpha, x_3) : x_\alpha \in A_i^{\overline{\varepsilon}},\ |x_3| < f_{\overline{\varepsilon}}(x_\alpha)\}$. Consider $\phi(\overline{\varepsilon}) \in C_0^\infty(A)$ such that

$$\begin{cases} 0 \le \phi(\overline{\varepsilon}) \le 1, & ||\nabla_\alpha \phi(\overline{\varepsilon})||_{L^\infty} \le 2K(\overline{\varepsilon}), \\ \phi(\overline{\varepsilon}) = \begin{cases} 1, & \text{if } \operatorname{dist}(x_\alpha, \partial A) > \frac{i(\overline{\varepsilon})+1}{K(\overline{\varepsilon})}, \\ \phi(\overline{\varepsilon}) = 0, & \text{if } \operatorname{dist}(x_\alpha, \partial A) \le \frac{i(\overline{\varepsilon})}{K(\overline{\varepsilon})}, \end{cases} \end{cases} \tag{2.0.12}$$

and set

$$w_{\overline{\varepsilon}} := \phi(\overline{\varepsilon}) v_{\overline{\varepsilon}} + (1 - \phi(\overline{\varepsilon}))u. \tag{2.0.13}$$

Note that $w_{\overline{\varepsilon}} = u$ in $\{(x_\alpha, x_3) : x_\alpha \in A \setminus K^{\overline{\varepsilon}}, |x_3| < f_{\overline{\varepsilon}}(x_\alpha)\}$, with $K^{\overline{\varepsilon}} := \left\{ x_\alpha \in A : \operatorname{dist}(x_\alpha, \partial A) \ge \frac{i(\overline{\varepsilon})}{K(\overline{\varepsilon})} \right\}$, and that $w_{\overline{\varepsilon}} \in W^{1,p}(A_{\overline{\varepsilon}}; \mathbb{R}^3)$. Furthermore, in view of (2.0.7),

$$(w_{\overline{\varepsilon}} - u)\chi_{A_{\overline{\varepsilon}}} \to 0 \text{ in } L^p(\Omega; \mathbb{R}^3). \tag{2.0.14}$$

Then, by (2.0.1), together with (2.0.12), (2.0.13),

$$\begin{aligned}
&J_{\{\varepsilon^{\mathcal{R}}\}}(u;A) \geq \\
&\limsup_{\overline{\varepsilon}\to 0^+} \int_{A_{\overline{\varepsilon}}\cap\left[\{x_\alpha :\, \mathrm{dist}(x_\alpha,\partial A)>\frac{i(\overline{\varepsilon})+1}{K(\overline{\varepsilon})}\}\times(-1,1)\right]} W_{\overline{\varepsilon}}\left(x_\alpha, x_3; \nabla_\alpha v_{\overline{\varepsilon}}\Big|\frac{1}{\overline{\varepsilon}}\nabla_3 v_{\overline{\varepsilon}}\right) dx \\
&\geqq \limsup_{\overline{\varepsilon}\to 0^+}\left\{\int_{A_{\overline{\varepsilon}}} W_{\overline{\varepsilon}}\left(x_\alpha, x_3; \nabla_\alpha w_{\overline{\varepsilon}}\Big|\frac{1}{\overline{\varepsilon}}\nabla_3 w_{\overline{\varepsilon}}\right) dx\right. \\
&-C\int_{A_{\overline{\varepsilon}}\cap\left[\{x_\alpha :\, \mathrm{dist}(x_\alpha,\partial A)<\frac{i(\overline{\varepsilon})}{K(\overline{\varepsilon})}\}\times(-1,1)\right]} (1+|\nabla_\alpha u|^p)\, dx \\
&-C\int_{(A^{\overline{\varepsilon}}_{i(\overline{\varepsilon})})_{\overline{\varepsilon}}} \left(1+|\nabla_\alpha v_{\overline{\varepsilon}}|^p+\frac{1}{\overline{\varepsilon}^p}|\nabla_3 v_{\overline{\varepsilon}}|^p\right) dx \\
&\left.-C|K(\overline{\varepsilon})|^p\int_{(A^{\overline{\varepsilon}}_{i(\overline{\varepsilon})})_{\overline{\varepsilon}}} |v_{\overline{\varepsilon}}-u|^p dx\right\} \\
&\geqq \limsup_{\overline{\varepsilon}\to 0^+} E_{\overline{\varepsilon}}(w_{\overline{\varepsilon}};A) - C\liminf_{\overline{\varepsilon}\to 0^+}\mathcal{L}^2(A(\overline{\varepsilon})) - C\liminf_{\overline{\varepsilon}\to 0^+}\frac{1}{M(\overline{\varepsilon})} \\
&-\liminf_{\overline{\varepsilon}\to 0^+} ||v_{\overline{\varepsilon}}-u||^{p/2}_{L^p(A_{\overline{\varepsilon}})} = \limsup_{\overline{\varepsilon}\to 0^+} E_{\overline{\varepsilon}}(w_{\overline{\varepsilon}};A),
\end{aligned} \tag{2.0.15}$$

where (2.0.10) and (2.0.11) have been used in deriving the last inequality in (2.0.15). But the definition (2.0.7) of $J_{\{\varepsilon^{\mathcal{R}}\}}(u;A)$, together with (2.0.14), imply that

$$J_{\{\varepsilon^{\mathcal{R}}\}}(u;A) \leq \liminf_{\overline{\varepsilon}\to 0^+} E_{\overline{\varepsilon}}(w_{\overline{\varepsilon}};A),$$

which, in view of (2.0.15), yields the desired result. □

Proof of Theorem 2.0.4- First Part. Let $u \in W^{1,p}(A;\mathbb{R}^3)$, let A be an open subset of ω. In order to prove that $J_{\{\varepsilon^{\mathcal{R}}\}}(u;A)$ is the Γ–limit of $E_{\varepsilon^{\mathcal{R}}}(v^{C^\delta}_{\varepsilon^{\mathcal{R}}};A)$ it suffices to assert the existence of a recovery sequence. To this end, fix $\delta > 0$ and choose a subset C^δ of A in $\mathcal{R}$ such that

$$\begin{cases} \overline{C^\delta} \subset A, \\ \displaystyle\int_{A\setminus C^\delta} (1+|Du_\alpha|)\, dx_\alpha < \frac{\delta}{C}. \end{cases}$$

Consider a sequence $v^{C^\delta}_{\varepsilon^{\mathcal{R}}}$ satisfying

$$\lim_{\varepsilon^{\mathcal{R}}\to 0^+} E_{\varepsilon^{\mathcal{R}}}(v^{C^\delta}_{\varepsilon^{\mathcal{R}}};C^\delta) = J_{\{\varepsilon^{\mathcal{R}}\}}(u;C^\delta).$$

In view of Lemma 2.0.8, we may assume $v_{\varepsilon^{\mathcal{R}}}^{C^\delta} = u$ on ∂C^δ and we extend $v_{\varepsilon^{\mathcal{R}}}^{C^\delta}$ as u outside $C^\delta_{\varepsilon^{\mathcal{R}}}$ so as to belong to $W^{1,p}(A_{\varepsilon^{\mathcal{R}}};\mathbb{R}^3)$, and since $J_{\{\varepsilon^{\mathcal{R}}\}}(u;C^\delta) \leq J_{\{\varepsilon^{\mathcal{R}}\}}(u;A)$ for all $\delta > 0$, we have

$$\begin{aligned}
&\limsup_{\delta\to 0^+}\limsup_{\varepsilon^{\mathcal{R}}\to 0^+} E_{\varepsilon^{\mathcal{R}}}(v_{\varepsilon^{\mathcal{R}}}^{C^\delta};A) \leq \\
&\quad \limsup_{\delta\to 0^+}\lim_{\varepsilon^{\mathcal{R}}\to 0^+}\left\{E_{\varepsilon^{\mathcal{R}}}(v_\varepsilon^{C^\delta} R;C^\delta) + C\int_{A\setminus C^\delta}(1+|\nabla_\alpha u|^p)dx\right\} \\
&\quad = \limsup_{\delta\to 0^+} J_{\{\varepsilon^{\mathcal{R}}\}}(u;C^\delta) \\
&\quad \leqq J_{\{\varepsilon^{\mathcal{R}}\}}(u;A) \\
&\quad \leqq \liminf_{\delta\to 0^+}\liminf_{\varepsilon^{\mathcal{R}}\to 0^+} E_{\varepsilon^{\mathcal{R}}}(v_{\varepsilon^{\mathcal{R}}}^{C^\delta};A).
\end{aligned}$$

Lemma 2.0.7 permits to conclude the existence of a decreasing sequence $\{\delta(\varepsilon^{\mathcal{R}})\} \searrow 0^+$ such that

$$\begin{cases}(v_{\varepsilon^{\mathcal{R}}}^{C^{\delta(\varepsilon^{\mathcal{R}})}} - u)\chi_{A_{\varepsilon^{\mathcal{R}}}} \to 0 \text{ in } L^p(\Omega;\mathbb{R}^3) \\ J_{\{\varepsilon^{\mathcal{R}}\}}(u;A) = \lim\limits_{\varepsilon^{\mathcal{R}}\to 0^+} E_{\varepsilon^{\mathcal{R}}}\left(v_{\varepsilon^{\mathcal{R}}}^{C^{\delta(\varepsilon^{\mathcal{R}})}};A\right),\end{cases}$$

which, together with (2.0.7), asserts that $J_{\{\varepsilon^{\mathcal{R}}\}}(\cdot;A)$ is the $\Gamma(L^p)$–limit of $E_{\varepsilon^{\mathcal{R}}}(\cdot;A)$. □

In order to establish (2.0.8) we use Theorem 2.0.5 applied to $J_{\{\varepsilon^{\mathcal{R}}\}}$, that maps any pair (u,A), $u \in W^{1,p}(\omega;\mathbb{R}^3)$, A an open subset of ω, into $\mathbb{R}$. Furthermore, and following the numbering in the statement of Theorem 2.0.5,

(i) $J_{\{\varepsilon^{\mathcal{R}}\}}(u;A) = J_{\{\varepsilon^{\mathcal{R}}\}}(v;A)$ whenever $u = v$, a.e. on $\mathbb{R}^2$,

(iii) $J_{\{\varepsilon^{\mathcal{R}}\}}(u;A) \leq C\int_A(1+|D_\alpha u|^p)dx_\alpha$,

(iv) $J_{\{\varepsilon^{\mathcal{R}}\}}(u+c;A) = J_{\{\varepsilon^{\mathcal{R}}\}}(u;A)$, $\quad c \in \mathbb{R}$.

It remains, therefore, to prove that

(ii) $J_{\{\varepsilon^{\mathcal{R}}\}}(u;\cdot)$ is a finite nonnegative Radon measure.

To this end we will use Lemma 2.0.6. The definition (2.0.7) of $J_{\{\varepsilon^{\mathcal{R}}\}}(u;\omega)$ implies the existence of a subsequence $\{\overline{\varepsilon}\}$ of $\{\varepsilon^{\mathcal{R}}\}$ and of an associated subsequence $\{v_{\overline{\varepsilon}}\}$ in $W^{1,p}(\omega_{\overline{\varepsilon}};\mathbb{R}^3)$ such that

$$\begin{cases}(v_{\overline{\varepsilon}} - u)\chi_{\omega_{\overline{\varepsilon}}} \to 0 \text{ in } L^p(\Omega;\mathbb{R}^3), \\ J_{\{\varepsilon^{\mathcal{R}}\}}(u;\omega) = \lim\limits_{\overline{\varepsilon}\to 0^+} E_{\overline{\varepsilon}}(v_{\overline{\varepsilon}};\omega).\end{cases} \tag{2.0.16}$$

Extracting a subsequence of $\{\overline{\varepsilon}\}$ if necessary, still denoted by $\{\overline{\varepsilon}\}$, there exists a finite Radon measure μ such that

$$W_{\overline{\varepsilon}}(x_\alpha, x_3; \nabla_\alpha v_{\overline{\varepsilon}} | 1/\overline{\varepsilon} \nabla_3 v_{\overline{\varepsilon}}) \chi_{\omega_{\overline{\varepsilon}}} \mathcal{L}^3 \overset{*}{\rightharpoonup} \mu \text{ weakly-* in the sense of measures .} \tag{2.0.17}$$

Set, for any Borel subset $X \subset \mathbb{R}^3$, $\hat{\mu}(X) := \mu(X \times [-1,1])$. Then, by virtue of (2.0.16), (2.0.17),

$$J_{\{\varepsilon^{\mathcal{R}}\}}(u;\omega) \geq \hat{\mu}(\mathbb{R}^2),$$

while, clearly for all open subsets $A \subset \omega$,

$$\begin{aligned} J_{\{\varepsilon^{\mathcal{R}}\}}(u;A) &\leq \liminf_{\overline{\varepsilon}\to 0^+} E_{\overline{\varepsilon}}(v_{\overline{\varepsilon}};A) \\ &= \liminf_{\overline{\varepsilon}\to 0^+} \int_{A_{\overline{\varepsilon}}} W_{\overline{\varepsilon}} \left(x_\alpha, x_3; \nabla_\alpha v_{\overline{\varepsilon}} \middle| \tfrac{1}{\overline{\varepsilon}} \nabla_3 v_{\overline{\varepsilon}} \right) dx \\ &\leqq \mu(\overline{A} \times [-1,1]) = \hat{\mu}(\overline{A}). \end{aligned}$$

This establishes properties (iii) and (iv) in Lemma 2.0.6. The bound from above in (2.0.1) immediately implies that it is absolutely continuous with respect to $\mathcal{L}^2 \lfloor \omega$, and, in turn, property (ii) is trivially satisfied.

We now show that $J_{\{\varepsilon^{\mathcal{R}}\}}$ is subadditive, that is that for every open subsets C, B, A of ω with $C \subset\subset B \subset A$,

$$J_{\{\varepsilon^{\mathcal{R}}\}}(u;A) \leq J_{\{\varepsilon^{\mathcal{R}}\}}(u;B) + J_{\{\varepsilon^{\mathcal{R}}\}}(u;A \setminus \overline{C}). \tag{2.0.18}$$

To this effect we consider, for any small enough $\delta > 0$, B^δ, D^δ two elements of $\mathcal{R}$ with $B^\delta \subset B, D^\delta \subset A \setminus \overline{C}$, such that

$$\int_{A\setminus(B^\delta \cup D^\delta)} (1 + |\nabla_\alpha u|)^p \, dx_\alpha < \delta. \tag{2.0.19}$$

Note that a small enough δ ensures that $B^\delta \cap D^\delta \neq \emptyset$. Then, there exist two sequences $\left\{v_{\varepsilon^{\mathcal{R}}}^{B^\delta}\right\}, \left\{v_{\varepsilon^{\mathcal{R}}}^{D^\delta}\right\}$ such that (2.0.7) is satisfied for B^δ and D^δ, respectively, and (see Lemma 2.0.8)

$$v_{\varepsilon^{\mathcal{R}}}^{B^\delta} = u \text{ on } \partial_t B^\delta_{\varepsilon^{\mathcal{R}}}, \quad v_{\varepsilon^{\mathcal{R}}}^{D^\delta} = u \text{ on } \partial_t D^\delta_{\varepsilon^{\mathcal{R}}}.$$

Consider the sequence of Radon measures

$$\begin{aligned} \lambda_{\varepsilon^{\mathcal{R}}} := \Bigg\{ 1 + \left|\nabla_\alpha v_{\varepsilon^{\mathcal{R}}}^{B^\delta}\right|^p + \left|\nabla_\alpha v_{\varepsilon^{\mathcal{R}}}^{D^\delta}\right|^p \\ + \left(\frac{1}{\varepsilon^{\mathcal{R}}}\right)^p \left(\left|\nabla_3 v_{\varepsilon^{\mathcal{R}}}^{B^\delta}\right|^p + \left|\nabla_3 v_{\varepsilon^{\mathcal{R}}}^{D^\delta}\right|^p \right) \Bigg\} \chi_{(B^\delta \cap D^\delta)_{\varepsilon^{\mathcal{R}}}} \, \mathcal{L}^3 \end{aligned}$$

where, as usual, $(B^\delta \cap D^\delta)_{\varepsilon^{\mathcal{R}}} = \{x_\alpha, x_3; x_\alpha \in B^\delta \cap D^\delta, |x_3| < f_{\varepsilon^{\mathcal{R}}}(x_\alpha)\}$. By virtue of the coercivity hypothesis in (2.0.1), $\{\lambda_{\varepsilon^{\mathcal{R}}}\}$ is a bounded sequence of finite nonnegative Radon measures on $\mathbb{R}^3$, hence there exists a finite nonnegative Radon measure λ such that a subsequence of $\{\lambda_{\varepsilon^{\mathcal{R}}}\}$ – denoted by $\{\lambda_{\overline{\varepsilon}}\}$ – satisfies

$$\lambda_{\overline{\varepsilon}} \overset{*}{\rightharpoonup} \lambda \text{ weakly-* in the sense of measures.} \tag{2.0.20}$$

Set $\hat{\lambda}(X) := \lambda(X \times [-1,1])$ for any Borel subset X of ω. Define, for $0 < \eta < 1$,

$$S^\delta_\eta := \{x \in B^\delta \cap D^\delta : \operatorname{dist}(x_\alpha, \partial B^\delta) = \eta\}.$$

The family $\{S^\delta_\eta\}_\eta$ is made up of pairwise disjoint elements, thus there exists $\eta_0 \in (0,1)$ such that

$$\hat{\lambda}(S^\delta_{\eta_0}) = 0. \tag{2.0.21}$$

For L^δ_ζ, a layer of thickness ζ around $S^\delta_{\eta_0}$, i.e.,

$$L^\delta_\zeta := \{x_\alpha \in B^\delta \cap D^\delta : \operatorname{dist}(x_\alpha, S^\delta_{\eta_0}) \le \zeta\},$$

consider a smooth cut-off function $\phi^\delta_\zeta \in C^\infty_0(\mathbb{R}^2)$ such that

$$\begin{cases} ||\phi^\delta_\zeta||_{L^\infty} \le 1, & ||\nabla_\alpha \phi^\delta_\zeta||_{L^\infty} \le C/\zeta, \\ \phi^\delta_\zeta = \begin{cases} 0 & \text{if } x_\alpha \in B^\delta \text{ and } \operatorname{dist}(x_\alpha, \partial B^\delta) \ge \eta_0 + \zeta, \\ 1 & \text{if } x_\alpha \notin B^\delta \text{ or } \operatorname{dist}(x_\alpha, \partial B^\delta) \le \eta_0 - \zeta. \end{cases} \end{cases} \tag{2.0.22}$$

Setting

$$v_{\zeta,\overline{\varepsilon}} := \phi^\delta_\zeta v^{D^\delta}_{\overline{\varepsilon}} + (1 - \phi^\delta_\zeta) v^{B^\delta}_{\overline{\varepsilon}} + \chi_{(A \setminus (B^\delta \cup D^\delta))_{\overline{\varepsilon}}} u,$$

then $v^\delta_{\zeta,\overline{\varepsilon}} \in W^{1,p}(A_{\overline{\varepsilon}}; \mathbb{R}^3)$, and

$$(v_{\zeta,\overline{\varepsilon}} - u)\chi_{A_{\overline{\varepsilon}}} \to 0 \text{ in } L^p(\Omega; \mathbb{R}^3).$$

Thus, by the very definition (2.0.7) and in view of (2.0.22),

$$\begin{aligned} J_{\{\varepsilon^{\mathcal{R}}\}}(u; A) &\le \liminf_{\overline{\varepsilon} \to 0^+} E_{\overline{\varepsilon}}(v_{\zeta,\overline{\varepsilon}}; A) \\ &\le J_{\{\varepsilon^{\mathcal{R}}\}}(u; B^\delta) + J_{\{\varepsilon^{\mathcal{R}}\}}(u; D^\delta) \\ &+ 2\beta \int_{A \setminus (B^\delta \cup D^\delta)} (1 + |\nabla_\alpha u|)\, dx_\alpha \\ &+ \beta \Big\{ \limsup_{\overline{\varepsilon} \to 0^+} \lambda_{\overline{\varepsilon}}(L^\delta_\zeta \times (-1,1)) \\ &+ \frac{C}{\zeta^p} \limsup_{\overline{\varepsilon} \to 0^+} \int_{(L^\delta_\zeta)_{\overline{\varepsilon}}} \left| v^{B^\delta}_{\overline{\varepsilon}} - v^{D^\delta}_{\overline{\varepsilon}} \right|^p dx \Big\}, \end{aligned} \tag{2.0.23}$$

where $(L^\delta_\zeta)_{\overline{\varepsilon}} := \{(x_\alpha, x_3) : x_\alpha \in L^\delta_\zeta \text{ and } |x_3| < f_{\overline{\varepsilon}}(x_\alpha)\}$. Since $\left|v^{B^\delta}_{\overline{\varepsilon}} - v^{D^\delta}_{\overline{\varepsilon}}\right| \chi_{(L^\delta_\zeta)_{\overline{\varepsilon}}} \leq \left|v^{B^\delta}_{\overline{\varepsilon}} - u\right| \chi_{B^\delta{}_{\overline{\varepsilon}}} + \left|v^{D^\delta}_{\overline{\varepsilon}} - u\right| \chi_{D^\delta{}_{\overline{\varepsilon}}}$, the last term in the last expression in (2.0.23) is 0, while, by virtue of (2.0.20),

$$\limsup_{\overline{\varepsilon} \to 0^+} \lambda_{\overline{\varepsilon}}(L_\zeta \times (-1,1)) \leq \hat{\lambda}(\overline{L}_\zeta).$$

But, as ζ tends to 0, $\hat{\lambda}(\overline{L^\delta_\zeta})$ goes to $\hat{\lambda}(S^\delta_{\eta_0}) = 0$ (cf. (2.0.21)), therefore, upon letting ζ tend to 0 in (2.0.23), and by (2.0.19), we obtain

$$\begin{aligned} J_{\{\varepsilon^{\mathcal{R}}\}}(u; A) &\leq \liminf_{\delta \to 0^+} \left[J_{\{\varepsilon^{\mathcal{R}}\}}(u; B^\delta) + J_{\{\varepsilon^{\mathcal{R}}\}}(u; D^\delta) + \beta\delta\right] \\ &\leq J_{\{\varepsilon^{\mathcal{R}}\}}(u; B) + J_{\{\varepsilon^{\mathcal{R}}\}}(u; A \setminus \overline{C}), \end{aligned}$$

and this proves (2.0.18).

The characterization of the effective energy density in the case of a spatially heterogeneous membrane has been obtained by Babadjian and Francfort in (7).

3 The periodic case

Let $W(x_\alpha, x_3; F)$ be a Carathéodory function from $Q' \times (-1,1) \times \mathbb{R}^9$ into $\mathbb{R}$ satisfying

$$\frac{1}{C}|F|^p - C \leq W(x_\alpha, x_3; F) \leq C(1 + |F|^p),$$

with $1 \leq p < \infty$ and $C > 0$. The function W is extended by Q'-periodicity to $\mathbb{R}^2 \times (-1,1) \times \mathbb{R}^9$ and we set

$$W_\varepsilon(x_\alpha, x_3; F) := W\left(\frac{x_\alpha}{\varepsilon}, x_3; F\right).$$

For more general multiscale periodic contexts we refer to the work of Babadjian and Baia in (8) and (9).

Assume that f is a continuous function from Q' into $[0,1]$ with $0 < \gamma \leq \min f$ and we set

$$f_\varepsilon(x_\alpha) := f\left(\frac{x_\alpha}{\varepsilon}\right).$$

Define, for any $\overline{F} \in \mathbb{R}^{3\times 2}$,

$$W_{\text{hom}}(\overline{F}) := \liminf_{t \nearrow \infty} g(t),$$

where, for any $t > 0$,

$$g(t) \quad := \quad \frac{1}{t^2} \inf_{\phi} \left\{ \int_{(tQ')^f} W(x_\alpha, x_3; \overline{F} + \nabla_\alpha \phi | \nabla_3 \phi) dx : \phi \in W^{1,p}((tQ')^f; \mathbb{R}^3), \right.$$
$$\left. \phi(x_\alpha, x_3) = 0 \text{ if } x_\alpha \in \partial(tQ'),\ |x_3| < f(x_\alpha) \right\},$$

and where, for $A \subset \mathbb{R}^2$, $A^f := \{(x_\alpha, x_3) : x_\alpha \in A, |x_3| < f(x_\alpha)\}$.

Theorem 3.0.9. *If $u \in W^{1,p}(\omega; \mathbb{R}^3)$ and if A is an open subset of ω, then*

$$\Gamma(L^p)\text{-}\lim E_\varepsilon(u; A) = \int_A W_{hom}(D_\alpha u)\, dx_\alpha.$$

Proof. Consider a sequence $\{\varepsilon\}$, with $\varepsilon \searrow 0^+$. By Theorem 2.0.4 there exist a subsequence $\{\varepsilon^{\mathcal{R}}\}$ of $\{\varepsilon\}$ and a Carathéodory function $W_{\{\varepsilon^{\mathcal{R}}\}}$ such that

$$J_{\{\varepsilon^{\mathcal{R}}\}}(u; A) = \int_A W_{\{\varepsilon^{\mathcal{R}}\}}(x_\alpha; \nabla_\alpha u)\, dx_\alpha.$$

We show that $W_{\{\varepsilon^{\mathcal{R}}\}}$ is independent of x_α.

Fix $\overline{F} \in \mathbb{R}^{3\times 2}$, and let $x_0, y_0 \in \omega$ be Lebesgue points for $W_{\{\varepsilon^{\mathcal{R}}\}}(\cdot; \overline{F})$, so that

$$\begin{aligned} W_{\{\varepsilon^{\mathcal{R}}\}}(x_0; \overline{F}) &= \lim_{\delta\to 0^+} \frac{1}{\delta^2} \int_{Q'(x_0,\delta)} W_{\{\varepsilon^{\mathcal{R}}\}}(x_\alpha; \overline{F})\, dx_\alpha \qquad (3.0.24) \\ &= \lim_{\delta\to 0^+} \frac{1}{\delta^2} J_{\{\varepsilon^{\mathcal{R}}\}}(\overline{F} x_\alpha; Q'(x_0, \delta)), \end{aligned}$$

– idem for y_0. Assume that δ is small enough. By Lemma 2.0.8 there exists a sequence $\{\psi^\delta_{\varepsilon^{\mathcal{R}}}\}$ with $\psi^\delta_{\varepsilon^{\mathcal{R}}} = 0$ on $\{(x_\alpha, x_3) : |x_3| < f(x_\alpha/\varepsilon^{\mathcal{R}}), x_\alpha \in \partial Q'(x_0, \delta)\}$ and $\psi^\delta_{\varepsilon^{\mathcal{R}}} \chi_{Q'(x_0,\delta)_{\varepsilon^{\mathcal{R}}}} \to 0$ in $L^p(\Omega; \mathbb{R}^3)$ (with, as usual, $Q'(x_0, \delta)_{\varepsilon^{\mathcal{R}}} := \{(x_\alpha, x_3) : |x_3| < f(x_\alpha/\varepsilon^{\mathcal{R}}), x_\alpha \in Q'(x_0, \delta)\}$), such that

$$J_{\{\varepsilon^{\mathcal{R}}\}}(\overline{F} x_\alpha; Q'(x_0, \delta)) = \lim_{\varepsilon^{\mathcal{R}}\to 0^+} E_{\varepsilon^{\mathcal{R}}}(\overline{F} x_\alpha + \psi^\delta_{\varepsilon^{\mathcal{R}}}; Q'(x_0, \delta)). \qquad (3.0.25)$$

Define the vector $\tau_{\varepsilon^{\mathcal{R}}} \in \varepsilon^{\mathcal{R}} \mathbb{Z}^N$ as

$$(\tau_{\varepsilon^{\mathcal{R}}})_i := \varepsilon^{\mathcal{R}} \left\| \left[\frac{(y_0 - x_0)_i}{\varepsilon^{\mathcal{R}}} \right] \right\|, \qquad \text{for } i = 1, \dots, N.$$

Clearly $\tau_{\varepsilon^{\mathcal{R}}} \to y_0 - x_0$ as $\varepsilon^{\mathcal{R}} \to 0^+$. Let

$$\phi^{\delta}_{\varepsilon^{\mathcal{R}}}(x_\alpha, x_3) := \psi^{\delta}_{\varepsilon^{\mathcal{R}}}(x_\alpha - \tau_{\varepsilon^{\mathcal{R}}}, x_3),$$

where we have extended $\psi^{\delta}_{\varepsilon^{\mathcal{R}}}$ by 0 to $[\mathbb{R}^2 \setminus Q'(x_0, \delta)]_{\varepsilon^{\mathcal{R}}}$.

Fix $r > 1$ and consider $\varepsilon^{\mathcal{R}}$ small enough so that

$$Q'(y_0 - \tau_{\varepsilon^{\mathcal{R}}}, \delta) \subset Q'(x_0, r\delta). \tag{3.0.26}$$

Since $\phi^{\delta}_{\varepsilon^{\mathcal{R}}} \chi_{Q'(y_0,\delta)_{\varepsilon^{\mathcal{R}}}} \to 0$ in $L^p(\Omega; \mathbb{R}^3)$, we have

$$\begin{aligned}
&J_{\{\varepsilon^{\mathcal{R}}\}}(\overline{F}x_\alpha; Q'(y_0, \delta)) \\
&\quad \leqq \liminf_{\varepsilon^{\mathcal{R}} \to 0^+} \int_{Q'(y_0,\delta)_{\varepsilon^{\mathcal{R}}}} W\left(\frac{x_\alpha}{\varepsilon^{\mathcal{R}}}, x_3; \overline{F} + \nabla_\alpha \phi^{\delta}_{\varepsilon^{\mathcal{R}}} \middle| \frac{1}{\varepsilon^{\mathcal{R}}} \nabla_3 \phi^{\delta}_{\varepsilon^{\mathcal{R}}}\right) dx \\
&\leqq \liminf_{\varepsilon^{\mathcal{R}} \to 0^+} \int_{Q'(y_0 - \tau_{\varepsilon^{\mathcal{R}}},\delta)_{\varepsilon^{\mathcal{R}}}} W\left(\frac{x_\alpha + \tau_{\varepsilon^{\mathcal{R}}}}{\varepsilon^{\mathcal{R}}}, x_3; \overline{F} + \nabla_\alpha \phi^{\delta}_{\overline{\varepsilon}}(x_\alpha + \tau_{\varepsilon^{\mathcal{R}}}, x_3) \right| \\
&\qquad\qquad \left. \frac{1}{\varepsilon^{\mathcal{R}}} \nabla_3 \phi^{\delta}_{\varepsilon^{\mathcal{R}}}(x_\alpha - \tau_{\varepsilon^{\mathcal{R}}}, x_3)\right) dx \\
&\leqq \liminf_{\varepsilon^{\mathcal{R}} \to 0^+} \int_{Q'(x_0,r\delta)_{\varepsilon^{\mathcal{R}}}} W\left(\frac{x_\alpha}{\varepsilon^{\mathcal{R}}}, x_3; \overline{F} + \nabla_\alpha \psi^{\delta}_{\overline{\varepsilon}}(x_\alpha, x_3) \right| \\
&\qquad\qquad \left. \frac{1}{\varepsilon^{\mathcal{R}}} \nabla_3 \psi^{\delta}_{\varepsilon^{\mathcal{R}}}(x_\alpha, x_3)\right) dx \\
&\leqq J_{\{\varepsilon^{\mathcal{R}}\}}(\overline{F}x_\alpha; Q'(x_0, \delta)) + C(1 + |\overline{F}|^p)\mathcal{L}^2(Q'(x_0, r\delta) \setminus Q'(x_0, \delta))
\end{aligned}$$

where we have used (3.0.25), (3.0.26) and the periodicity of $W(\cdot, x_3)$. Letting $r \to 1$, we finally obtain

$$J_{\{\varepsilon^{\mathcal{R}}\}}(\overline{F}x_\alpha; Q'(y_0, \delta)) \leq J_{\{\varepsilon^{\mathcal{R}}\}}(\overline{F}x_\alpha; Q'(x_0, \delta)),$$

hence, in view of (3.0.24),

$$W_{\{\varepsilon^{\mathcal{R}}\}}(y_0; \overline{F}) = \lim_{\delta \to 0^+} \frac{1}{\delta^2} J_{\{\varepsilon^{\mathcal{R}}\}}(\overline{F}x_\alpha; Q'(y_0, \delta)) \leq W_{\{\varepsilon^{\mathcal{R}}\}}(x_0; \overline{F}).$$

Given the arbitrariness of x_0 and y_0 we conclude that

$$W_{\{\varepsilon^{\mathcal{R}}\}}(y_0; \overline{F}) = W_{\{\varepsilon^{\mathcal{R}}\}}(x_0; \overline{F}) =: W_{\{\varepsilon^{\mathcal{R}}\}}(\overline{F}).$$

We now identify $W_{\{\varepsilon^{\mathcal{R}}\}}(\overline{F})$. Assuming, without loss of generality, that $0 \in \omega$ and $Q' \subset \omega$, by Lemma 2.0.8 there exists a sequence $\{\psi_{\varepsilon^{\mathcal{R}}}\}$ with

$\psi_{\varepsilon^{\mathcal{R}}} = 0$ on $\{(x_\alpha, x_3) : |x_3| < f(x_\alpha/\varepsilon^{\mathcal{R}}), x_\alpha \in \partial Q'\}$ and $\psi_{\varepsilon^{\mathcal{R}}} \chi_{Q'_{\varepsilon^{\mathcal{R}}}} \to 0$ in $L^p(\Omega; \mathbb{R}^3)$, such that

$$W_{\{\varepsilon^{\mathcal{R}}\}}(\overline{F}) = J_{\{\varepsilon^{\mathcal{R}}\}}(\overline{F} x_\alpha; Q') = \lim_{\varepsilon^{\mathcal{R}} \to 0^+} E_{\varepsilon^{\mathcal{R}}}(\overline{F} x_\alpha + \psi_{\varepsilon^{\mathcal{R}}}; Q').$$

Define

$$\phi_{\varepsilon^{\mathcal{R}}}(x_\alpha, x_3) := \frac{1}{\varepsilon^{\mathcal{R}}} \psi_{\varepsilon^{\mathcal{R}}}(\varepsilon^{\mathcal{R}} x_\alpha, x_3).$$

Then $\phi_{\varepsilon^{\mathcal{R}}} \in W^{1,p}\left(\left[\left(0, 1/\varepsilon^{\mathcal{R}}\right)^2\right]^f; \mathbb{R}^3\right)$ and it is equal to 0 if $x_\alpha \in \partial(0, 1/\varepsilon^{\mathcal{R}})^2$; thus it is an admissible test function in the definition of $g(1/\varepsilon^{\mathcal{R}})$ and so

$$\begin{aligned} &\limsup_{\varepsilon^{\mathcal{R}} \to 0^+} g\left(\frac{1}{\varepsilon^{\mathcal{R}}}\right) \leqq \\ &\limsup_{\varepsilon^{\mathcal{R}} \to 0^+} \varepsilon^{\mathcal{R}2} \int_{\left[\left(0, \frac{1}{\varepsilon^{\mathcal{R}}}\right)^2\right]^f} W\left(x_\alpha, x_3; \overline{F} + \nabla_\alpha \phi_{\varepsilon^{\mathcal{R}}} | \nabla_3 \phi_{\varepsilon^{\mathcal{R}}}\right) dx \\ &= \limsup_{\varepsilon^{\mathcal{R}} \to 0^+} \int_{Q'_{\varepsilon^{\mathcal{R}}}} W\left(\frac{x_\alpha}{\varepsilon^{\mathcal{R}}}, x_3; \overline{F} + \nabla_\alpha \psi_{\varepsilon^{\mathcal{R}}} \middle| \frac{1}{\varepsilon^{\mathcal{R}}} \nabla_3 \psi_{\varepsilon^{\mathcal{R}}}\right) dx \\ &= W_{\{\varepsilon^{\mathcal{R}}\}}(\overline{F}). \end{aligned} \tag{3.0.27}$$

Conversely, consider $\lambda_n \nearrow \infty$ such that

$$g(\lambda_n) \to \liminf_{t \nearrow \infty} g(t).$$

For each n, take $\phi_n \in W^{1,p}(\{(0, \lambda_n)^2 \times (-1, 1) : |x_3| < f(x_\alpha)\}; \mathbb{R}^3)$ with $\phi_n = 0$ if $x_\alpha \in \partial(0, \lambda_n)^2$, and such that

$$g(\lambda_n) + \frac{1}{\lambda_n^3} \geq \frac{1}{\lambda_n^2} \int_{[(0,\lambda_n)^2]^f} W(x_\alpha, x_3; \overline{F} + \nabla_\alpha \phi_n | \nabla_3 \phi_n) dx. \tag{3.0.28}$$

Set $\psi^n_{\varepsilon^{\mathcal{R}}} := \varepsilon^{\mathcal{R}} \phi_n\left(\frac{x_\alpha}{\varepsilon^{\mathcal{R}}}, x_3\right)$, where ϕ_n has been extended by zero to $([\![\lambda_n]\!] +$

$1)^2$, and then to the whole of $\mathbb{R}^2$ by $([\![\lambda_n]\!]+1)^2$-periodicity. Then,

$$J_{\{\varepsilon^{\mathcal{R}}\}}(\overline{F}x_\alpha;Q') \le \liminf_{\varepsilon^{\mathcal{R}}\to 0^+} E_{\varepsilon^{\mathcal{R}}}(\overline{F}x_\alpha + \psi^n_{\varepsilon^{\mathcal{R}}};Q')$$

$$= \liminf_{\varepsilon^{\mathcal{R}}\to 0^+} \int_{Q'_{\varepsilon^{\mathcal{R}}}} W\left(\frac{x_\alpha}{\varepsilon^{\mathcal{R}}}, x_3; \overline{F} + \nabla_\alpha \psi^n_{\varepsilon^{\mathcal{R}}} \middle| \frac{1}{\varepsilon^{\mathcal{R}}}\nabla_3 \psi^n_{\varepsilon^{\mathcal{R}}}\right) dx$$

$$= \liminf_{\varepsilon^{\mathcal{R}}\to 0^+} \int_{Q'} \left[\int_{-f(\frac{x_\alpha}{\varepsilon^{\mathcal{R}}})}^{f(\frac{x_\alpha}{\varepsilon^{\mathcal{R}}})} W\left(\frac{x_\alpha}{\varepsilon^{\mathcal{R}}}, x_3; \overline{F} + \nabla_\alpha\phi_n\left(\frac{x_\alpha}{\varepsilon^{\mathcal{R}}}, x_3\right) \middle| \nabla_3\phi_n\left(\frac{x_\alpha}{\varepsilon^{\mathcal{R}}}, x_3\right)\right) dx_3\right] dx_\alpha$$

$$= \frac{1}{([\![\lambda_n]\!]+1)^2}\int_{(0,\lambda_n)^2}\left[\int_{-f(x_\alpha)}^{f(x_\alpha)} W(x_\alpha, x_3; \overline{F} + \nabla_\alpha\phi_n(x_\alpha,x_3)| \nabla_3\phi_n(x_\alpha,x_3))\, dx_3\right] dx_\alpha$$

$$+\frac{1}{([\![\lambda_n]\!]+1)^2}\int_{[(0,[\![\lambda_n]\!]+1)^2\setminus(0,\lambda_n)^2]^f} W(x_\alpha,x_3;\overline{F}|0)\, dx_\alpha$$

$$\le \frac{\lambda_n^2}{([\![\lambda_n]\!]+1)^2}\left(g(\lambda_n) + \frac{1}{\lambda_n^3}\right) + O\left(\frac{1}{n}\right),$$

where we have used (3.0.28) as well as the $([\![\lambda_n]\!]+1)^2$-periodic character of

$$\int_{f(\cdot)}^{-f(\cdot)} W(\cdot, x_3; \overline{F} + \nabla_\alpha\phi^n(\cdot,x_3)|\nabla_3\phi_n(\cdot,x_3))\, dx_3.$$

Thus, letting n tend to ∞,

$$J_{\{\varepsilon^{\mathcal{R}}\}}(\overline{F}x_\alpha;Q') \le \liminf_{t\nearrow\infty} g(t),$$

or still

$$W_{\{\varepsilon^{\mathcal{R}}\}}(\overline{F}) \le \liminf_{t\nearrow\infty} g(t). \tag{3.0.29}$$

Recalling (3.0.27), (3.0.29), we obtain

$$\liminf_{t\nearrow\infty} g(t) \le \limsup_{\varepsilon^{\mathcal{R}}\to 0^+} g\left(\frac{1}{\varepsilon^{\mathcal{R}}}\right) \le W_{\{\varepsilon^{\mathcal{R}}\}}(\overline{F}) \le \liminf_{t\nearrow\infty} g(t),$$

which proves the desired result. Since the $\Gamma(L^p)$–limit of $E_{\varepsilon^{\mathcal{R}}}(u;A)$ is independent of the specific sequence $\{\varepsilon^{\mathcal{R}}\}$, in light of Proposition 7.11 in (16) we conclude that $E_\varepsilon(u;A)$ $\Gamma(L^p)$-converges to $\int_A W_{\text{hom}}(\nabla_\alpha u)\, dx_\alpha$. □

4 Optimal design of a thin film

We refer to (27) and references therein for a detailed motivation of the problem considered below and for relevant results in the so-called cylindrical case (see Remark 4.0.12 below).

Let $W_i(F)$, $i = 1, 2$, be continuous real-valued functions on $\mathbb{R}^{3\times 3}$ such that

$$\frac{1}{C}|F|^p - C \leq W_i(F) \leq C(1 + |F|^p), 0 < C, \, 1 \leq p < \infty.$$

For any $v \in W^{1,p}(\omega; \mathbb{R}^3)$, $\theta \in L^\infty(\omega \times (-1,1); [0,1])$ and any open subdomain $A \subset \mathbb{R}^2$, the goal is to characterize

$$J(v; \theta; A) := \liminf_{\{\varepsilon\} \searrow 0^+} J_{\{\varepsilon\}}(v; \theta; A),$$

where

$$J_{\{\varepsilon\}}(v; \theta; A) :=$$
$$\inf_{\{\chi_\varepsilon\},\{v_\varepsilon\}} \left\{ \liminf_{\varepsilon \to 0^+} \int_{A\times(-1,1)} (\chi_\varepsilon W_1 + (1-\chi_\varepsilon) W_2) \left(\nabla_\alpha v_\varepsilon \Big| \frac{1}{\varepsilon} \nabla_3 v_\varepsilon \right) dx : \right.$$
$$v_\varepsilon \in W^{1,p}(A \times (-1,1); \mathbb{R}^3), \chi_\varepsilon \in L^\infty(A \times (-1,1); \{0,1\}),$$
$$\left. v_\varepsilon \to v \text{ in } L^p(A \times (-1,1); \mathbb{R}^3), \chi_\varepsilon \overset{*}{\rightharpoonup} \theta \text{in } L^\infty(A \times (-1,1); [0,1]) \right\}.$$

Define, for any $\overline{F} \in \mathbb{R}^{3\times 2}$, $\theta \in [0,1]$,

$$\widehat{W}(\theta, \overline{F}) := \tag{4.0.30}$$
$$\inf_{\lambda>0} \inf_{\phi,\chi} \left\{ \frac{1}{2} \int_{Q'\times(-1,1)} (\chi W_1 + (1-\chi) W_2)(\overline{F} + \nabla_\alpha \phi | \lambda \nabla_3 \phi) dx : \right.$$
$$\phi \in W^{1,p}(Q' \times (-1,1); \mathbb{R}^3), \phi = 0 \text{ on } \partial Q' \times (-1,1),$$
$$\left. \chi \in L^\infty(Q' \times (-1,1); \{0,1\}), \; \frac{1}{2} \int_{Q'\times(-1,1)} \chi dx = \theta \right\}$$
$$= \inf_{k>0} \inf_{\phi,\chi} \left\{ \frac{1}{2k} \int_{Q'\times(-k,k)} (\chi W_1 + (1-\chi) W_2)(\overline{F} + \nabla_\alpha \phi | \nabla_3 \phi) dx : \right.$$
$$\phi \in W^{1,p}(Q' \times (-k,k); \mathbb{R}^3), \phi = 0 \text{ on } \partial Q' \times (-k,k),$$
$$\left. \chi \in L^\infty(Q' \times (-k,k); \{0,1\}), \; \frac{1}{2k} \int_{Q'\times(-k,k)} \chi dx = \theta \right\}.$$

Remark 4.0.10. *It is easily proved that* $\widehat{W}$ *is an upper-semicontinuous function of* $(\theta, \overline{F}) \in [0,1] \times \mathbb{R}^{3\times 2}$ *(see the proof of Proposition 2.9 in (27)).*

Theorem 4.0.11.

$$J(v;\theta;A) \geq 2\int_A \widehat{W}\left(\frac{1}{2}\int_{-1}^{1}\theta(x_\alpha,s)\,ds, Dv(x_\alpha)\right)\,dx_\alpha, \tag{4.0.31}$$

and equality holds if $\theta \in L^\infty(\omega;[0,1])$ *and if* W *satisfies the symmetry property:*

$$W(\overline{F}|F_3) = W(\overline{F}|-F_3),\ \overline{F}\in\mathbb{R}^{3\times 2}, F_3\in\mathbb{R}^3. \tag{4.0.32}$$

Remark 4.0.12. *In contrast with the setting investigated in (27) the material distribution – the characteristic functions* χ_ε *–is not restricted to cylindrical geometries; in other words* χ_ε *may also depend on the transverse variable* x_3. *This is the only difference between Theorem 4.0.11 and Theorem 2.3 in (27). Note that the equality in Theorem 4.0.11 holds if* θ *is independent of* x_3, *that is if we are truly in a plate-like setting.*

Remark 4.0.13. *Theorem 4.0.11 is not a mere application of Theorem 2.0.4 because of the presence of the additional field* χ_ε, *but the main argument can be carried out with natural modifications where needed.*

Lemma 4.0.14. *For any sequence* $\{\varepsilon\}\searrow 0^+$, *there exists a subsequence* $\{\varepsilon^{\mathcal{R}}\}$ *such that*

$$J_{\{\varepsilon^{\mathcal{R}}\}}(v;\theta;A) \geq 2\int_A \widehat{W}\left(\frac{1}{2}\int_{-1}^{1}\theta(x_\alpha,s)\,ds, Dv(x_\alpha)\right)\,dx_\alpha,$$

thus, in particular,

$$J(v;\theta;A) \geq 2\int_A \widehat{W}\left(\frac{1}{2}\int_{-1}^{1}\theta(x_\alpha,s)\,ds, Dv(x_\alpha)\right)\,dx_\alpha.$$

Proof. To establish (4.0.31) we show that for any sequence $\{\varepsilon\}\searrow 0^+$ there exists a subsequence $\{\varepsilon^{\mathcal{R}}\}$ such that

$$J_{\{\varepsilon^{\mathcal{R}}\}}(v;\theta;A) \geq 2\int_A \widehat{W}\left(\frac{1}{2}\int_{-1}^{1}\theta(x_\alpha,s)\,ds, Dv(x_\alpha)\right)\,dx_\alpha.$$

The subsequence $\{\varepsilon^{\mathcal{R}}\}$ is defined exactly as in Section 2(see argument leading to (2.0.7)). The proof is then divided into two steps: The first step is devoted to asserting that $J_{\{\varepsilon^{\mathcal{R}}\}}(v;\theta;\cdot)$ is a finite nonnegative Radon measure which is absolutely continuous with respect to $\mathcal{L}^2\lfloor\omega$, and the second step establishes that the Radon-Nykodim derivative $\frac{dJ_{\{\varepsilon^{\mathcal{R}}\}}(v;\theta;\cdot)}{d\mathcal{L}^2}(x_0)$ is, for suitable x_0's in ω, greater than or equal to $2\widehat{W}\left(1/2\int_{-1}^{1}\theta(x_0,s)\,ds, \nabla_\alpha v(x_0)\right)$.

We now address the second part of Theorem 4.0.11 and thus assume, from now onward, that θ is independent of x_3.

Step 1: Here it is assumed that $\theta \equiv \theta_\infty$ and $v \equiv v_\infty$ are, respectively, constant and affine functions and we claim: Let T be a triangle on the plane. Then there exists a sequence $\{v_n, \chi_n\} \in W^{1,p}(T \times (-1,1); \mathbb{R}^3) \times L^\infty(T \times (-1,1); \{0,1\})$ with $v_n = v_\infty$ on $\partial T \times (-1,1)$ such that, as $n \nearrow \infty$,

$$\begin{cases} \chi_n \overset{*}{\rightharpoonup} \theta_\infty & \text{in } L^\infty(T \times (-1,1); [0,1]), \\ v_n \to v_\infty & \text{in } L^p(T \times (-1,1); \mathbb{R}^3), \end{cases}$$

and

$$\begin{aligned} J(v_\infty; \theta_\infty; T) &= \lim_{n \to +\infty} \int_{T \times (-1,1)} (\chi_n W_1 + (1-\chi_n) W_2)(\nabla_\alpha v_n | n \nabla_3 v_n) dx \\ &= 2\mathcal{L}^2(T)\, \widehat{W}(\theta_\infty, \nabla_\alpha v_\infty). \end{aligned}$$

The proof is a blow-up argument in the spirit of (29). From the definition (4.0.30) of $\widehat{W}$, together with the density of $W^{1,\infty}(Q' \times (-1,1); \mathbb{R}^3)$ into $W^{1,p}(Q' \times (-1,1); \mathbb{R}^3)$, we deduce that, for any $\eta > 0$, there exist $\lambda^\eta > 0$, $\phi^\eta \in W^{1,\infty}(Q' \times (-1,1); \mathbb{R}^3)$, with $\phi^\eta = 0$ on $\partial Q' \times (-1,1)$, and $\chi^\eta \in L^\infty(Q' \times (-1,1); \{0,1\})$ with $1/2 \int_{Q' \times (-1,1)} \chi^\eta dx = \theta_\infty$ such that

$$2\widehat{W}(\theta_\infty, \nabla_\alpha v_\infty) \geq \int_{Q' \times (-1,1)} (\chi^\eta W_1 + (1-\chi^\eta) W_2)(\nabla_\alpha v_\infty + \nabla_\alpha \phi^\eta | \lambda^\eta \nabla_3 \phi^\eta) dx - \eta. \tag{4.0.33}$$

Extend ϕ^η, χ^η to be $Q' \times (-2,2)$-periodic by setting

$$\overline{\phi}^\eta, \overline{\chi}^\eta(x_\alpha, x_3) := \begin{cases} \phi^\eta, \chi^\eta(x_\alpha, x_3), & -1 \leq x_3 \leq 1, \\ \phi^\eta, \chi^\eta(x_\alpha, -2 - x_3), & -2 \leq x_3 \leq -1, \\ \phi^\eta, \chi^\eta(x_\alpha, 2 - x_3), & 1 \leq x_3 \leq 2, \end{cases}$$

then extend $\overline{\phi}^\eta, \overline{\chi}^\eta$ by $Q' \times (-2,2)$-periodicity to $\mathbb{R}^3$. Note that, in view of (4.0.32), (4.0.33) also reads as

$$4\widehat{W}(\theta_\infty, \nabla_\alpha v_\infty) \geq \int_{Q' \times (-2,2)} (\overline{\chi}^\eta W_1 + (1-\overline{\chi}^\eta) W_2)(\nabla_\alpha v_\infty + \nabla_\alpha \overline{\phi}^\eta | \lambda^\eta \nabla_3 \overline{\phi}^\eta) dx - 2\eta. \tag{4.0.34}$$

Remark 4.0.15. *This is the only instance where assumption (4.0.32) is used in this study.*

Set

$$\begin{cases} \chi_n^\eta(x_\alpha, x_3) & := \overline{\chi}^\eta(n^2 x_\alpha, n x_3), \\ v_n^\eta(x_\alpha, x_3) & := v_\infty(x_\alpha) + \frac{1}{n^2}\overline{\phi}^\eta(n^2 x_\alpha, n\lambda^\eta x_3). \end{cases}$$

If $Q'(a,r)$ is a square on $\mathbb{R}^2$, then it is easily checked that, as $n \nearrow \infty$,

$$\begin{cases} \chi_n^\eta \overset{*}{\rightharpoonup} \theta_\infty & \text{in } L^\infty(Q'(a,r)\times(-1,1);[0,1]), \\ v_n^\eta \to v_\infty & \text{in } L^p(Q'(a,r)\times(-1,1);\mathbb{R}^3). \end{cases} \tag{4.0.35}$$

Further, in view of (4.0.34), together with the periodic character of v_n^η, χ_n^η,

$$\begin{aligned} &J(v_\infty;\theta_\infty;Q'(a,r)) \leq \\ &\liminf_{\eta\to0^+}\lim_{n\to+\infty}\int_{Q'(a,r)\times(-1,1)}(\chi_n^\eta W_1 + (1-\chi_n^\eta)W_2)(\nabla_\alpha v_n^\eta | n\nabla_3 v_n^\eta)dx \\ &\leqq \limsup_{\eta\to0^+}\lim_{n\to+\infty}\int_{Q'(a,r)\times(-1,1)}(\chi_n^\eta W_1 + (1-\chi_n^\eta)W_2)(\nabla_\alpha v_n^\eta | n\nabla_3 v_n^\eta)dx \\ &\leqq 2\mathcal{L}^2(Q'(a,r))\,\widehat{W}(\theta_\infty, \nabla_\alpha v_\infty). \end{aligned}$$

But, in view of (4.0.31),

$$J(v_\infty;\theta_\infty;Q'(a,r)) \geq 2\mathcal{L}^2(Q'(a,r))\,\widehat{W}(\theta_\infty, \nabla_\alpha v_\infty),$$

so that

$$\begin{aligned} &J(v_\infty;\theta_\infty;Q'(a,r)) = \\ &\lim_{\eta\to0^+}\lim_{n\to+\infty}\int_{Q'(a,r)\times(-1,1)}(\chi_n^\eta W_1 + (1-\chi_n^\eta)W_2)(\nabla_\alpha v_n^\eta | n\nabla_3 v_n^\eta)dx \\ &= 2\mathcal{L}^2(Q'(a,r))\,\widehat{W}(\theta_\infty, \nabla_\alpha v_\infty), \end{aligned}$$

hence, by Lemma 2.0.7 there exists a sequence $\{\eta(n)\}_n$ such that $v_n := v_n^{\eta(n)}, \chi_n := \chi_n^{\eta(n)}$ satisfy (4.0.35) as well as

$$\begin{aligned} J(v_\infty;\theta_\infty;Q'(a,r)) &= \lim_{n\to+\infty}\int_{Q'(a,r)\times(-1,1)}(\chi_n W_1 + (1-\chi_n)W_2)(\nabla_\alpha v_n | n\nabla_3 v_n)dx \\ &= 2\mathcal{L}^2(Q'(a,r))\,\widehat{W}(\theta_\infty, \nabla_\alpha v_\infty). \end{aligned} \tag{4.0.36}$$

Consider now a triangle T covered with squares of the type $Q'(a,r), a \in \mathbb{R}^2, r > 0$, up to a set of measure $1/m$, i.e.,

$$T_m := \cup_{i=1}^{N(m)} Q'(a_i^m, r_i^m) \subset\subset T,$$

with $\mathcal{L}^2(T \setminus T_m) \leq 1/m$. An easy construction, identical to that of Step 2 in Lemma 4.2 in (27) (see Remark 4.4 in (27)) yields (4.0.36) for T in lieu of $Q'(a,r)$.

Step 2: The second step addresses the case of general domains $A \subset \omega$, general θ's and general v's, and yields Theorem 4.0.11.

In view of Step 1 – and in particular of the boundary condition $v_n = v_\infty$ for the corresponding sequence –, the conclusions on this step also hold true if T is replaced by a triangulation of the plane on which v_∞ is continuous and piecewise affine and θ_∞ piecewise constant. Then, let v, θ be arbitrary elements of $W^{1,p}(\omega;\mathbb{R}^3)$ and $L^\infty(\omega;[0,1])$, respectively, and consider $\{w_k, \theta_k\}$ a piecewise affine/piecewise constant pair defined on triangulations of the plane such that

$$\begin{cases} \theta_k \to \theta & \text{a.e. in } \omega, \\ w_k \to v & \text{in } W^{1,p}(\omega;\mathbb{R}^3). \end{cases}$$

For each pair (w_k, θ_k) there exists a pair-sequence $\{v_k^n, \theta_k^n\}$ defined on the same triangulation satisfying the properties of Step 1 for that triangulation and for $v_\infty := w_k, \theta_\infty := \theta_k$. A diagonalization process (with n replaced by $n(k)$) immediately yields a sequence $\{v_k, \chi_k\} \in W^{1,p}(A \times (-1,1);\mathbb{R}^3) \times L^\infty(A \times (-1,1);\{0,1\})$ such that

$$\begin{cases} \chi_k \overset{*}{\rightharpoonup} \theta & \text{in } L^\infty(A \times (-1,1);[0,1]), \\ v_k \to v & \text{in } L^p(A \times (-1,1);\mathbb{R}^3), \end{cases} \tag{4.0.37}$$

and

$$\begin{aligned} J(v;\theta;A) &\leq \liminf_{k \to +\infty} \int_{A \times (-1,1)} (\chi_k W_1 + (1-\chi_k) W_2)(\nabla_\alpha v_k | n(k) \nabla_3 v_k) dx \\ &= \liminf_{k \to +\infty} J(v_k;\theta_k;A) \\ &= \liminf_k 2 \int_A \widehat{W}(\theta_k, \nabla_\alpha v_k)\, dx_\alpha. \end{aligned} \tag{4.0.38}$$

The bound from above for W_i, the first convergence in (4.0.37), Fatou's lemma, and the upper semicontinuity property of $\widehat{W}$ (see Remark 4.0.10) imply that

$$\liminf_{k \to +\infty} \int_A \{\beta(1 + |\nabla_\alpha v_k|^p) - \widehat{W}(\theta_k, \nabla_\alpha v_k)\} dx_\alpha \geq$$

$$\int_A \{\beta(1 + |\nabla_\alpha v|^p) - \widehat{W}(\theta, \nabla_\alpha v)\} dx_\alpha,$$

thus

$$\int_A \widehat{W}(\theta, \nabla_\alpha v)\, dx_\alpha \geq \limsup_{k\to+\infty} \int_A \widehat{W}(\theta_k, \nabla_\alpha v_k)\, dx_\alpha,$$

which, together with (4.0.38), yields

$$J(v;\theta;A) \leq 2 \int_A \widehat{W}(\theta, \nabla_\alpha v)\, dx_\alpha.$$

We recall that (4.0.31) provides the other inequality and the proof of Theorem 4.0.11 is complete. □

Bibliography

[1] Acerbi, E., Buttazzo, G., Percivale, D., A variational definition for the strain energy of an elastic string. *J. Elasticity*, **25** (1991), 137–148.

[2] Acerbi, E., Fusco, N., Semicontinuity problems in the calculus of variations. *Arch. Rat. Mech. Anal.*, **86** (1984), 125–145.

[3] Acerbi, E., Chiadò Piat, V., Dal Maso, G., Percivale, D., An extension theorem from connected sets, and homogenization in general periodic domains. *Nonlinear Analysis*, **18** (1992), 481-496.

[4] Anzelotti, E., Baldo, S., Percivale, D., Dimensional reduction in variational problems, asymptotic developments in Γ- convergence, and thin structures in elasticity. *Asymptot. Anal.*, **9** (1994), 61–100.

[5] Ansini, N., Braides, A., Homogenization of oscillating boundaries and applications to thin films. *J. Anal. Math.*, **83** (2001), 151–182.

[6] Anzelotti, E., Baldo, S., Percivale, D., Dimensional reduction in variational problems, asymptotic developments in Γ–convergence, and thin structures in elasticity. *Asymptot. Anal.*, **9** (1994), 61–100.

[7] Babaddjian, J.-F., Francfort, G., Spatial heterogeneity in 3D-2D dimensional reduction. *ESAIM Control Optim. Calc. Var.*, **11** (2005), 139-160.

[8] Babaddjian, J.-F., Baia, M., 3D-2D analysis of a thin film with periodic microstructure. *Proc. Roy. Soc. Edinburgh, Sect A*, **136** (2006), 223-243.

[9] Babaddjian, J.-F., Baia, M., Multiscale nonconvex relaxation and application to thin films. *Asymptot. Anal.*, **48** (2006), 173-218.

[10] Bhattacharya, K., Braides, A., Thin films with many small cracks. *R. Soc. Lond. Proc. Ser. A Math. Phys. Eng. Sci.*, **458** (2002), 823–840.

[11] Bhattacharya, K., Fonseca, I., Francfort, G., An asymptotic study of the debonding of thin films. *Arch. Rat. Mech. Anal.*, **161** (2002) 205–229.

[12] Bhattacharya, K., James, R. D., A theory of thin films of martensitic materials with applications to microactuators. *J. Mech. Phys. Solids*, **47** (1999), 531-576.

[13] Blanchard, D., Francfort, G., Asymptotic thermoelastic behaviour of flat plates. *Quart. Appl. Math.*, **45**(4) (1987), 645–667.

[14] Bocea, M., Fonseca, I., Equi-integrability results for 3D-2D dimension reduction problems. *ESAIM: Control, Optimisation and Calculus of Variations*, **7** (2002), 443–470.

[15] Bouchitté, G., Fonseca, I., Mascarenhas, M. L., Bending moment in membrane theory. *J. Elasticity*, **73** (2003), 75–99 (2004).

[16] Braides, A., Defranceschi, A., *Homogenization of Multiple Integrals.* Oxford Lecture Series in Mathematics and its Applications, **12**, Oxford, 1998.

[17] Braides, A., Fonseca, I., Francfort, G., 3D-2D asymptotic analysis for inhomogeneous thin films. *Indiana Univ. Math. J.*, **49** (2000), 1367–1404.

[18] Braides, A., Fonseca, I., Brittle thin films. *Applied Math. and Optimization*, **44** (2001), 299–323.

[19] Buttazzo, G., *Semicontinuity, Relaxation and Integral Representation in the Calculus of Variations.* Pitman, London, (1989).

[20] Caillerie, D., Thin elastic and periodic plates. *Math. Methods in Appl. Sci.*, **6** (1984), 159–191.

[21] P.G. Ciarlet, A justification of the von Kàrmàn equations. *Arch. Rat. Mech. Anal.*, **73**, (1980), 349–389.

[22] Ciarlet, P. G., *Introduction to Linear Shell Theory.* North Holland, Amsterdam, (1998).

[23] Ciarlet, P. G., Destuynde, Ph., A justification of the two–dimensional linear plate model. *J. Mécanique*, **18** (1979), 315–344.

[24] Ciarlet, P. G., Destuynder, Ph., A justification of a nonlinear model in plate theory. *Comput. Methods Appl. Mech. Engrg.*, **7/18** (1979), 227–258.

[25] Cimetière, A., Geymonat, G., Le Dret, H., Raoult,A., Tutek, Z., Asymptotic theory and analysis for diplacements and stress distribution in nonlinear elastic straight slender rods. *J. Elasticity*, **19**(II) (1988), 111–161.

[26] De Giorgi, E., Letta, G., Une notion générale de convergence faible pour des fonctions croissantes d'ensemble. *Ann. Sc. Norm. Sup. Pisa Cl. Sci.*, **4** (1977), 61–99.

[27] Fonseca, I., Francfort, G., 3D–2D asymptotic analysis of an optimal design problem for thin films. *J. Reine Angew. Math.*, **505** (1998), 173–202.

[28] Fonseca, I., Francfort, G., On the inadequacy of the scaling of linear elasticity for 3D–2D asymptotics in a nonlinear setting. *J. Math. Pures Appl.*, **80** (2001), 547-562.

[29] Fonseca, I., Müller, S., Quasiconvex integrands and lower semicontinuity in L^1. *SIAM J. Math. Anal.*, **23**, (1992), 1081–1098.

[30] Fox, D., Raoult, A., Simo, J.C., A justification of nonlinear properly invariant plate theories. *Arch. Rat. Mech. Anal.*, **124** (1993), 157–199.

[31] Freddi, L., Paroni, R., The energy density of martensitic thin films via dimension reduction. *Interfaces Free Bound.*, **6** (2004), 439–459.

[32] Geymonat, G., Krasucki, F., Marigo, J. J., Stress distribution in anisotropic elastic composite beams. In: P.G. Ciarlet, E. Sanchez–Palencia, editors, *Applications of multiple scalings in mechanics*, Masson, Paris (1987) 118–133.

[33] Geymonat, G., Krasucki, F., Marigo, J. J., Sur la commutativité des passages à la limite en théorie asymptotique des poutres composites. *C.R. Acad. Sci. Paris*, **305**(II) (1987), 225–228.

[34] Kohn, R. V., Vogelius, M., A new model for thin plates with rapidly varying thickness II, III. *Quart. Appl. Math.*, **43, 44** (1985), (1986), 1–22, 35–48.

[35] Le Dret, H., Raoult, A., The nonlinear membrane model as variational limit of nonlinear three–dimensional elasticity. *J. Math. Pures Appl.*, **74** (1995), 549–578.

[36] Le Dret, H., Raoult, A., The membrane shell model in nonlinear elasticity: a variational asymptotic derivation. *J. Nonlinear Sc.*, **6** (1996), 59–84.

[37] Le Dret, H., Raoult, A., Variational convergence for nonlinear shell models with directors and related semicontinuity and relaxation results. *Arch. Rat. Mech. Anal.*, **154** (2000), 101–134.

[38] Murat, F., Sili, A., Problèmes monotones dans des cylindres de faible diamètre. *C. R. Acad. Sci. Paris*, **319**(I) (1995), 567–572.

[39] Shu, Y., Heterogeneous thin films of martensitic materials. *Arch. Rat. Mech. Anal.*, **153** (2000), 39–90.

[40] Trabucho, L., Viaño, J. M., Mathematical Modelling of Rods. In: Ciarlet, P. G., Lions, J. L., editors, *Handbook of Numerical Analysis*, North Holland, Amsterdam, (1996), 487–974.

GPSR Compliance

The European Union's (EU) General Product Safety Regulation (GPSR) is a set of rules that requires consumer products to be safe and our obligations to ensure this.

If you have any concerns about our products, you can contact us on ProductSafety@springernature.com

In case Publisher is established outside the EU, the EU authorized representative is:

Springer Nature Customer Service Center GmbH
Europaplatz 3
69115 Heidelberg, Germany

Batch number: 10577876

Printed by Printforce, the Netherlands